Excel Data Analysis

MW00445937

Hector Guerrero

Excel Data Analysis

Modeling and Simulation

 Springer

Dr. Hector Guerrero
Mason School of Business
College of William & Mary
Williamsburg, VA 23189
USA
hector.guerrero@mason.wm.edu

ISBN 978-3-642-10834-1   e-ISBN 978-3-642-10835-8
DOI 10.1007/978-3-642-10835-8
Springer Heidelberg Dordrecht London New York

Library of Congress Control Number: 2010920153

© Springer-Verlag Berlin Heidelberg 2010
This work is subject to copyright. All rights are reserved, whether the whole or part of the material is concerned, specifically the rights of translation, reprinting, reuse of illustrations, recitation, broadcasting, reproduction on microfilm or in any other way, and storage in data banks. Duplication of this publication or parts thereof is permitted only under the provisions of the German Copyright Law of September 9, 1965, in its current version, and permission for use must always be obtained from Springer. Violations are liable to prosecution under the German Copyright Law.
The use of general descriptive names, registered names, trademarks, etc. in this publication does not imply, even in the absence of a specific statement, that such names are exempt from the relevant protective laws and regulations and therefore free for general use.

Cover design: WMXDesign GmbH, Heidelberg

Printed on acid-free paper

Springer is part of Springer Science+Business Media (www.springer.com)

To my wonderful parents . . . Paco and Nena

Preface

Why does the World Need—*Excel Data Analysis, Modeling, and Simulation?*

When spreadsheets first became widely available in the early 1980s, it spawned a revolution in teaching. What previously could only be done with arcane software and large scale computing was now available to the *common-man*, on a desktop. Also, before spreadsheets, most substantial analytical work was done outside the classroom where the tools were; spreadsheets and personal computers moved the work into the classroom. Not only did it change how the analysis curriculum was taught, but it also empowered students to venture out on their own to explore new ways to use the tools. I can't tell you how many phone calls, office visits, and/or emails I have received in my teaching career from ecstatic students crowing about what they have just done with a spreadsheet model.

I have been teaching courses related to spreadsheet based analysis and modeling for about 25 years and I have watched and participated in the spreadsheet revolution. During that time, I have been a witness to the following observations:

- Each year has led to more and more demand for Excel based analysis and modeling skills, both from students, practitioners, and recruiters
- Excel has evolved as an ever more powerful suite of tools, functions, and capabilities, including the recent iteration and basis for this book—Excel 2007
- The ingenuity of Excel users to create applications and tools to deal with complex problems continues to amaze me
- Those students that preceded the spreadsheet revolution often find themselves at a loss as to where to go for an introduction to what is commonly taught to most many undergraduates in business and sciences.

Each of one these observations have motivated me to write this book. The first suggests that there is no foreseeable end to the demand for the skills that Excel enables; in fact, the need for continuing productivity in all economies guarantees that an individual with proficiency in spreadsheet analysis will be highly prized by an organization. At a minimum, these skills permit you freedom from *specialists* that can delay or hold you captive while waiting for a solution. This was common in the early days of information technology (IT); you requested that the IT group provide you with a solution or tool and you waited, and waited, and waited. Today if you need a solution you can do it yourself.

The combination of the 2nd and 3rd observations suggests that when you couple bright and energetic people with powerful tools and a good learning environment, wonderful things can happen. I have seen this throughout my teaching career, as well as in my consulting practice. The trick is to provide a teaching vehicle that makes the analysis accessible. My hope is that this book is such a teaching vehicle. I believe that there are three simple factors that facilitate learning—select examples that contain interesting questions, methodically lead students through the rationale of the analysis, and thoroughly explain the Excel tools to achieve the analysis.

The last observation has fueled my desire to lend a hand to the many students that passed through the educational system *before* the spreadsheet analysis revolution; to provide them with a book that points them in the right direction. Several years ago, I encountered a former MBA student in a Cincinnati Airport bookstore. He explained to me that he was looking for a good Excel-based book on Data analysis and modeling—"You know it's been more than 20 years since I was in a Tuck School classroom, and I desperately need to understand what my interns seem to be able to do so easily." By providing a broad variety of exemplary problems, from *graphical/statistical analysis* to *modeling/simulation* to *optimization*, and the Excel tools to accomplish these analyses, most readers should be able to achieve success in their self-study attempts to master spreadsheet analysis. Besides a good compass, students also need to be made aware of *the possible*. It is not usual to hear from students "Can you use Excel to do *this*?" or "I didn't know you could do *that* with Excel!"

Who Benefits from this Book?

This book is targeted at the student or practitioner that is looking for a *single* introductory Excel-based resource that covers three essential business skills—Data Analysis, Business Modeling, and Simulation. I have successfully used this material with undergraduates, MBAs, Executive MBAs and in Executive Education programs. For my students, the book has been the main teaching resource for both semester and half-semester long courses. The examples used in the books are sufficiently flexible to guide teaching goals in many directions. For executives, the book has served as a compliment to classroom lectures, as well as an excellent post-program, self-study resource. Finally, I believe that it will serve practitioners, like that former student I met in Cincinnati, that have the desire and motivation to refurbish their understanding of data analysis, modeling, and simulation concepts through self-study.

Key Features of this Book

I have used a number of examples in this book that I have developed over many years of teaching and consulting. Some are brief and to the point; others are more complex and require considerable effort to digest. I urge you to not become frustrated with the more complex examples. There is much to be learned from these examples, not only the analytical techniques, but also *approaches* to solving complex problems. These examples, as is always the case in real-world, messy problems, require making reasonable assumptions and some concession to simplification if a solution is to be obtained. My hope is that the approach will be as valuable to the reader as the analytical techniques. I have also taken great pains to provide an abundance of Excel screen shots that should give the reader a solid understanding of the chapter examples.

But, let me vigorously warn you of one thing—this is not an Excel *how-to* book. Excel *how-to* books concentrate on the Excel tools and not on analysis—it is assumed that you will fill in the analysis blanks. There are many excellent Excel *how-to* books on the market and a number of excellent websites (e.g. MrExcel.com) where you can find help with the details of specific Excel issues. I have attempted to write a book that is about analysis, analysis that can be easily and thoroughly handled with Excel. Keep this in mind as you proceed. So in summary, remember that the analysis is the primary focus and that Excel simply serves as an excellent vehicle by which to achieve the analysis.

Acknowledgements

I would like to thank the editorial staff of Springer for their invaluable support—Dr. Niels Peter Thomas, Ms. Alice Blanck, and Ms. Ulrike Stricker. Thanks to Ms. Elizabeth Bowman for her excellent editing effort over many years. Special thanks to the countless students I have taught over the years, in particular Bill Jelen, the world-wide-web's Mr. Excel that made a believer out of me. Finally, thanks to my family and friends that took a back seat to the book over the years of development—Tere, Rob, Brandy, Mac, Lili, PT, and Scout.

Contents

About the Author

Dr. Guerrero is a professor at Mason School of Business at the College of William and Mary, in Williamsburg, Virginia. He teaches in the areas of decision making, statistics, operations and business quantitative methods. He has previously taught at the Amos Tuck School of Business at Dartmouth College, and the College of Business of the University of Notre Dame. He is well known among his students for his quest to bring clarity to complex decision problems.

He earned a Ph.D. Operations and Systems Analysis, University of Washington and a BS in Electrical Engineering and an MBA at the University of Texas. He has published scholarly work in the areas of operations management, product design, and catastrophic planning.

Prior to entering academe, he worked as an engineer for Dow Chemical Company and Lockheed Missiles and Space Co. He is also very active in consulting and executive education with a wide variety of clients— U.S. Government, International firms, as well as many small and large U.S. manufacturing and service firms.

It is not unusual to find him relaxing on a quiet beach with a challenging Excel workbook and an excellent cabernet.

Chapter 1
Introduction to Spreadsheet Modeling

Contents

1.1 Introduction

Spreadsheets have become as commonplace as calculators in analysis and decision making. In this chapter we explore the importance of creating decision making models with Excel. We also consider the characteristics that make spreadsheets useful, not only for ourselves, but for others with whom we collaborate. As with any tool, learning to use them effectively requires carefully conceived planning and repeated practice; thus, we will terminate the chapter with an example of a poorly planned spreadsheet that is rehabilitated into a shining example of what a spreadsheet *can* be.

Some texts provide you with very detailed, in depth explanations of the intricacies of Excel; this text opts to concentrate on the types of analysis and model building you can perform with Excel. The ultimate goal of this book is to provide you with an Excel-centric approach to solving problems and to do so with *relatively*

H. Guerrero, *Excel Data Analysis*, DOI 10.1007/978-3-642-10835-8_1,
© Springer-Verlag Berlin Heidelberg 2010

simple and *abbreviated* examples. In other words, this book is for the individual that shouts—"I'm not interested in a 900 page text, full of *Ctl-Shift-F4-R key stroke shortcuts*. What I need is a good and instructive example so I can solve this problem before I leave the office tonight."

Finally, for many texts the introductory chapter is a "throw-away", to be read casually before getting to substantial material in the chapters that follow, but that is not the case for this chapter. It sets the stage for some important guidelines for constructing worksheets and workbooks that will be essential throughout the remaining chapters. I urge you to read this material carefully and to consider the content seriously.

Let's begin by considering the following encounter between two graduate school classmates of the class of 1990. In it, we begin to answer the question that decision makers face as Excel becomes the standard for analysis and collaboration—How can I quickly and effectively learn the capabilities of this powerful tool?

1.2 What's an MBA to do?

It was late Friday afternoon when Julia Lopez received an unexpected phone call from an MBA classmate, Ram Das, whom she had not heard from in years. They both work in Washington, DC and agreed to meet at a coffee shop on Wisconsin Avenue to catch up on their careers.

Ram: Julia, it's great to see you. I don't remember you looking as prosperous when we were struggling with our quantitative and computer classes in school.

Julia: No kidding! In those days I was just trying to keep up and survive. You don't look any worse for wear yourself. Still doing that rocket-science analysis you loved in school?

Ram: Yes, but it's getting tougher to defend my status as a rocket scientist. This summer we hired an undergraduate intern that just blew us away. This kid could do any type of analysis we asked, and do it on one software platform, Excel. Now my boss expects the same from me, but many years out of school, there is no way I have the training to equal that intern's skills.

Julia: Join the club. We had an intern we called the Excel Wonder Woman. I don't know about you, but in the last few years, people are expecting more and better analytical skills from MBAs. As a product manager, I'm expected to know as much about complex business analysis as I do about understanding my customers and markets. I even bought 5 or 6 books on business decision making with Excel. It's just impossible to get through hundreds of pages of detailed keystrokes and tricks for using Excel, much less simultaneously understand the basics of the analysis. Who has the time to do it?

Ram: I'd be satisfied with a brief, readable book that gives me a clear view of the *kinds* of things you can do with Excel, and just one straightforward example. Our intern was doing things that I would never have believed possible— analyzing qualitative data, querying databases, simulations, optimization, statistical analysis, collecting data on web pages, you name it. It used to take me six separate software packages to do all those things. I would love to do it all in Excel, and I know that to some degree you can.

Julia: Just before I came over here my boss dumped another project on my desk that he wants done in Excel. The Excel Wonder Woman convinced him that we ought to be building all our important analytical tools on Excel—*Decision Support Systems* she calls them. And if I hear the term *collaborative* one more time, I'm going to explode.

Ram: Julia, I have to go, but let's talk more about this. Maybe we can help each other learn more about the capabilities of Excel.

Julia: This is exciting. Reminds me of our study group work in the MBA.

This brief episode is occurring with uncomfortable frequency for many people in decision making roles. Technology, in the form of desktop software and hardware, is becoming as much a part of day-to-day business analysis as the concepts and techniques that have been with us for years. Although sometimes complex, the difficulty has not been in *understanding* these concepts and techniques, but more often, how to put them to use. For many individuals, if software were available for modeling problems, it could be unfriendly and inflexible; if software were not available, then we were limited to solving *baby* problems that were generally of little practical interest.

1.3 Why Model Problems?

It may appear to be trivial to ask why we model problems, but it is worth considering. Usually, there are at least two reasons for modeling problems—(1) if a problem has important financial and organizational implications, then it deserves serious consideration, and modeling permits serious analysis, and (2) on a very practical level, often we are directed by superiors to model a problem because *they* believe it is worthwhile. For a subordinate decision maker and analyst, important problems generally call for more than a gratuitous "I think..." or "I feel..." to satisfy a superior's questions. Increasingly, superiors are asking questions about decisions that require careful investigation of assumptions, and that question the sensitivity of decision outcomes to changes in environmental conditions and the assumptions. To deal with these questions, formality in decision making is a must; thus, we build models that can accommodate this higher degree of scrutiny. Ultimately, modeling can, and should, lead to better overall decision making.

1.4 Why Model Decision Problems with Excel?

So, if the modeling of decision problems is important and necessary in our work, then what modeling tool(s) do we select? In recent years there has been little doubt as to the answer of this question for most decision makers: Microsoft Excel. Excel is the most pervasive, all-purpose modeling tool on the planet due to its ease of use. It has a wealth of internal capability that continues to grow as each new version is introduced. Excel also resides in Microsoft Office, a suite of similarly popular tools that permit interoperability. Finally, there are tremendous advantages to "one-stop shopping" in the selection of a modeling tool, that is, a tool with many capabilities. There is so much power and capability built into Excel, that unless you have received very recent training in its latest capabilities, you might be unaware of the variety of modeling that is possible with Excel. Herein lies the first layer of important questions for decision makers who are considering a decision tool choice:

1. What forms of analysis are possible with Excel?
2. If my modeling effort requires multiple forms of analysis, can Excel handle the various techniques required?
3. If I commit to using Excel, will it be capable of handling new forms of analysis and a potential increase in the scale and complexity of my models?

The general answer to these questions is—just about any analytical technique that you can conceive that fits in the row-column structure of spreadsheets can be modeled with Excel. Note that this is a very broad and bold statement. Obviously, if you are modeling phenomena related to high energy physics or theoretical mathematics, you are very likely to choose other modeling tools. Yet, for the individual looking to model business problems, Excel is a must, and that is why this book will be of value to you. More specifically, Table 1.1 provides a partial list of the types of analysis this book will address.

When we first conceptualize and plan to solve a decision problem, one of the first considerations we face is which modeling approach to use. There are business problems that are sufficiently unique and complex that they will require a much more targeted and specialized modeling approach than Excel. Yet, most of us are involved with business problems that span a variety of problem areas—e.g. marketing issues that require qualitative database analysis, finance problems that require simulation of financial statements, and risk analysis that requires the determination of risk profiles. Spreadsheets permit us to unify these analyses on a single modeling platform. This makes our modeling effort: (1) *durable*—a robust structure that can anticipate varied use, (2) *flexible*—capable of adaptation as the problem changes and evolves, and (3) *shareable*—models that can be shared by a variety of individuals at many levels of the organization, all of whom are collaborating in the solution

Table 1.1 Types of analysis this book will undertake

Quantitative Data Presentation—Graphs and Charts
Quantitative Data Analysis—Summary Statistics and Data Exploration and Manipulation
Qualitative Data Presentation—Pivot Tables and Pivot Charts
Qualitative Data Analysis—Data Tables, Data Queries, and Data Filters
Advanced Statistical Analysis—Hypothesis testing, Correlation Analysis, and Regression Model
   Sensitivity Analysis—One-way, Two-way, Data Tables, Graphical Presentation
Optimization Models and Goal Seeking—Solver for Constrained Optimization, Scenarios
Models with Uncertainty—Monte Carlo Simulation

---

of the problem. Additionally, the standard programming required for spreadsheets is easier to learn than other forms of sophisticated programming languages found in many modeling systems. Even so, Excel has anticipated the occasional need for more formal programming by providing a powerful programming language, VBA (Visual Basic for Applications).

The ubiquitous nature of Excel spreadsheets has led to serious academic research and investigation into their use and misuse. Under the general title of **spreadsheet engineering**, academics have begun to apply many of the important principles of software engineering to spreadsheets, attempting to achieve better modeling results: more useful models, fewer mistakes in programming, and a greater impact on decision making. The growth in the importance of this topic is evidence of the potentially high costs associated with poorly designed spreadsheets.

In the next section, I address some **best practices** that will lead to superior everyday spreadsheet and workbook designs, or *good spreadsheet engineering*. Unlike some of the high level concepts of spreadsheet engineering, I provide very simple and specific guidance for spreadsheet development. My recommendations are aimed at the day-to-day users, and just as the ancient art of **Feng Shui** provides a sense of order and wellbeing in a building, public space, or home, these best practices can do the same for frequent users of spreadsheets.

1.5 Spreadsheet Feng Shui[1]/ Spreadsheet Engineering

The initial development of a spreadsheet project should focus on two areas—(1) planning and organizing the problem to be modeled, and (2) some general practices of good spreadsheet engineering. In this section we focus on the latter. In succeeding chapters we will deal with the former by presenting numerous forms of analysis that can be used to model business decisions. The following are five best practices to consider when designing a spreadsheet model:

Think workbooks not worksheets—Spare the worksheet; spoil the workbook. When spreadsheets were first introduced, a workbook consisted of a single worksheet. Over time spreadsheets have evolved into multi-worksheet workbooks, with interconnectivity between worksheets and even other workbooks and files. In workbooks that represent serious analytical effort, you should be conscious of not attempting to place too much information, data, or analysis on a single worksheet. Thus, I always include on separate worksheets: (1) an *introductory* or *cover page* with documentation that identifies the purpose, authors, contact information, and intended use of the spreadsheet model and, (2) a *table of contents* providing users with a glimpse of how the workbook will proceed. In deciding on whether or not to include additional worksheets, it is important to ask yourself the following question—Does the addition of a worksheet make the workbook easier to view and

---

[1] The ancient Chinese study of arrangement and location in one's physical environment, currently very popular in fields of architecture and interior design.

use? If the answer is *yes,* then your course of action is clear. Yet, there is a cost to adding worksheets—extra worksheets lead to the use of extra computer memory for a workbook. Thus, it is always a good idea to avoid the inclusion of gratuitous worksheets, which regardless of their memory overhead cost can be annoying to users. When in doubt, I generally decide in favor of adding a worksheet.

Place variables and parameters in a central location—Every workbook needs a Brain. I define a workbook's *Brain* as a central location for variables and parameters. Call it what you like—data center, variable depot, etc.—these values generally do not belong in cell formulas hidden from easy viewing. Why? If it is necessary to change a value that is used in the individual cell formulas of a worksheet, the change must be made in every cell containing the value. This idea can be generalized in the following concept: if you have a value that is used in numerous cell locations and you anticipate the possibility of changing that value, then you should have the cells that utilize the value, reference the value at some central location (*Brain*). For example, if a specific interest or discount rate is used in many cell formulas and/or in many worksheets you should locate that value in a single cell in the *Brain* to make a change in the value easier to manage. As we will see later, a *Brain* is also quite useful in conducting the sensitivity analysis for a model.

Design workbook layout with users in mind—User friendliness and designer control. As the lead designer of the workbook, you should consider how you want others to interact with your workbook. User interaction should consider not only the ultimate end use of the workbook, but also the collaborative interaction by others involved in the workbook design and creation process. Here are some specific questions to consider that facilitate **user friendliness** and designer control:

1. What areas of the workbook will the *end* user be allowed to access when the design becomes fixed?
2. Should certain worksheets or ranges be hidden from *users*?
3. What specific level of design interaction will *collaborators* be allowed?
4. What specific worksheets and ranges will *collaborators* be allowed to access?

Remember that your authority as lead designer extends to testing the workbook and determining how end users will employ the workbook. Therefore, not only do you need to exercise direction and control for the development process of the workbook, but also how it will be used.

Document workbook content and development—Insert text and comments liberally. There is nothing more annoying than viewing a workbook that is incomprehensible. This can occur even in carefully designed spreadsheets. What leads to spreadsheets that are difficult to comprehend? From the user perspective, the complexity of a workbook can be such that it may be necessary to provide explanatory documentation; otherwise, worksheet details and overall analytical approach can bewilder the user. Additionally, the designer often needs to provide users and collaborators with perspective on how and why a workbook developed as it did—e.g.

why were certain analytical approaches incorporated in the design, what assumptions were made, and what were the alternatives considered? You might view this as justification or defense of the workbook design.

There are a number of choices available for documentation: (1) text entered directly into cells, (2) naming cell ranges with descriptive titles (e.g. Revenue, Expenses, COGS, etc.), (3) explanatory text placed in text boxes, and (4) comments inserted into cells. I recommend the latter three approaches—text boxes for more detailed and longer explanations, range names to provide users with descriptive and understandable formulas since these names will appear in cell formulas that reference them, and cell comments for quick and brief explanations. In late chapters, I will demonstrate each of these forms of documentation.

Provide convenient workbook navigation— Beam me up Scotty! The ability to easily navigate around a well designed workbook is a must. This can be achieved through the use of **hyperlinks.** Hyperlinks are convenient connections to cell locations within a worksheet, to other worksheets in the same workbook, or to other workbooks or other files.

Navigation is not only a convenience, but also it provides a form of control for the workbook designer. Navigation is integral to our discussion of *"Design workbook layout with users in mind."* It permits control and influence over the user's movement and access to the workbook. For example, in a serious spreadsheet project it is essential to provide a table of contents on a single worksheet. The table of contents should contain a detailed list of the worksheets, a brief explanation of what is contained in the worksheet, and hyperlinks the user can use to access the various worksheets.

Organizations that use spreadsheet analysis are constantly seeking ways to incorporate best practices into operations. By standardizing the five general practices, you provide valuable guidelines for designing workbooks that have a useful and enduring life. Additionally, standardization will lead to a common "structure and look" that allows decision makers to focus more directly on the modeling content of a workbook, rather than the *noise* often caused by poor design and layout. The five best practices are summarized in Table 1.2.

Table 1.2 Five best practices for workbook deign

Think workbooks not worksheets—Spare the worksheet; spoil the workbook
Place variables and parameters in a central location—Every workbook needs a Brain
Design workbook layout with users in mind—User friendliness and designer control
Document workbook content and development—Insert text and comments liberally
Provide convenient workbook navigation—Beam me up Scotty

1.6 A Spreadsheet Makeover

Now let's consider a specific problem that will allow us to apply the best practices we have discussed. Our friends Julia and Ram are meeting several weeks after their initial encounter. It is early Sunday afternoon and they have just returned from running a 10 k race. The following discussion takes place after the run.

Julia: Ram, you didn't do badly on the run.
Ram: Thanks, but you're obviously being kind. I feel exhausted.
Julia: Speaking of exhaustion, remember that project I told you my boss dumped on my desk? Well, I have a spreadsheet that I think does a pretty good job of solving the problem. Can you take a look at it?
Ram: Sure. By the way, do you know that Prof. Gomez from our MBA has written a book on spreadsheet analysis? The old guy did a pretty good job of it too. I brought along a copy for you.
Julia: Thanks. I remember him as being pretty good at simplifying some tough concepts.
Ram: His first chapter discusses a simple way to think about spreadsheet structure and workbook design—workbook *feng shui* as he puts it. It's actually 5 best practices to consider in workbook design.
Julia: Maybe we can apply it to my spreadsheet?
Ram: Let's do it.

1.6.1 Julia's Business Problem—A Very Uncertain Outcome

Julia works for a consultancy, Market Focus International (MFI), which advises firms on marketing to American, ethic markets—Hispanic Americans, Armenian Americans, Chinese Americans, etc. One of her customers, Mid-Atlantic Foods Inc., a prominent food distributor in the Mid-Atlantic of the US, is considering the addition of a new product to their ethnic foods line—flour *tortillas*.[2] The firm is interested in a forecast of the financial effect of adding tortillas to their product lines. This is considered a controversial product line extension by some of the Mid-Atlantic's management, so much so, that one of the executives has dubbed the project *A Very Uncertain Outcome*.

Julia has decided to perform a **pro forma** (forecasted or projected) profit or loss analysis, with a relatively simple structure. (The profit or loss statement is one of the most important financial statements in business.) After interviews with the relevant individuals at the client firm, Julia assembles the important variables and relationships that she will incorporate into her spreadsheet analysis. These relationships are shown in Exhibit 1.1. The information collected reveals the considerable uncertainty involved in forecasting the success of the flour tortilla introduction. For example, the *Sales Revenue* (*Sales Volume* ∗ *Average Unit Selling Price*) forecast is based on three possible values of *Sales Volume* and three possible values of *Average Unit Selling Price*. This leads to nine (3 × 3) possible combinations of *Sales Revenue*. One combination of values leading to *Sales Revenue* is volume of 3.5 million units in

---

[2]A tortilla is a form of flat, unleavened bread popular in Mexico, parts of Latin America, and the U.S.

Sales Revenue— Sales Volume * Average Selling Price
      Sales Volume— (low- 2,000,000 / high- 5,000,000 / most likely- 3,500,000)
      Probability of Sales Volume— (low- 17.5% / high- 17.5% / most likely- 65%)
      Average Selling Price— (4, 5, or 6 with equal probability)

Cost of Goods Sold Expense— assumed to be a percent of the Sales Revenue- either 40% **or** 80% with
      equal probability

        Gross Margin—   **Sales Revenue- Cost of Goods Sold Expense**

Variable Operating Expenses—
      Sales Volume Driven (VOESVD)—
            Sales Revenue * VOESVD%
            VOESVD% is 10% if sales volume is low or most likely; 20% otherwise

      Sales Revenue Driven (VOESRD)— Sales Revenue * VOESRD%
            If Sales Volume is =2,000,000 VOESRD% is 15%
            If Sales Volume is =3,500,000 VOESRD% is 10%
            If Sales Volume is =5,000,000 VOESRD% is 7.5%

        Contribution Margin—   **Gross Margin - Variable Operating Expenses**

Fixed Expenses—
      Operating Expenses— $300,000

      Depreciation Expense— $250,000

        Operating Earnings (EBIT)— **Contribution Margin - Fixed Expenses**
                                       (Earnings before interest and taxes)

Interest Expense— $170,000

      Earnings before income tax (EBT)—   **Operating Earnings - Interest Expense**

Income Tax expense—   Progressive
                      23% Marginal tax rate for 1-5,000,000 EDT
                      34% Marginal tax rate >5,000,000 EBT

        Net Income—   **Earnings before income tax - Income Tax**
                                  (Bottom-line Profit)

Exhibit 1.1   A very uncertain outcome

sales and a selling unit price of $5, or total price of $16.5 million. Another source of
uncertainty is the percentage of the *Sales Revenue* used to calculate *Costs of Goods
Sold Expense*, either 40 or 80% with equal probability of occurrence. **Uncertainty**
in sales volume and sales price also affects the variable expenses. Volume driven
and revenue driven variable expenses are dependent on the uncertain outcomes of
Sales Revenue and *Sales Volume*.

    Julia's workbook appears in Exhibits 1.2 and 1.3. These exhibits provide details
on the cell formulas used in the calculations. Note that Exhibit 1.2 consists of a
single worksheet comprised of a *single* forecasted Profit or Loss scenario; that is,
she has selected a single value for the uncertain variables (the most likely) for her
calculations. The *Sales Revenue* in Exhibit 1.3 is based on sales of 3.5 million units,
the most likely value for volume, and a unit price of $5, the mid-point of equally
possible unit sales prices.

	A	B	C	D	E	F
			D3 ▾ (fx =3500000*5*0.6			
1						
2		Sales Revenue		17500000		
3		Costs of Goods Sold		10500000		
4			Gross Margin	7000000		
5		Variable Operating Expenses				
6			Sales Volume Driven	2625000		
7			Sales Revenue Driven	1750000		
8			Contribution Margin	2625000		
9		Fixed Expenses				
10			Operating Expenses	300000		
11			Depreciation Expenses	250000		
12			Operating Earnings (EBIT)	2075000		
13		Interest Expenses		170000		
14			Earnings Before Income Taxes	1905000		
15		Income Tax Expense		647700		
16			Net Income	1257300		

Exhibit 1.2   Julia's initial workbook

	A	B	C	D	E
		A1 ▾ (fx			
1					
2		Sales Revenue		=3500000*5	
3		Costs of Goods Sold		=3500000*5*0.6	
4			Gross Margin	=D2-D3	
5		Variable Operating Expenses			
6			Sales Volume Driven	=3500000*5*0.15	
7			Sales Revenue Driven	=3500000*5*0.1	
8			Contribution Margin	=D4-SUM(D6:D7)	
9		Fixed Expenses			
10			Operating Expenses	=300000	
11			Depreciation Expenses	=250000	
12			Operating Earnings (EBIT)	=D8-SUM(D10:D11)	
13		Interest Expenses		=170000	
14			Earnings Before Income Taxes	=D12-D13	
15		Income Tax Expense		=D14*0.34	
16			Net Income	=D14-D15	
17					
18					

Exhibit 1.3   Julia's initial workbook with cell formulas

Her calculation of *Cost of Goods Sold Expense* (COGS) is not quite as simple to determine. There are two equally possible percentages, 40 or 80%, that can be multiplied times the *Sales Revenue* to determine COGS. Rather than select one, she has decided to use a percentage value that is at the midpoint of the range, 60%. Thus, she has made some assumptions in her calculations that may need explanation to the client, yet there is no documentation of her reasons for this or any other assumption.

Additionally, in Exhibit 1.3 the inflexibility of the workbook is apparent—all parameters and variables are imbedded in the workbook formulas; thus, if Julia wants to make changes to these assumed values, it will be difficult to undertake. To make these changes quickly and accurately, it would be wiser to place these parameters in a central location—in a *Brain*—and have the cell formulas refer to this location. It is quite conceivable that the client will want to ask some **what-if** questions about her analysis. For example, what if the unit price range is changed from 4, 5 and 6 dollars to 3, 4, and 5 dollars; what if the most likely *Sales Volume* is raised to 4.5 million. Obviously, there are many more questions that could be asked and Ram will provide a formal critique of Julia's workbook and analysis that is organized around the 5 best practices. Julia hopes that by sending the workbook to Ram he will suggest changes to improve the workbook.

1.6.2 Ram's Critique

After considerable examination of the worksheet, Ram gives Julia his recommendations for a "spreadsheet makeover" in Table 1.3. He also makes some general analytical recommendations that he believes will improve the usefulness of the

Table 1.3 Makeover recommendations

General Comment—I don't believe that you have adequately captured the uncertainty associated with the problem. In most cases you have used a single value of a set, or distribution, of possible values—e.g. you use 3,500,000 as the Sales Volume. Although this is the most likely value, 2,000,000 and 5,000,000 have a combined probability of occurrence of 35% (a non-trivial probability of occurrence). By using the full range of possible values, you can provide the user with a view of the *variability* of the resulting "bottom line value-Net Income" in the form of a *risk profile*. This requires randomly selecting (random sampling) values of the uncertain parameters from their stated distributions. You can do this through the use of the RAND() function in Excel, and repeating these experiments many times, say 100 times. This is known as **Monte Carlo Simulation**. (Chaps. 7 and 8 are devoted to this topic.)

P1—The Workbook is simply a single spreadsheet. Although it is possible that an analysis would only require a single spreadsheet, I don't believe that it is sufficient for this complex problem, and certainly the customer will expect a more complete and sophisticated analysis.—*Modify the workbook to include more analysis, more documentation, and expanded presentation of results on separate worksheets.*

P2—There are many instances where variables in this problem are imbedded in cell formulas (see Exhibit 1.2 cell G3). The variables should have a separate worksheet location for quick access and presentation—a *Brain*. The cell formulas can then reference the cell location in the Brain to access the value of the variable or parameter. This will allow you to easily make changes in a single location and note the sensitivity of the model to these changes. If the client asks *what if* questions during your presentation of results, the current spreadsheet will be very difficult to use.—*Create a Brain worksheet.*

P3—The new layout that results from the changes I suggest, should include a number of user friendliness considerations—(1) *create a table of contents, (2) place important analysis on separate worksheets, and (3) place the results of the analysis into a graph that provides a "risk profile" of the problem results (see Exhibit 1.7).* Number (3) *is related to a larger issue of appropriateness of analysis (see General Comment).*

P4—Document the workbook to provide the user with information regarding the assumptions and form of analysis employed—*Use text boxes to provide users with information on assumed values (Sales Volume, Average Selling Price, etc.), use cell comments to guide users to cells where the input of data can be performed, and name cell ranges so formulas reflect directly the operation being performed in the cell.*

P5—Provide the user with navigation from the table of content to, and within, the various worksheets of the workbook—*Insert hypertext links throughout.*

workbook. Ram has serious misgivings about her analytical approach. It does not, in his opinion, capture the substantial *uncertainty* of her *A Very Uncertain Outcome* problem. Although there are many possible avenues for improvement, it is important to provide Julia with rapid and actionable feedback; she has a deadline that must be met for the presentation of her analytical findings. His recommendations are organized in terms of the 5 best practices (P1 = practice 1, etc):

1.6.3 Julia's New and Improved Workbook

Julia's initial reaction to Ram's critique is a bit guarded. She wonders what added value will result from applying the best practices to workbook and how the sophisticated analysis that Ram is suggesting will help the client's decision making. More importantly, she also wonders if she is capable of making the changes. Yet, she understands that the client is quite interested in the results of the analysis, and anything she can do to improve her ability to provide insight to this problem and, of course, sell future consulting services are worth considering carefully. With Ram's critique in mind, she begins the process of rehabilitating the spreadsheet she has constructed by concentrating on three issues: reconsideration of the overall analysis to provide greater insight of the uncertainty, structuring and organizing the analysis within the new multi-worksheet structure, and incorporating the 5 best practices to improve spreadsheet functionality.

In reconsidering the analysis, Julia agrees that a single-point estimate of the P/L statement is severely limited in its potential to provide Mid-Atlantic Foods with a broad view of the uncertainty associated with the extension of the product line. A **risk profile**, a distribution of the net income outcomes associated with the uncertain values of volume, price, and expenses, is a far more useful tool for this purpose. Thus, to create a risk profile it will be necessary to perform the following:

1. place important input data on a single worksheet that can be referenced ("*Brain*")
2. simulate the possible P/L outcomes on a single worksheet ("*Analysis*") by randomly selecting values of uncertain factors
3. repeat the process numerous times—100 (an arbitrary choice in this example)
4. collect the data on a separate worksheet ("*Data Collection Area*")
5. present the data in a graphical format on another worksheet ("*Graph-Risk Profile*")

This suggests three worksheets associated with the analysis ("*Analysis*", "*Data Collection Area*", and "*Graph-Risk Profile*"). If we consider the additional worksheet for the location of important parameter values ("*Brain*") and a location from which the user can navigate the multiple worksheets ("*Table of Contents*"), we are now up to a total of five worksheets. Additionally, Julia realizes that she has to avoid the issues of inflexibility we discussed above in her initial workbook (Exhibit 1.3). Finally, she is aware that she will have to automate the data collection process by creating a simple **macro** that generates simulated outcomes, captures the results, and stores 100 such results in worksheet. A macro is a computer program written in a simple language (**VBA**) that performs specific Excel programming tasks for the user, and it is beyond Julia's capabilities. Ram has skill in creating macros and has volunteered to help her.

Exhibit 1.4 presents the new five worksheet structure that Julia has settled on. Each of the colored tabs, a feature available in the Office XP version of Excel, represents a worksheet. The worksheet displayed, *T of C*, is the Table of Contents. Note that the underlined text items in the table are *hyperlinks* that transfer you to the various worksheets. Moving the cursor over the link will permit you to click the link and then automatically transfer you to the specified location. Insertion of a hyperlink is performed by selecting the icon in the menu bar that is represented by a globe and three links of a chain (see the Insert menu tab in Exhibit 1.4). When this Globe icon is selected, a dialog box will ask you where you would like the link to transfer the cursor, including questions regarding whether the transfer will be to this or other worksheets, or even other workbooks or files. This worksheet also provides documentation describing the project in a text box.

In Exhibit 1.5 Julia has created a Brain, which she has playfully entitled Señor (Mr.) Brain. We can see how data from her earlier spreadsheet (see Exhibit 1.1) is carefully organized to permit direct and simple referencing by formulas in the *Analysis* worksheet. If the client should desire a change to any of the assumed parameters, the Brain is the place to perform the change. Observing the sensitivity of the P/L outcomes to these changes is simply a matter of adjusting the relevant data elements in the Brain, and noting the new outcomes. Thus, Julia is prepared for the clients *what if* questions. In later chapters we will refer to this process as **Sensitivity Analysis**.

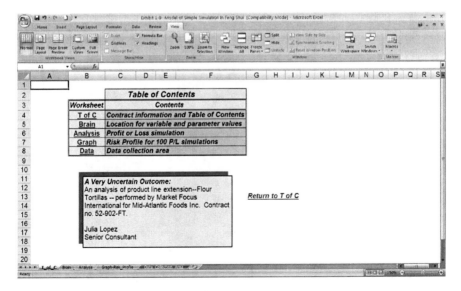

Exhibit 1.4 Improved workbook—table of contents

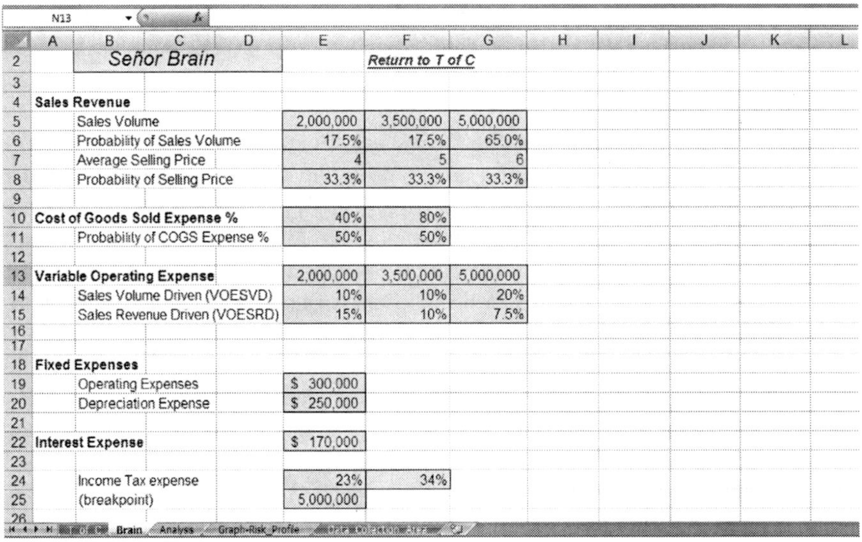

	N13		ƒx								
	A	B	C	D	E	F	G	H	I	J	K	L
2		*Señor Brain*				*Return to T of C*						
3												
4	Sales Revenue											
5		Sales Volume			2,000,000	3,500,000	5,000,000					
6		Probability of Sales Volume			17.5%	17.5%	65.0%					
7		Average Selling Price			4	5	6					
8		Probability of Selling Price			33.3%	33.3%	33.3%					
9												
10	Cost of Goods Sold Expense %				40%	80%						
11		Probability of COGS Expense %			50%	50%						
12												
13	Variable Operating Expense				2,000,000	3,500,000	5,000,000					
14		Sales Volume Driven (VOESVD)			10%	10%	20%					
15		Sales Revenue Driven (VOESRD)			15%	10%	7.5%					
16												
17												
18	Fixed Expenses											
19		Operating Expenses			$ 300,000							
20		Depreciation Expense			$ 250,000							
21												
22	Interest Expense				$ 170,000							
23												
24		Income Tax expense			23%	34%						
25		(breakpoint)			5,000,000							
26												

Brain / Analysis / Graph-Risk_Profile

Exhibit 1.5 Improved workbook—brain

The heart of the workbook, the *Analysis* worksheet in Exhibit 1.6, simulates individual scenarios of P/L *Net Income* based on randomly generated values of uncertain parameters. The determination of these uncertain values occurs off the screen image in columns N, O, and P. The values of sales volume, sales price, and COGS percentage are selected fairly (randomly) and used to calculate a *Net Income*. This can be thought of as a single scenario: a result based on a specific set of randomly selected

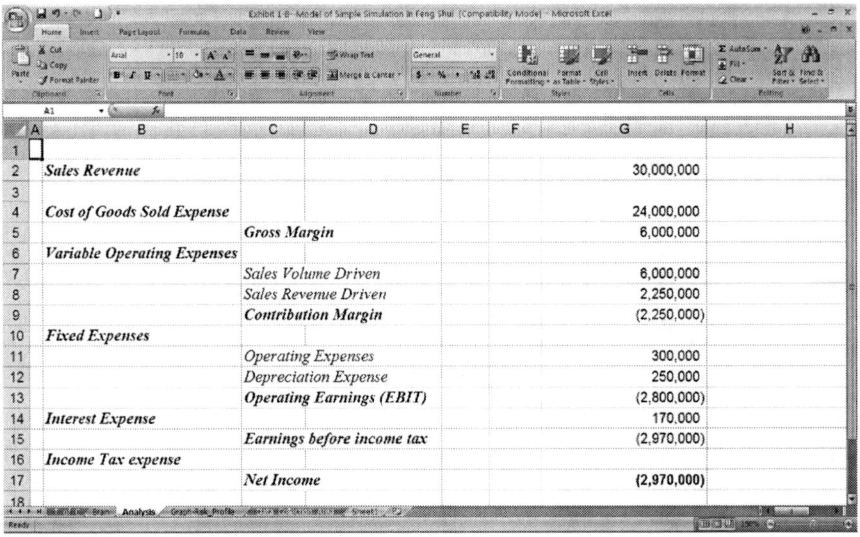

	A	B	C	D	E	F	G	H
1								
2		*Sales Revenue*					30,000,000	
3								
4		*Cost of Goods Sold Expense*					24,000,000	
5			*Gross Margin*				6,000,000	
6		*Variable Operating Expenses*						
7			*Sales Volume Driven*				6,000,000	
8			*Sales Revenue Driven*				2,250,000	
9			*Contribution Margin*				(2,250,000)	
10		*Fixed Expenses*						
11			*Operating Expenses*				300,000	
12			*Depreciation Expense*				250,000	
13			*Operating Earnings (EBIT)*				(2,800,000)	
14		*Interest Expense*					170,000	
15			*Earnings before income tax*				(2,970,000)	
16		*Income Tax expense*						
17			*Net Income*				(2,970,000)	
18								

Brain / Analysis / Graph-Risk_Profile

Exhibit 1.6 Improved workbook—analysis

variables. Then the process is repeated to generate new P/L outcome scenarios. All of this is managed by the macro that automatically makes the random selection, calculates new *Net Income*, and records the *Net Income* to a worksheet called *Data Collection Area*. The appropriate number of scenarios, or iterations, for this process is a question of simulation design. It is important to select a number of scenarios that reflect accurately the full behavior of the *Net Income*. Too few scenarios may lead to unrepresentative results, and too many scenarios can be costly and tedious to collect. Note that the particular scenario in Exhibit 1.6 shows a loss of 2.97 million dollars. This is a very different result from her simple analysis in Exhibit 1.2, where a profit of over $1,000,000 was presented. (More discussion of the proper number of scenarios can be found in Chaps. 7 and 8.)

In Exhibit 1.7, *Graph-Risk Profile*, simulation results (recorded in the Data Collection Area shown in Exhibit 1.8) are arranged into a frequency distribution by using the Data Analysis tool (more on this tool in Chaps. 2, 3, 4, and 5) available in the Data Tab. A frequency distribution is determined from a sample of variable values and provides the number of scenarios that fall into a relatively narrow range of *Net Income* performance; for example, a range from $1,000,000 to $1,500,000. By carefully selecting these ranges, also known as bins, and counting the scenarios falling in each, a profile of outcomes can be presented graphically. We often refer to these graphs as *Risk Profiles*. The title is appropriate given that the client is presented with both the positive (higher net income) and negative (lower net income) risk associated with the adoption of the flour tortilla product line.

It is now up to the client to take this information and apply some decision criteria to either accept or reject the product line. Those executives that are *not* predisposed to adopting the product line might concentrate on the negative potential outcomes. Note that in 46 of the 100 simulations the P/L outcome is a loss, with a substantial down side risk—31 observations are losses of more than 2 million dollars. This

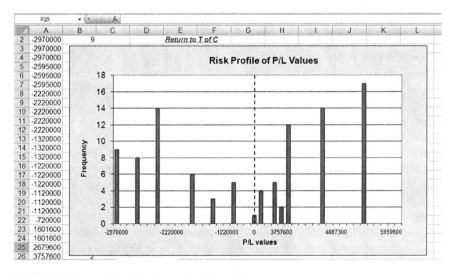

Exhibit 1.7   Improved workbook—graph-risk profile

	A	B	C	D	E	F	G	H	I	J	K	L
1			100 Simulations of P/L	hector.guerrero: Step 1								
2			$ 2,140,600	Place cursor here and go to "Analysis"								
3			$ (2,970,000)	work sheet to begin simulation for 100								
4			$ 5,959,800	experiments (P/L). Results will be placed								
5			$ 4,887,300	in the column below "100 Simulations of								
6			$ 1,601,600	P/L".								
7			$ (2,595,000)									
8			$ (720,000)									
9			$ 2,140,600									
10			$ 3,814,800		Return to T of C							
11			$ 2,140,600									
12			$ (2,970,000)									
13			$ 3,814,800									
14			$ 5,959,800									
15			$ (2,970,000)									
16			$ (2,970,000)									
17			$ 4,144,800									
18			$ (2,220,000)									
19			$ (2,970,000)									
20			$ 4,887,300									
21			$ (2,970,000)									

Exhibit 1.8 Improved workbook—data collection area

information can be gleaned from the risk profile or the frequency distribution that underlies the risk profile. Clearly the information content of the risk profile is far more revealing than Julia's original calculation of a single profit of $1,257,300, based on her selective use of specific parameter values. As a manager seeking as thorough an analysis as possible, there is little doubt that I would prefer the risk profile to the single scenario that Julia initially produced.

1.7 Summary

This example is one that is relatively sophisticated for the casual or first time user of Excel. Do not worry if you do not understand every detail of the simulation. It is presented here to help us focus on how a simple analysis can be extended and how our best practices can improve the utility of a spreadsheet analysis. In later chapters we will return to these types of models and you will see how such models can be constructed.

It is easy to convince oneself of the lack of importance of an introductory chapter of a textbook, especially one that in later chapters focuses on relatively complex analytical issues. Most readers often skip an introduction or skim the material in a casual manner, preferring instead to get the "real meat of the book." Yet, in my opinion this chapter may be one of the most important chapters of this book. With an understanding of the important issues in spreadsheet design, you can turn an ineffective, cumbersome, and unfocused analysis into one that users will hail as an "analytical triumph." Remember that spreadsheets are used by a variety of individuals in the organization, some at higher levels and some at lower levels. The design effort required to create a workbook that can easily be used by others and serve as a collaborative document by numerous colleagues is not an impossible goal to achieve, but it does require thoughtful planning and the application of a few simple, best practices. As we saw in our example, even the analysis of a relatively simple problem can be greatly enhanced by applying the five practices in Table 1.2. Of course, the significant change in the analytical approach is also important, and the remaining chapters of the book are dedicated to these analytical topics.

In the coming chapters we will continue to apply the five practices and explore the numerous analytical techniques that are contained in Excel. For example, in the next four chapters we examine the data analysis capabilities of Excel with quantitative (numerical—e.g. 2345.81 or 53%) and qualitative (categorical—e.g. male or Texas) data. We will also see how both quantitative and qualitative data can be presented in charts and tables to answer many important business questions; graphical data analysis can be very persuasive in decision making.

Key Terms

Spreadsheet Engineering
Best Practices
Feng Shui
User Friendliness
Hyperlinks
Pro Forma
Uncertainty
What-if
Monte Carlo Simulation
Risk Profile
Macro
VBA
Sensitivity Analysis

Problems and Exercises

1. Consider a workbook project that you or a colleague have developed in the past and apply the best practices of the *Feng Shui of Spreadsheets* to your old work book.
2. Create a workbook that has four worksheets—Table of Contents, Me, My Favorite Pet, and My Least Favorite Pet. Place hyperlinks on the Table of Contents to permit you to link to each of the pages and return to the Table of Contents. Insert a picture of yourself on the Me page and a picture of pets on the My Favorite Pet and My Least Favorite Pet page. Be creative and insert any text you like in text boxes explaining who you are and why these pets are your favorite and least favorite.
3. What is a *risk profile*? How can it be used for decision making?

4. Explain to a classmate or colleague why Best Practices in creating workbooks and worksheets are important.

5. *Advanced Problem*—An investor is considering the purchase of one to three condominiums in the tropical paradise of Costa Rica. The investor has no intention of using the condo for her personal use and is only concerned with the income producing capability that it will produce. After some discussion with a long time and real estate savvy resident of Costa Rica, the investor decides to perform a simple analysis of the operating profit/loss based on the following information:

	A	B	C
Variable Property Cost*	Based on:	Based on:	Based on:
	Most likely monthly occupancy of 20 day	Most likely monthly occupancy of 25 day	Most likely monthly occupancy of 15 day
	12 months per year operation	12 months per year operation	10 months per year operation
	2000 Colones per occupancy day cost	1000 Colones per occupancy day cost	3500 Colones per occupancy day cost
Fixed Property Cost*	3,000,000	2,500,000	4,500,000
Daily Revenue*	33,800	26,000	78,000

* All *Cost* and *Revenues* in **Colones**–520 Costa Rican **Colones/US Dollar**.

Additionally, the exchange rate may vary ±15%, and the most likely occupancy days can vary from a low and high of 15–25, 20–30, and 10–20 for A, B, and C, respectively. Based on this information create a workbook that determines the best case, most likely, and worse case annual cash flows for each of the properties.

Chapter 2
Presentation of Quantitative Data

Contents

2.1 Introduction

We often think of data as being strictly numerical values, and in business, those values are often stated in terms of dollars. Although data in the form of dollars are ubiquitous, it is quite easy to imagine other numerical units: percentages, counts in categories, units of sales, etc. This chapter, and Chap. 3, discusses how we can best use Excel's graphics capabilities to effectively present quantitative data (**ratio**

H. Guerrero, *Excel Data Analysis*, DOI 10.1007/978-3-642-10835-8_2,
© Springer-Verlag Berlin Heidelberg 2010

and **interval**), whether it is in dollars or some other quantitative measure, to inform and influence an audience. In Chaps. 4 and 5 we will acknowledge that not all data are numerical by focusing on qualitative (**categorical/nominal** or **ordinal**) data. The process of data gathering often produces a combination of data types, and throughout our discussions it will be impossible to ignore this fact: quantitative and qualitative data often occur together.

Unfortunately, the scope of this book does not permit in depth coverage of the data collection process, so I strongly suggest you consult a reference on data research methods before you begin a significant data collection project. I will make some brief remarks about the planning and collection of data, but we will generally assume that data has been collected in an efficient and effective manner. Now, let us consider the essential ingredients of good data presentation and the issues that can make it either easy or difficult to succeed. We will begin with a general discussion of data: how to classify it and the context or orientation within which it exists.

2.2 Data Classification

Skilled data analysts spend a great deal of time and effort in planning a data collection effort. They begin by considering the type of data they *can* and *will* collect in light of their goals for the use of the data. Just as carpenters are careful in selecting their tools, so are analysts in their choice of data. You cannot ask a *low precision* tool to perform *high precision* work. The same is true for data. A good analyst is cognizant of the types of analyses they can perform on various categories of data. This is particularly true in statistical analysis, where there are often rules for the types of analyses that can be performed on various types of data.

The standard characteristics that help us categorize data are presented in Table 2.1. Each successive category permits greater *measurement* precision and also permits more extensive statistical analysis. Thus, we can see from Table 2.1 that ratio data measurement is more precise than nominal data measurement. It is important to remember that all these forms of data, regardless of their classification, are valuable, and we collect data in different forms by considering availability and our analysis goals. For example, nominal data are used in many marketing studies, while ratio data are more often the tools of finance, operations, and economics; yet, all business functions collect data in each of these categories.

For nominal and ordinal data, we use non-metric measurement scales in the form of categorical properties or attributes. Interval and ratio data are based on metric measurement scales allowing a wide variety of mathematical operations to be performed on the data. The major difference between interval and ratio measurement scales is the existence of an absolute zero for ratio scales and arbitrary zero points for interval scales. For example, consider a comparison of the Fahrenheit and Celsius temperature scales. The zero points for these scales are arbitrarily set and do not indicate an "absolute absence" of temperature. Similarly, it is incorrect to suggest that 40° Celsius is half as hot as 80° Celsius. By contrast, it can be said that 16

Table 2.1 Data categorization

Data	Description	Properties	Examples
Nominal or Categorical Data	Data that can be placed into mutually exclusive categories	Quantitative relationships among and between data are meaningless and descriptive statistics are meaningless	Country in which you were born, a geographic region, your gender—these are either/or categories
Ordinal Data	Data are ordered or often ranked according to some characteristic	Categories can be compared to one another, but the difference in categories is generally meaningless and calculating averages is suspect	Ranking breakfast cereals—preferring cereal X *more* than Y implies nothing about *how much more* you like one versus the other
Interval Data	Data characterized and ordered by a specific distance between each observation, but having no natural zero	Ratios are meaningless, thus 15 degrees Celsius is not half as warm as 30 degrees Celsius	The Fahrenheit (or Celsius) temperature scale or consumer survey scales that are *specified* to be interval scales
Ratio data	Data that have a natural zero	These data have both ratios and differences that are meaningful	Sales revenue, time to perform a task, length, or weight

ounces of coffee is, in fact, twice as heavy as 8 ounces. Ultimately, the ratio scale has the highest information content of any of the measurement scales.

Just as thorough problem definition is essential to problem solving, careful selection of appropriate data categories is essential in a data collection effort. Data collection is an arduous and often costly task, so why not carefully plan for the use of the data prior to its collection? Additionally, remember that there are few things that will anger a cost conscious superior more than the news that you have to repeat a data collection effort.

2.3 Data Context and Data Orientation

The data that we collect and assemble for presentation purposes exists in a particular **data context** : a set of conditions or an environment related to the data. This context is important to our understanding of the data. We relate data to time (e.g. daily, quarterly, yearly, etc.), to categorical treatment (e.g. an economic downturn, sales in Europe, etc.), and to events (e.g. sales promotions, demographic changes, etc.). Just as we record the values of quantitative data, we also record the context of data— e.g. revenue generated by product A, in quarter B, due to salesperson C, in sales territory D. Thus, associated with the quantitative data element that we record are numerous other important data elements that may, or may not, be quantitative.

Sometimes the context is obvious, sometimes the context is complex and difficult to identify, and often, there is more than a single context that is essential to consider. Without an understanding of the data context, important insights related to the data can be lost. To make matters worse, the context related to the data may change or reveal itself only after substantial time has passed. For example, consider data which indicates a substantial loss of value in your stock portfolio, recorded from 1990 to 2008. If the only context that is considered is time, it is possible to ignore a host of important contextual issues—e.g. the bursting of the dot-com *bubble* of the late 1990s. Without knowledge of this event context, you may simply conclude that you are a poor stock picker.

It is impossible to anticipate all the elements of data context that should be collected, but whatever data we collect should be sufficient to provide a context that suits our needs and goals. If I am interested in promoting the idea that the revenues of my business are growing over time and growing only in selected product categories, I will assemble time oriented revenue data for the various products of interest. Thus, the related dimensions of my revenue data are time and product. There may also be an economic context, such as demographic conditions that may influence particular types of sales. Determining the contextual dimensions that are important will influence what data we collect and how we present it. Additionally, you can save a great deal of effort and *after the fact* data adjustment by carefully considering in advance the various dimensions that you will need.

Consider the owner of a small business that is interested in recording expenses in a variety of accounts for cash flow management, income statement preparation, and tax purposes. This is an important activity for any small business. Cash flow is the life blood of these businesses, and if it is not managed well, the results can be catastrophic. Each time the business owner incurs an expense, he either collects a receipt (upon final payment) or an invoice (a request for payment). Additionally, suppliers to small businesses often request a deposit that represents a form of partial payment and a commitment to the services provided by the supplier.

An example of these data is shown in the worksheet in Table 2.2. Each of the primary data entries, referred to as **records**, contain important and diverse dimensions referred to as **fields**—date, amount, nature of the expense, names, addresses, and an occasional hand entered comment, etc. A record represents a single observation of the collected data fields, as in item 3 (printing on 1/5/2004) of Table 2.2. This record contains 7 fields—Printing, $2,543.21, 1/5/2004, etc.—and each record is a row in the worksheet.

Somewhere in our business owner's office is an old shoebox that is the final resting place for his primary data. It is filled with scraps of paper: invoices and receipts. At the end of each week our businessperson empties the box and records what he believes to be the important elements of each receipt or invoice. Table 2.2 is an example of the type of data that the owner might collect from the receipts and invoices over time. The receipts and invoices can contain more data than needs to be recorded or used for analysis and decision making. The dilemma the owner faces

Table 2.2 Payment example

Item	Account	$ Amount	Date Rcvd.	Deposit	Days to Pay	Comment
1	Office Supply	$123.45	1/2/2004	$10.00	0	Project X
2	Office Supply	$54.40	1/5/2004	$0.00	0	Project Y
3	Printing	$2,543.21	1/5/2004	$350.00	45	Feb. Brochure
4	Cleaning Service	$78.83	1/8/2004	$0.00	15	Monthly
5	Coffee Service	$56.92	1/9/2004	$0.00	15	Monthly
6	Office Supply	$914.22	1/12/2004	$100.00	30	Project X
7	Printing	$755.00	1/13/2004	$50.00	30	Hand Bills
8	Office Supply	$478.88	1/16/2004	$50.00	30	Computer
9	Office Rent	$1,632.00	1/19/2004	$0.00	15	Monthly
10	Fire Insurance	$1,254.73	1/22/2004	$0.00	60	Quarterly
11	Cleaning Service	$135.64	1/22/2004	$0.00	15	Water Damage
12	Orphan's Fund	$300.00	1/27/2004	$0.00	0	Charity*
13	Office Supply	$343.78	1/30/2004	$100.00	15	Laser Printer
14	Printing	$2,211.82	2/4/2004	$350.00	45	Mar. Brochure
15	Coffee Service	$56.92	2/5/2004	$0.00	15	Monthly
16	Cleaning Service	$78.83	2/10/2004	$0.00	15	Monthly
17	Printing	$254.17	2/12/2004	$50.00	15	Hand Bills
18	Office Supply	$412.19	2/12/2004	$50.00	30	Project Y
19	Office Supply	$1,467.44	2/13/2004	$150.00	30	Project W
20	Office Supply	$221.52	2/16/2004	$50.00	15	Project X
21	Office Rent	$1,632.00	2/18/2004	$0.00	15	Monthly
22	Police Fund	$250.00	2/19/2004	$0.00	15	Charity
23	Printing	$87.34	2/23/2004	$25.00	0	Posters
24	Printing	$94.12	2/23/2004	$25.00	0	Posters
25	Entertaining	$298.32	2/26/2004	$0.00	0	Project Y
26	Orphan's Fund	$300.00	2/27/2004	$0.00	0	Charity
27	Office Supply	$1,669.76	3/1/2004	$150.00	45	Project Z
28	Office Supply	$1,111.02	3/2/2004	$150.00	30	Project W
29	Office Supply	$76.21	3/4/2004	$25.00	0	Project W
30	Coffee Service	$56.92	3/5/2004	$0.00	15	Monthly
31	Office Supply	$914.22	3/8/2004	$100.00	30	Project X
32	Cleaning Service	$78.83	3/9/2004	$0.00	15	Monthly
33	Printing	$455.10	3/12/2002	$100.00	15	Hand Bills
34	Office Supply	$1,572.31	3/15/2002	$150.00	45	Project Y
35	Office Rent	$1,632.00	3/17/2002	$0.00	15	Monthly
36	Police Fund	$250.00	3/23/2002	$0.00	15	Charity
37	Office Supply	$642.11	3/26/2002	$100.00	30	Project W
38	Office Supply	$712.16	3/29/2002	$100.00	30	Project Z
39	Orphan's Fund	$300.00	3/29/2002	$0.00	0	Charity

is the amount and type of data to record in the worksheet: recording too much data can lead to wasted effort and neglect of other important activities, and recording too little data can lead to overlooking important business issues.

What advice can we provide our businessperson that might make their efforts in collecting, assembling, and recording data more useful and efficient? Below I provide a number of guidelines that can make the process of planning for a data collection effort straightforward.

2.3.1 Data Preparation Advice

1. *Not all data are created equal*—Spend some time and effort considering the category of data (nominal, ratio, etc.) that you will collect and how you will use it. Do you have choices in the categorical type of data you can collect? How will you use the data in analysis and presentation?
2. *More is better*—If you are uncertain of the specific dimensions of a data observation that you will need for analysis, err on the side of recording a greater number of dimensions (more information on the context). It is easier not to use collected data than to add the un-collected data later. Adding data later can be costly and assumes that you will be able to locate it, which may be difficult or impossible.
3. *More is **not** better*—If you can communicate what you need to communicate with less data, then by all means do so. Bloated databases can lead to distractions and misunderstanding. With new computer memory technology the cost of data storage is declining rapidly, but there is still a cost to data entry, storage, and of archiving records for long periods of time.
4. *Keep it simple and columnar*—Select a simple, unique title for each data dimension or field (e.g. Revenue, Address, etc.) and record the data in a column, with each row representing a record, or observation, of recorded data. Each column or field represents a different dimension of the data. Table 2.2 is a good example of columnar data entry for seven data fields.
5. *Comments are useful*—It may be wise to place a *miscellaneous* dimension or field reserved for written observations—a **comment field**. Be careful, because of their unique nature, comments are often difficult, if not impossible, to query via structured database query languages. Try to pick key words for entry (*overdue*, *lost sale*, etc.) if you plan to later query the field.
6. *Consistency in category titles*—Although you may not consider a significant difference between the category titles *Deposit* and *$Deposit*, Excel will view them as completely distinct field titles. Excel is not capable of understanding that the terms may be synonymous in your mind.

Let's examine Table 2.2 in light of the data preparation advice we have just received. But first, let's take a look at a typical invoice and the data that it might contain. Exhibit 2.1 shows an invoice for office supply items purchased at Hamm

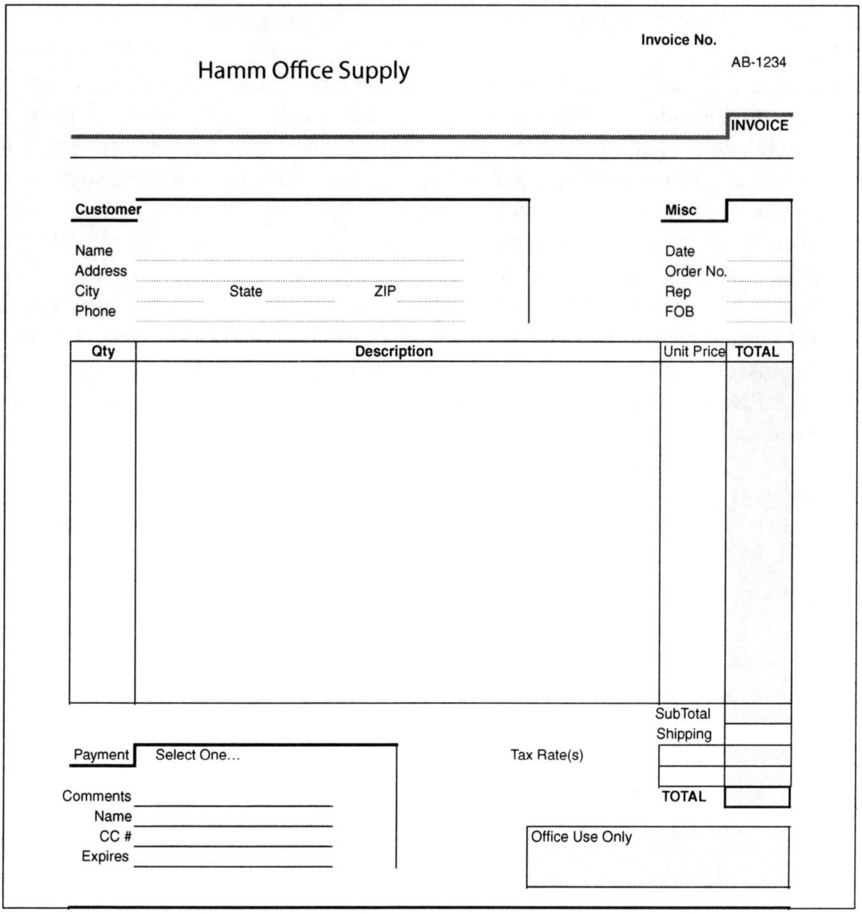

Invoice No.
Hamm Office Supply
AB-1234
INVOICE

Customer

Name
Address
City State ZIP
Phone

Misc

Date
Order No.
Rep
FOB

Qty	Description	Unit Price	TOTAL
		SubTotal	
		Shipping	
	Tax Rate(s)		
		TOTAL	

Payment Select One...

Comments
Name
CC #
Expires

Office Use Only

Exhibit 2.1 Generic invoice

Office Supply, Inc. Note the amount of data that this generic invoice (an MS Office Template) contains is quite substantial: approximately 20 fields. Of course, some of the data are only of marginal value, such as our address—we know that the invoice was intended for our firm and we know where we are located. Yet, it is verification that the Hamm invoice is in fact intended for our firm. Notice that each line item in the invoice will require *multiple* item entries—qty (quantity), description, unit price, and total. Given the potential for large quantities of data, it would be wise to consider a **relational database**, such as MS Access, to optimize data entry effort. Of course, even if the data are stored in a relational database, that does not restrict us from using Excel to analyze the data by downloading data from Access to Excel; in fact, this is a wonderful advantage of the Office suite.

Now for our examination of the data in Table 2.2 in light of our advice:

1. *Not all data are created equal*— Our businessperson has assembled a variety of
   data dimensions or fields to provide the central data element ($ Amount) with
   ample context and orientation. The 7 fields that comprise each record appear to
   be sufficient for the businessperson's goal of recording the expenses and describing the context associated with his business operation. This includes recording
   each expense to ultimately calculate annual profit or loss, tracking particular
   expenses associated with projects or other uses of funds (e.g. charity), and the
   timing of expenses (Date Rcvd., Days to Pay, etc.) and subsequent cash flow. If
   the businessperson expands his examination of the transactions, some data may
   be missing, for example Order Number or Shipping Cost. Only the future will
   reveal if these data elements will become important, and for now, these data are
   not collected.
2. *More is better*—The data elements that our businessperson has selected may not
   all be used in our graphical presentation, but this could change in the future.
   Better to collect a little too much data initially than to perform an extensive
   collection of data at a later date. Those invoices and scraps of paper representing
   primary data may be difficult to find or identify in 3 months.
3. *More is **not** better*—Our businessperson has carefully selected the data that he
   feels is necessary without creating excessive data entry effort.
4. *Keep it simple and columnar*—Unique and simple titles for the various data
   dimensions (e.g. Account, Date Rcvd., etc.) have been selected and arranged
   in columnar fashion. Adding, inserting, or deleting a column is virtually costless
   for even an unskilled Excel user.
5. *Comments are useful*—The Comment field has been designated for the specific
   project (e.g. Project X), source item (e.g. Computer), or other important information (e.g. Monthly charge). If any criticism can be made here, it is that maybe
   these data elements deserve a title other than Comment. For example, entitle this
   data element *Project/Sources of Expense* and use the Comment title as a less
   structured data category. These could range from comments relating to customer
   service experiences, to information on possible competitors that provide similar
   services.
6. *Consistency in category titles*—Although you may not consider there to be
   a significant difference between the account titles Office Supply and Office
   Supplies, Excel will view them as completely distinct accounts. Our businessperson appears to have been consistent in the use of account types and comment
   entries. It is not unusual for these entries to be converted to numerical codes, for
   example, replacing Printing with account code 351.

2.4 Types of Charts and Graphs

There are literally hundreds of types of **charts** and **graphs** (these are synonymous
terms) available in Excel. Thus, the possibilities for selecting a presentation format are both interesting and daunting. What graph type is best for my needs? Often

the answer is that more than one type of graph will perform the presentation goal required; thus, the selection is a matter of your taste or that of your audience. Therefore, it is convenient to divide the problem of selecting a presentation format into two parts: the actual data presentation and the *embellishment* that will surround it. In certain situations we choose to do as little embellishment as possible; in others, we find it necessary to dress the data presentation in lovely colors, backgrounds, and labeling. To determine how to blend these two parts, ask yourself few simple questions:

1. What is the purpose of the data presentation? Is it possible to show the data without embellishment or do you want to attract attention through your presentation *style*? In a business world where people are exposed to many, many presentations, it may be necessary to do something extraordinary to gain attention or simply conform to the norm.
2. At what point does my embellishment of the data become distracting? Does the embellishment cover or conceal the data? Don't forget that from an information perspective it is *all* about the data, so don't detract from its presentation by adding superfluous and distracting adornment.
3. Am I being true to my taste and style of presentation? This author's taste in formatting is guided by some simple principles that can be stated in a number of familiar laws: *less is more, small is beautiful*, and *keep it simple*. As long as you are able to deliver the desired information and achieve your presentation goal, there is no problem with our differences in taste.
4. Formatting should be consistent among graphs in a workbook.

2.4.1 Ribbons and the Excel Menu System

So how do we put together a graph or chart? In pre-2007 Excel an ingenious tool called a **Chart Wizard** is available to perform these tasks. As the name implies, the Chart Wizard guides you through standardized steps, 4 to be exact, that take the guesswork out of creating charts. If you follow the 4 steps it is almost fool proof, and if you read all the options available to you for each of the 4 steps it will allow you to create charts very quickly. In Excel 2007 the wizard has been replaced because of a major development in the Excel 2007 user interface—**ribbons**. Ribbons replace the old hierarchical pull-down menu system that was the basis for user interaction with Excel. Ribbons are menus and commands organized in **tabs** that provide access to the functionality for specific uses. Some of these will appear familiar to pre-Excel 2007 users and others will not—Home, Insert, Page Layout, Formulas, Data, Review, and View. Within each tab you will find **groups** of related functionality and commands. Additionally, some menus specific to an activity, for example the creation of a graph or chart, will appear as the activity is taking place. For those just beginning to use Excel 2007 and with no previous exposure to Excel, you will probably find the menu system quite easy to use; for those with prior experience with Excel, the transition may be a bit frustrating at times. I have found the new

system quite useful, in spite of the occasional difficulty of finding functionality that
I was accustomed to before Excel 2007. Exhibit 2.2 shows the Insert tab where the
Charts group is found.

In this exhibit, a very simple graph of six data points for two data series,
data1 and *data2*, is shown as two variations of the column graph. One also displays the data used to create the graph. Additionally, since the leftmost graph
has been selected, indicated by the border that surrounds the graph, a group
of menus appear at the top of the ribbon—**Chart Tools**. These tools contain
menus for Design, Layout, and Format. This group is relevant to the creation of
a chart or graph. Ultimately, ribbons lead to a flatter, or less hierarchical, menu
system.

Our first step in chart creation is to organize our data in a worksheet. In Exhibit
2.2 the six data points for the two series have a familiar columnar orientation and
have titles, *data1* and *data2*. By capturing the **data range** containing the data that
you intend to chart before engaging the charts group in the Insert tab, you automatically identify the data to be graphed. Note that this can, but need not, include
the column title of the data specified as text. By capturing the title, the graph will
assume that you want to name the data series the same as title selected. If you place
alphabetic characters, a through f, in the first column of the captured data, the graph
will use these characters as the x-axis of the chart.

If you prefer not to capture the data prior to engaging the charts group, you can
either: (1) open and capture a blank chart type and *copy* the data and *paste* the
data to the blank chart type, or (2) use a right click of your mouse to *select data*.
Obviously, there will be numerous detailed steps to capturing data and labeling the

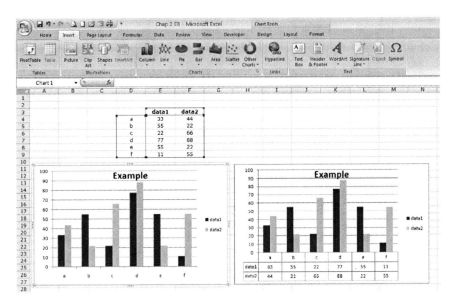

Exhibit 2.2   Insert tab and excel chart group

graph appropriately. We defer a detailed example of creating graphs using the chart group for the next section.

2.4.2 Some Frequently Used Charts

It is always dangerous to make bold assertions, but it is generally understood that the *mother of all graphs* is the **Column** or **Bar chart**. They differ only in their vertical and horizontal orientation, respectively. They easily represent the most often occurring data situation: some observed numerical variable that is measured in a single dimension (often time). Consider a simple set of data related to five products (A–E) and their sales over a two year period of time, measured in millions of dollars. The first four quarters represent year 1 and the second four year 2. These data are shown in Table 2.3. Thus, in quarter 1 of the second year, sales for product B results in sales of $49,000,000.

A quick visual examination of the data in Table 2.3 reveals that the product sales are relatively similar in magnitude (less than 100), but with differences in quarterly increases and decreases within the individual products. For example, product A varies substantially over the 8 quarters, while product D shows relatively little variation. Additionally, it appears that when product A shows high sales in early quarters (1 and 2), product E shows low sales in early quarters—they appear to be somewhat negatively correlated, although a graph may reveal more conclusive information. **Negative correlation** implies that one data series moves in the opposite direction from another; **positive correlation** suggests that both series move in the same direction. In later chapters we will discuss statistical correlation in greater detail.

Let's experiment with a few chart types to examine the data and tease out insights related to product A–E sales. The first graph, Exhibit 2.3, displays a simple Column chart of sales for the 5 product series in each of 8 quarters. The relative magnitude of the 5 products in a quarter is easily observed, but note that the 5 product series are difficult to follow through time, despite the color coding. It is difficult to concentrate solely on a single series, for example Product A, through time.

Table 2.3   Sales* data for products A–E

Quarter	A	B	C	D	E
1	98	45	64	21	23
2	58	21	45	23	14
3	23	36	21	31	56
4	43	21	14	30	78
1	89	49	27	35	27
2	52	20	40	40	20
3	24	43	58	37	67
4	34	21	76	40	89

* in millions of dollars

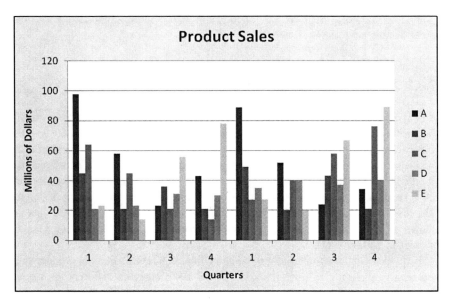

Exhibit 2.3 Column chart for products A–E

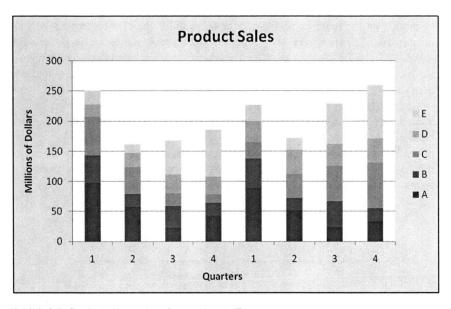

Exhibit 2.4 Stacked column chart for products A–E

In Exhibit 2.4 the chart type used is a **Stacked Column**. This graph provides a view not only of the individual product sales, but also of the quarterly totals. By observing the absolute height of each stacked column, one can see that total product sales in quarter 1 of year 1 (horizontal value 1) are greater that quarter

2 of year 1 (horizontal value 5). The relative size of each color within a column provides information of the sales quantities for each product in the quarter. For our data, the Stacked Column chart provides visual information about quarterly totals that is easier to discern. Yet, it still remains difficult to track the quarterly changes within products and among products over time. For example, it would be difficult to determine if product D is greater or smaller in quarter 3 or 4 of year 1, or to determine the magnitude of each.

Next, Exhibit 2.5 demonstrates a **3-D Column** (3 dimensional) chart. This is a visually impressive graph due to the 3-D effect, but much of the information relating to time based behavior of the products is lost due to the inability to clearly view columns hidden by other columns. The angle of perspective for 3-D graphs can be changed to remedy this problem partially, but if a single graph is used to chart many data series, they can still be difficult, or impossible, to view.

Now, let us convert the chart type to a **Line chart** and determine if there is an improvement or difference in the visual interpretation of the data. Before we begin, we must be careful to consider what we mean by an improvement, because an improvement is only an improvement relative to a goal that we establish for data presentation. For example, consider the goal that the presentation portrays the changes in each product's sales over quarters. Thus, we will want to use a chart that easily permits the viewer's eye to follow the quarterly related change in each specific series. Line charts will probably provide a better visual presentation of the data than Column charts, especially in **time series data,** if this is our goal.

Exhibit 2.6 shows the 5 product data in a simple and direct format. Note that the graph provides information in the three areas we have identified as important: (1) the relative value of a product's sales to other products *within* each quarter, (2) the relative value of a product's sales to other products *across* quarters, and

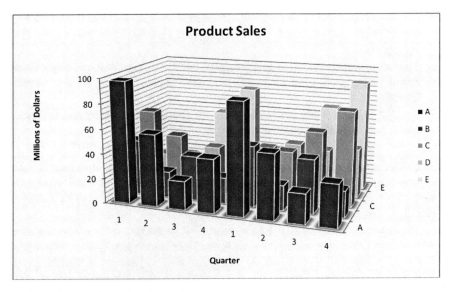

Exhibit 2.5   3-D column chart for products A–E

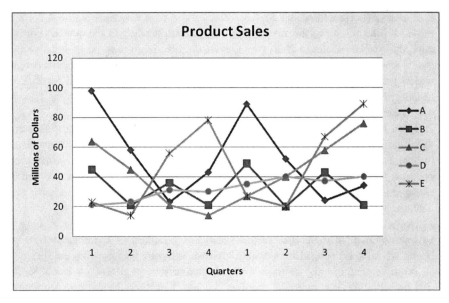

Exhibit 2.6 Line chart for products A–E

(3) the behavior of the individual product's sales over quarters. The line graph provides some interesting insights related to the data. For example:

1. Products A and E both appear to exhibit seasonal behavior that achieves highs and lows approximately every 4 quarters (e.g. highs in quarter 1 for A and quarter 4 for E).
2. The high for product E is offset approximately one quarter from that of the high for A. Thus, the peak in sales for Product A **lags** (occurs later) the peak for E by one quarter.
3. Product D seams to show little seasonality, but does appear to have a slight **linear trend** (increases at a constant rate). The trend is positive; that is, sales increase over time.
4. Product B has a stable pattern of quarterly alternating increases and decreases, and it may have slight positive trend from year 1 to year 2.

Needless to say, line graphs can be quite revealing, even if the behavior is based on scant data. Yet, we must also be careful not to convince ourselves of **systematic behavior** (regular or predictable) based on little data; more data may be needed to convince ourselves of true systematic behavior.

Finally, Exhibit 2.7 is also a Line graph, but in 3-D. It suffers from the same visual obstructions that we experienced in the 3-D Column graph—possibly appealing from a visual perspective, but providing less information content than the simple line graph in Exhibit 2.6 due to the obstructed view. It is difficult to see values of

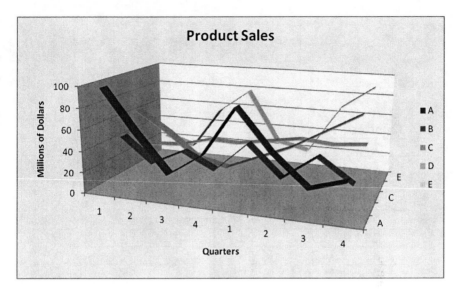

Exhibit 2.7 3-D line chart for products A–E

product E (the rear-most line) in early quarters. As I stated earlier, *simple* graphs are often *better* from a presentation perspective.

2.4.3 Specific Steps for Creating a Chart

We have seen the results of creating a chart in Exhibits 2.3, 2.4, 2.5, 2.6, and 2.7. Now let us create a chart, from beginning to end, for Exhibit 2.6, the Line chart for all products. The process we will use includes the following steps: (1) select a chart type, (2) identify the data to be charted including the x-axis, and (3) provide titles for the axes, series, and chart. For step 1, Exhibit 2.8 shows the selection of the Line chart format within the Charts group of the Insert tab. Note that there are also custom charts that are available for specialized circumstances. In pre-2007 Excel, these were a separate category of charts, but in 2007, they have been incorporated into the Format options.

The next step, selection of the data, has a number of possible options. The option shown in Exhibit 2.9 is one in which a blank chart type is selected and the chart is engaged. A right click of the mouse produces a set of options, including *Select Data*. The chart type can be selected and the data range *copied* into the blank chart (one with a border appearing around the chart as in Exhibit 2.9). By capturing the data range, including the titles (B1:F9), a series title (A, B, etc.) is automatically provided. Alternatively, if the data range had been selected prior to selecting the chart type, the data also would have been automatically captured in the line chart.

Note that the X-axis (horizontal) for Exhibit 2.9 is represented by the quarters of each of the two years, 1–4 for each year. In order to reflect this in the chart, you

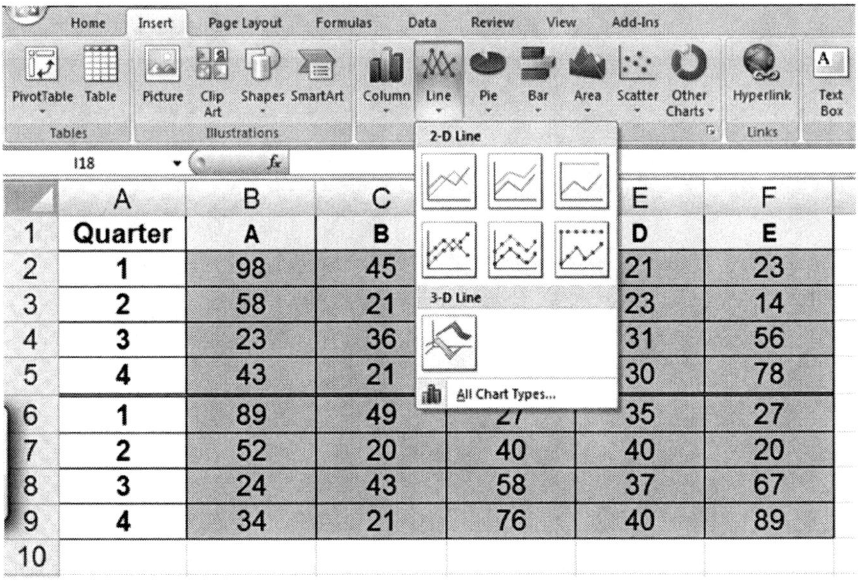

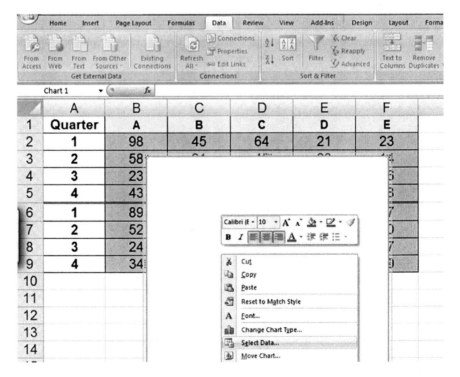

Exhibit 2.8 Step 1-selection of line chart from charts group

Exhibit 2.9 Step 1-selection of data range for product data

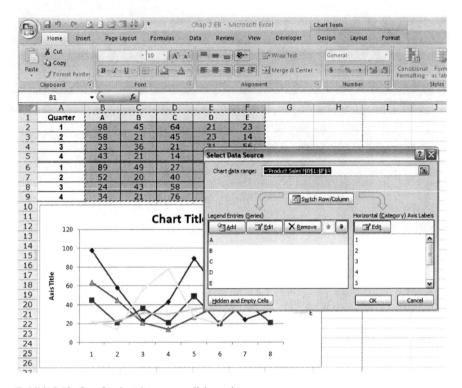

Exhibit 2.10 Step 2-select data source dialogue box

must specify the range where the axis labels are located. You can see that our axis labels are located in range A2:A9.

In Exhibit 2.10 we can see the partially completed chart. A right click on the chart area permits you to once again use the Select Data function. The dialogue box that appears permits you to select the new **Horizontal (Category) Axis Labels**. By depressing the Edit button in the Horizontal Axis Labels window, you can capture the appropriate range (A2:A9) to change the x-axis. This is shown in Exhibit 2.11.

Step 3 of the process permits titles for the chart and axes. Exhibit 2.12 shows the selection of layout in the Design tab Charts Layout group. The Layout and the Format tabs provide many more options for customizing the look of the chart to your needs.

As we mentioned earlier, many details for a chart can be handled by pointing and right clicking; for example, **Select Data**, **Format Chart Area**, and **Chart Type** changes. Selecting a particular part of the graph or chart with a left click, for example an axis or a **Chart Title**, then right clicking, also permits changes to the look of the chart or changes in the axes scale, pattern, font, etc. I would suggest that you take a simple data set, similar to the one I have provided, and experiment with all the options available. Also, try the various chart types to see how the data can be displayed.

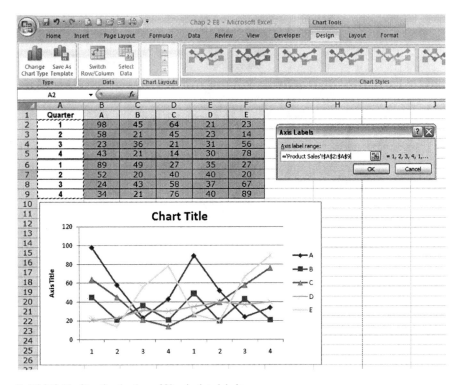

Exhibit 2.11 Step 3-selection of X-axis data labels

2.5 An Example of Graphical Data Analysis and Presentation

Before we begin a full scale example of graphical data analysis and presentation, let's consider the task we have before us. We are engaging in an exercise, **Data Analysis**, which can be organized into 4 basic activities: *collecting, summarizing, analyzing,* and *presenting* data. Our example will be organized into each of these steps, all of which are essential to successful graphical data analysis.

Collecting data not only involves the act of gathering, but also includes careful planning for the types of data to be collected (interval, ordinal, etc.). Data collection can be quite costly; thus, if we gather the wrong data or omit necessary data, we may have to make a costly future investment to repeat this activity. Some important questions we should ask before collecting are:

1. What data are necessary to achieve our analysis goals?
2. Where and when will we collect the data?
3. How many and what types of data elements related to an observation (e.g. customer name, date, etc.) are needed to describe the context or orientation? For example, each record of the 39 total in Table 2.2 represents an invoice or receipt observation with 7 data fields with nominal, interval, and ratio data elements.

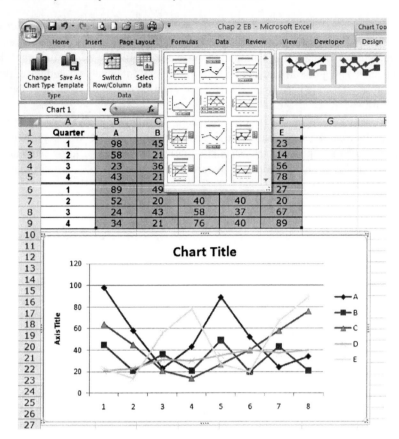

Exhibit 2.12   Step 3-chart design, layout, and format

Summarizing data can be as simple as placing primary data elements in a worksheet, but also it can include a number of modifications that make the data more useable. For example, if we collect data related to a date (1/23/2013), should the date also be represented as a day of the week (Monday, etc.)? This may sound redundant since a date implies a day of the week, but the data collector must often make these conversions of the data. Summarizing prepares the data for the analysis that is to follow. It is also possible that during analysis the data will need further summarization or modification to suit our goals.

There are many techniques for **analyzing** data. Not surprisingly, valuable analysis can be performed by simply **eyeballing** (careful visual examination) the data. We can place the data in a table, make charts of the data, and look for patterns of behavior or movement in the data. Of course, there are also formal mathematical techniques of analyzing data with descriptive or inferential statistics. Also, we can use powerful modeling techniques like Monte Carlo simulation and constrained optimization for analysis. We will see more of these topics in later chapters.

Once we have collected, summarized, and analyzed our data we are ready for **presenting** data results. Much of what we have discussed in this chapter is related to graphical presentation of data and represents a distillation of our understanding of the data. The goal of presentation is to inform and influence our audience. If our preliminary steps are performed well, the presentation of results should be relatively straight forward.

With this simple model in mind—collect, summarize, analyze, and present— let's apply what we have learned to an example problem. We will begin with the collection of data, proceed to a data summarization phase, perform some simple analysis, and then select various graphical presentation formats that will highlight the insights we have gained.

2.5.1 Example—Tere's Budget for the 2nd Semester of College

This example is motivated by a concerned parent, Dad, monitoring the second semester college expenditures for his daughter, Tere. Tere is in her first year of university. In the 1st semester, Tere's expenditures far exceeded Dad's planned budget. Therefore, Dad has decided to monitor how much she spends during the 2nd semester. The 2nd semester will constitute a data collection period to study expenditures. Dad is skilled in data analysis and what he learns from this semester will become the basis of his advice to Tere regarding her future spending. Table 2.4 provides a detailed breakdown of the expenditures that result from the 2nd semester, specifically 60 expenditures that Tere incurred. Dad has set as his goal for the analysis the determination of how and why expenditures occur over time. The following sections take us step by step through the data analysis process, with special emphasis on the presentation of results.

Table 2.4 2nd semester university student expenses

Obs.	Week	Date	Weekday	Amount	Cash/CRedit Card	Food/Personal/School
1	Week 01	6-Jan	Sn	111.46	R	F
2		7-Jan	M	43.23	C	S
3		8-Jan	T	17.11	C	S
4		10-Jan	Th	17.67	C	P
5	Week 02	13-Jan	Sn	107.00	R	F
6		14-Jan	M	36.65	C	P
7		14-Jan	M	33.91	C	P
8		17-Jan	Th	17.67	C	P
9		18-Jan	F	41.17	R	F
10	Week 03	20-Jan	Sn	91.53	R	F
11		21-Jan	M	49.76	C	P
12		21-Jan	M	32.97	C	S
13		22-Jan	T	14.03	C	P
14		24-Jan	Th	17.67	C	P
15		24-Jan	Th	17.67	C	P

Table 2.4   (continued)

Obs.	Week	Date	Weekday	Amount	Cash/CRedit Card	Food/Personal/ School
16	*Week 04*	27-Jan	Sn	76.19	R	F
17		31-Jan	Th	17.67	C	P
18		31-Jan	Th	17.67	C	P
19		1-Feb	F	33.03	R	F
20	*Week 05*	3-Feb	Sn	66.63	R	F
21		5-Feb	T	15.23	C	P
22		7-Feb	Th	17.67	C	P
23	*Week 06*	10-Feb	Sn	96.19	R	F
24		12-Feb	T	14.91	C	P
25		14-Feb	Th	17.67	C	P
26		15-Feb	F	40.30	R	F
27	*Week 07*	17-Feb	Sn	96.26	R	F
28		18-Feb	M	36.37	C	S
29		18-Feb	M	46.19	C	P
30		19-Feb	T	18.03	C	P
31		21-Feb	Th	17.67	C	P
32		22-Feb	F	28.49	R	F
33	*Week 08*	24-Feb	Sn	75.21	R	F
34		24-Feb	Sn	58.22	R	F
35		28-Feb	Th	17.67	C	P
36	*Week 09*	3-Mar	Sn	90.09	R	F
37		4-Mar	M	38.91	C	P
38		8-Mar	F	39.63	R	F
39	*Week 10*	10-Mar	Sn	106.49	R	F
40		11-Mar	M	27.64	C	S
41		11-Mar	M	34.36	C	P
42		16-Mar	S	53.32	R	S
43	*Week 11*	17-Mar	Sn	111.78	R	F
44		19-Mar	T	17.91	C	P
45		23-Mar	S	53.52	R	P
46	*Week 12*	24-Mar	Sn	69.00	R	F
47		28-Mar	Th	17.67	C	P
48	*Week 13*	31-Mar	Sn	56.12	R	F
49		1-Apr	M	48.24	C	S
50		4-Apr	Th	17.67	C	P
51		6-Apr	S	55.79	R	S
52	*Week 14*	7-Apr	Sn	107.88	R	F
53		8-Apr	M	47.37	C	P
54		13-Apr	S	39.05	R	P
55	*Week 15*	14-Apr	Sn	85.95	R	F
56		16-Apr	T	22.37	C	S
57		16-Apr	T	23.86	C	P
58		18-Apr	Th	17.67	C	P
59		19-Apr	F	28.60	R	F
60		20-Apr	S	48.82	R	S

2.5.2 Collecting Data

Dad meets with Tere to discuss the data collection effort. Dad convinces Tere that she should keep a detailed log of data regarding second semester expenditures, either paid for with a credit card or cash. Although Tere is reluctant, Dad convinces her that he will be fair in his analysis. They agree on a list of the most important issues and concerns he wants to address regarding expenditures:

1. What types of purchases are being made?
2. Are there interesting patterns occurring during the week, month, and semester?
3. How are the payments of expenditures divided among the credit card and cash?
4. Can some of the expenditures be identified as unnecessary?

To answer these questions, Dad assumes that each time an expenditure occurs, with either cash or credit card, an *observation* is generated. Next, he selects 6 data fields to describe each observation: (1) the number of the week (1–15) for the 15 week semester in which the expenditure occurs, (2) the date, (3) the weekday (Sunday = Sn, Monday = M, etc.) corresponding to the date, (4) the amount of the expenditure in dollars, (5) whether cash (C) or credit card (R) was used for payment, and finally, (6) one of three categories of expenditure types-food (F), personal (P), and school (S). Note that these data elements represent a wide variety of data types, from ratio data related to Amount, to categorical data representing food/personal/school, to ordinal data for the date. In Table 2.4 we see that the first observation in the first week was made by credit card on Sunday, January 6th for food in the amount $111.46. Thus, we have collected our data and now we can begin to consider summarization.

2.5.3 Summarizing Data

Let's begin the process of data analysis with some basic exploration; what is often referred to as a **fishing expedition**. It is called a fishing expedition, because we simply want to perform a cursory examination of the expenditures with no particular analytical direction in mind, other than becoming acquainted with the data. This initial process should then lead to more explicit directions for the analysis; that is, we will go where the fishing expedition leads us. Summarization of the data will be important to us at this stage. Exhibit 2.13 displays the data in a *loose* chronological order, but it does not provide a great deal of information for a number of reasons. First, each successive observation does not correspond to a *strict* chronological order. For example, the first seven observations in Exhibit 2.13 represent Sunday, Monday, Tuesday, Thursday, Sunday, Monday, and Monday expenditures, respectively. Thus, there are situations where several expenditures occur on the same day and there are days where no expenditures occur. If Dad's second question about patterns of expenditures is to be answered, we will have to modify the data to include all days of the week and impose strict chronological order; thus, our chart should

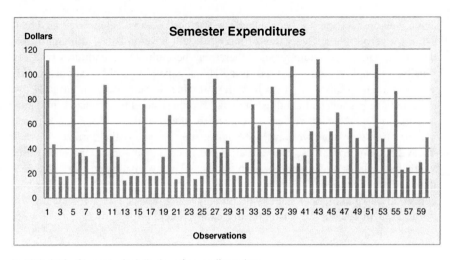

Exhibit 2.13 Chronological display of expenditure data

include days where there are no expenditures and multiple daily expenditures may have to be aggregated.

Table 2.5 displays a small portion of our expenditure data in this more rigid format which has inserted days for which there are no expenditures. Note, for example,

Table 2.5 Portion of modified expenditure data including no expenditure days

Obs.	Week	Date	Weekday	Amount	Cash/CRedit Card	Food/Personal/ School
1	Week 01	6-Jan	1	111.46	R	F
2		7-Jan	2	43.23	C	S
3		8-Jan	3	17.11	C	S
		9-Jan	4	0.00		
4		10-Jan	5	17.67	C	P
		11-Jan	6	0.00		
		12-Jan	7	0.00		
5	Week 02	13-Jan	1	107	R	F
6		14-Jan	2	36.65	C	P
7		14-Jan	2	33.91	C	P
		15-Jan	3	0.00		
		16-Jan	4	0.00		
8		17-Jan	5	17.67	C	P
9		18-Jan	6	41.17	R	F
		19-Jan	7	0.00		
10	Week 03	20-Jan	1	91.53	R	F
11		21-Jan	2	49.76	C	P
12		21-Jan	2	32.97	C	S
13		22-Jan	3	14.03	C	P
		23-Jan	4	0.00		

that a new observation has been added for Wednesday (now categorized as day 4), 9-Jan for zero dollars. Every day of the week will have an entry, although it may be zero dollars in expenditures, and there may be multiple expenditures on a day. Finally, although we are interested in individual expenditure observations, weekly, and even daily, totals could also be quite valuable. In summary, the original data collected needed substantial adjustment and summarization to organize it into more meaningful and informative data to achieve our stated goals.

Let us assume that we have reorganized our data into the format shown in Table 2.5. As before, these data are arranged in columnar format and each observation has 6 fields plus an observation number. We have made one more change to the data in anticipation of the analysis we will perform. The Weekday field has been converted into a numerical value, with Sunday being replaced with 1, Monday with 2, etc. We will discuss the reason for this change later.

2.5.4 Analyzing Data

Now we are ready to look for insights in the data we have collected and summarized; that is, perform analysis. First, focusing on the dollar value of the observations, we see considerable variation in amounts of expenditures. This is not unexpected given the relatively small number of observations in the semester. If we want to graphically analyze the data by type of payment (credit card or cash payments) *and* the category of expenditure (F, P, S), then we will have to further reorganize the data to provide this information. We will see that this can be managed with the **Sort tool** in the Data tab. The Sort tool permits us to *rearrange* our overall spreadsheet table of data observations into the observations of particular interest for our analysis.

Dad suspects that the expenditures for particular days of the week are higher than others from the data in Table 2.5. He begins by organizing the data according to day of the week—all Sundays (1), all Mondays (2), etc. To Sort the data by day, we first capture the entire data range we are interested in sorting, including the header row that contains column titles (Weekday, Amount, etc.), then we select the Sort tool in the Sort and Filter group of the Data tab. Sort permits us to set **sort keys** (the titles in the header row) that can then be selected, as well as an option for executing ascending or descending sorts. An ascending sort of text arranges data in ascending alphabetical order (a to z) and an ascending sort of numerical data is analogous. Now we can see that converting the Weekday field to a numerical value insures a Sort that places weekdays in ascending order. If the field values had remained Sn, M, etc., the sort would lead to an alphabetic sort and loss of the consecutive order of days—Friday as day 1 and Wednesday as day 7.

Exhibit 2.14 shows the data sort procedure for our original data. We begin by capturing the spreadsheet range of interest that includes the observed data and titles, now containing more than 60 observations due to our data summarization. In the Sort and Filter group we select the Sort tool. Exhibit 2.14 shows the dialog boxes

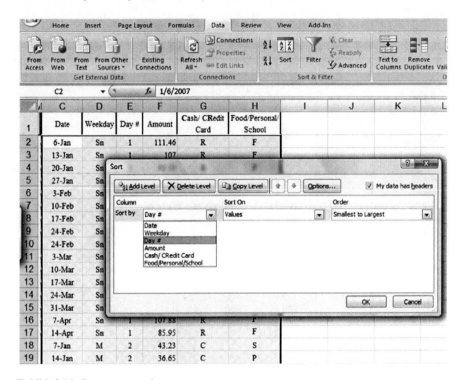

Exhibit 2.14 Data sort procedure

that ask the user for the key for sorting the data. The key used is Day #. As you can see in Exhibit 2.14, the first 16 sorted observations are for Sunday (Day 1). The complete sorted data are shown in Table 2.6.

At this point our data have come a long way from 60 basic observations and are ready to reveal some expenditure behavior. First, notice in Table 2.6 that all expenditures on Sunday are for food (F), they are made with a credit card, and are generally the highest $ values. This pattern occurs every Sunday of every week in the data. Immediately, Dad is alerted to this curious behavior—is it possible that Tere reserves grocery shopping for Sundays? Also, note that Monday's cash expenditures are of lesser value and never for food. Additionally, there are several multiple Monday expenditures and they occur irregularly over the weeks of the semester. Exhibit 2.15 provides a graph of this Sunday and Monday data comparison and Exhibit 2.16 compares Sunday and Thursday. In each case Dad has organized the data series by the specific day of each week. Also, he has aggregated multiple expenditures for a single day, such as Monday, Jan-14 expenditures of $33.91 and $36.65 (total $70.56). The Jan-14 quantity can be seen in Exhibit 2.15 in week 2 for Monday, and this has required manual summarization of the data in Table 2.6. Obviously, there are many other possible daily comparisons that can be performed and they, too, will require manual summarization.

Table 2.6　Modified expenditure data sorted by weekday and date

Date	Weekday	Amount	Cash/ cRedit Card	Food/ Personal/ School
6-Jan	1	111.46	R	F
13-Jan	1	107	R	F
20-Jan	1	91.53	R	F
27-Jan	1	76.19	R	F
3-Feb	1	66.63	R	F
10-Feb	1	96.19	R	F
17-Feb	1	96.26	R	F
24-Feb	1	58.22	R	F
24-Feb	1	75.21	R	F
3-Mar	1	90.09	R	F
10-Mar	1	106.49	R	F
17-Mar	1	111.78	R	F
24-Mar	1	69	R	F
31-Mar	1	56.12	R	F
7-Apr	1	107.88	R	F
14-Apr	1	85.95	R	F
7-Jan	2	43.23	C	S
14-Jan	2	33.91	C	P
14-Jan	2	36.65	C	P
21-Jan	2	32.97	C	S
21-Jan	2	49.76	C	P
28-Jan	2	0		
4-Feb	2	0		
11-Feb	2	0		
18-Feb	2	36.37	C	S
18-Feb	2	46.19	C	P
25-Feb	2	0		
4-Mar	2	38.91	C	P
11-Mar	2	27.64	C	S
11-Mar	2	34.36	C	P
18-Mar	2	0		
25-Mar	2	0		
1-Apr	2	48.24	C	S
8-Apr	2	47.37	C	P
15-Apr	2	0		
8-Jan	3	17.11	C	S
15-Jan	3	0		
22-Jan	3	14.03	C	P
29-Jan	3	0		
5-Feb	3	15.23	C	P
12-Feb	3	14.91	C	P
19-Feb	3	18.03	C	P
26-Feb	3	0		
5-Mar	3	0		
12-Mar	3	0		
19-Mar	3	17.91	C	P
26-Mar	3	0		
2-Apr	3	0		
9-Apr	3	0		

Table 2.6 (continued)

Date	Weekday	Amount	Cash/ cRedit Card	Food/ Personal/ School
16-Apr	3	22.37	C	S
16-Apr	3	23.86	C	P
9-Jan	4	0		
16-Jan	4	0		
23-Jan	4	0		
30-Jan	4	0		
6-Feb	4	0		
13-Feb	4	0		
20-Feb	4	0		
27-Feb	4	0		
6-Mar	4	0		
13-Mar	4	0		
20-Mar	4	0		
27-Mar	4	0		
3-Apr	4	0		
10-Apr	4	0		
17-Apr	4	0		
10-Jan	5	17.67	C	P
17-Jan	5	17.67	C	P
24-Jan	5	17.67	C	P
24-Jan	5	17.67	C	P
31-Jan	5	17.67	C	P
31-Jan	5	17.67	C	P
7-Feb	5	17.67	C	P
14-Feb	5	17.67	C	P
21-Feb	5	17.67	C	P
28-Feb	5	17.67	C	P
7-Mar	5	0		
14-Mar	5	0		
21-Mar	5	0		
28-Mar	5	17.67	C	P
4-Apr	5	17.67	C	P
11-Apr	5	0		
18-Apr	5	17.67	C	P
11-Jan	6	0		
18-Jan	6	41.17	R	F
25-Jan	6	0		
1-Feb	6	33.03	R	F
8-Feb	6	0		
15-Feb	6	40.3	R	F
22-Feb	6	28.49	R	F
1-Mar	6	0		
8-Mar	6	39.63	R	F
15-Mar	6	0		
22-Mar	6	0		
29-Mar	6	0		
5-Apr	6	0		
12-Apr	6	0		
19-Apr	6	28.6	R	F

Table 2.6 (continued)

Date	Weekday	Amount	Cash/ cRedit Card	Food/ Personal/ School
12-Jan	7	0		
19-Jan	7	0		
26-Jan	7	0		
2-Feb	7	0		
9-Feb	7	0		
16-Feb	7	0		
23-Feb	7	0		
2-Mar	7	0		
9-Mar	7	0		
16-Mar	7	53.32	R	S
23-Mar	7	53.52	R	P
30-Mar	7	0		
6-Apr	7	55.79	R	S
13-Apr	7	39.05	R	P
20-Apr	7	48.82	R	S

Now let's summarize some of Dad's early findings. Below are some of the most obvious results:

1) All Sunday expenditures (16 observations) are high dollar value, Credit Card, Food, and occur consistently on every Sunday.
2) Monday expenditures (12) are Cash, School, and Personal, and occur frequently, but occur less frequently than Sunday expenditures.
3) Tuesday expenditures (8) are Cash and predominantly Personal.
4) There are no Wednesday (0) expenditures.

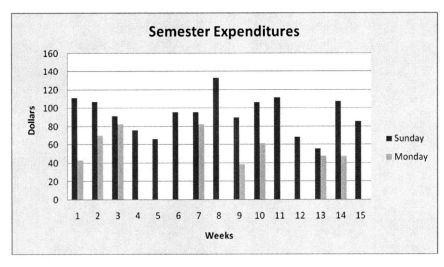

Exhibit 2.15 Modified expenditure data sorted by Sunday and Monday

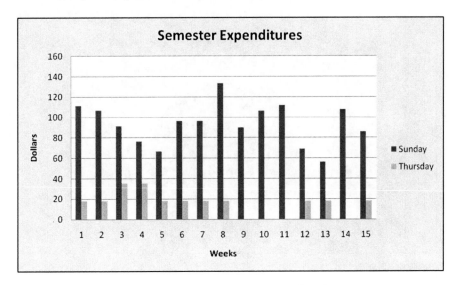

Exhibit 2.16 Modified expenditure data sorted by Sunday and Thursday

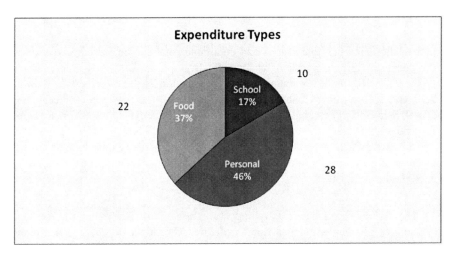

Exhibit 2.17 Number of expenditure types

5) Thursday expenditures (13) are all Personal, Cash, and exactly the same value ($17.67), although there are multiple expenditures on some Thursdays.
6) Friday expenditures (6) are all for Food and paid with Credit Card.
7) Saturday expenditures (5) are Credit Card and a mix of School and Personal.
8) The distribution of the *number* of expenditure types (Food, Personal, and School) is not proportional to the *dollars spent* on each type. (See Exhibits 2.17 and 2.18). Food accounts for fewer numbers of expenditures (37% of total) than personal, but for a greater percentage (60%) of the total dollar of expenditures.

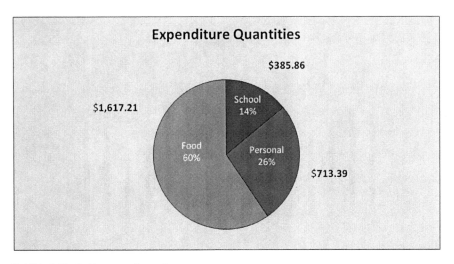

Exhibit 2.18 Dollar expenditures by type

2.5.5 Presenting Data

Exhibits 2.13, 2.14, 2.15, 2.16, 2.17, and 2.18 and Tables 2.4, 2.5, and 2.6 are examples of the many possible graphs and data tables that can be presented to explore the questions originally asked by Dad. Each graph requires data preparation to fit the analytical goal. For example, the construction of the pie charts in Exhibits 2.17 and 2.18 required that we count the number of expenditures of each type (Food, School, and Personal) and that we sum the dollar expenditures for each type, respectively.

Dad is now able to examine Tere's buying patterns more closely, and through discussion with Tere he finds some interesting behavior related to the data he has assembled:

1) The $17.67 Thursday expenditures are related to Tere's favorite personal activity—having a manicure and pedicure. The duplicate charge on a single Thursday represent a return to have her nails redone once she determines she is not happy with the first outcome.
2) Sunday expenditures are dinners (not grocery shopping) at her favorite sushi restaurant. The dollar amount of each expenditure is always high because she treats her friends, Dave and Suzanne, to dinner. Dad determines that this is a magnanimous, but fiscally irresponsible, gesture. She agrees to stop paying for her friends.
3) There are no expenditures on Wednesday because she has class all day long and is able to do little else, but study and attend class.
4) To avoid carrying a lot of cash, Tere generally prefers to use a credit card for larger expenditures. She is adhering to a bit of advice received from Dad for her own personal security.

5) She makes fewer expenditures near the end of the week because she is generally exhausted by her school work. Sunday dinner is a form of self-reward that she has established as a start to a new week. Of course, she wants to share her reward with her friends Dave and Suzanne.
6) Friday food expenditures, she explains to Dad, are due to grocery shopping.

Once Dad has obtained this information, he negotiates several money saving concessions. First, she agrees to not treat Dave and Suzanne to dinner every Sunday; every other Sunday is sufficiently generous. She also agrees to reduce her manicure visits to every other week, and she also agrees that cooking for her friends is equally entertaining as eating out.

We have not gone into great detail on the preparation of data to produce Exhibits 2.15, 2.16, 2.17, and 2.18, other than the sorting exercise we performed. Later in Chap. 4 we will learn to use the Filter and Advanced Filter capabilities of Excel. This will provide a simple method for preparing our data for graphical presentation.

2.6 Some Final Practical Graphical Presentation Advice

This chapter has presented a number of topics related to graphical presentation of quantitative data. Many of the topics are an introduction to data analysis which we will visit in far greater depth in later chapters. Before we move on, let me leave you with a set of suggestions that might guide you in your presentation choices. Over time you will develop a sense of your own presentation style and preferences for presenting data in effective formats. Don't be afraid to experiment as you explore your own style and taste.

1. *Some charts and graphs deserve their own worksheet*—Often a graph fits nicely on a worksheet that contains the data series that generate the graph. But also, it is quite acceptable to dedicate a separate worksheet to the graph if the data series make viewing the graph difficult or distracting. This is particularly true when the graph represents the important *results* presentation of a worksheet. (Later we will discuss static versus dynamic graphs, which make the choice relatively straightforward.)
2. *Axis labels are essential*—Some creators of graphs are lax in the identification of graph axes, both the units associated with the axis scale and the verbal description of the axis dimension. Because they are intimately acquainted with the data generating the graph, they forget that the viewer may not be similarly acquainted. Always provide clear and concise identification of axes, and remember that you are not the only one who will view the graph.
3. *Scale differences in values can be confusing*—Often graphs are used as tools for visual comparison. Sometimes this is done by plotting multiple series of interest on a single graph or by comparing individual series on separate graphs. In doing so, we may not be able to note series behavior due to scale differences for the graphs. This suggests that we may want to use multiple scales on a single graph to compare several series. For Excel 2003 see Custom Types in Step 1 of the

Chart Wizard; for Excel 2007 select the series, right click and select Format Data series where a Secondary Axis is available. Additionally, if we display series on separate graphs, we can impose similar scales on the multiple graphs to facilitate equitable comparison. Being alert to these differences can change our assessment of results.

4. *Fit the Chart Type by considering the graph's purpose*—The choice of the chart type should invariably be guided by one principle—*keep it simple*. There are often many ways to display data, whether the data are cross-sectional or time series. Consider the following ideas and questions relating to chart type selection.

Time Series Data (data related to a time axis)

a. Will the data be displayed over a chronological time horizon? If so, it is considered time series data.
b. In business or economics, time series data are invariably displayed with time on the horizontal axis.
c. With time series we can either display data discretely (bars) or continuously (lines and area). If the flow or continuity of data is important then Line and Area graphs are preferred. Be careful that viewers not assume that they can locate values between time increments, if these intermediate values are not meaningful.

Cross-sectional Data Time Snap-shot or (time dimension is not of primary importance)

a. For data that is a single snap-shot of time or time is not our focus, column or bar graphs are used most frequently. If you use column or bar graphs, it is important to have category titles on axes (horizontal or vertical). If you do not use a column or bar graph, then a Pie, Doughnut, Cone, or Pyramid graph may be appropriate. Line graphs are usually not advisable for cross-sectional data.
b. Flat Pie graphs are far easier to read and interpret than 3-D Pie graphs. Also, when data result in several very small pie segments, relative to others, then precise comparisons can be difficult.
c. Is the categorical order of the data important? There may be a natural order in categories that should be preserved in the data presentation—e.g. the application of chronologically successive marketing promotions in a sales campaign.
d. Displaying multiple series in a Doughnut graph can be confusing. The creation of Doughnuts within Doughnuts can lead to implied proportional relationships which do not exit.

Co-related Data

a. Scatter diagrams are excellent tools for viewing the co-relationship (correlation statistical jargon) of one variable to another. They represent two associated data items on a two dimensional surface—e.g. the number of housing starts in a time period and the corresponding purchase of plumbing supplies.

b. Bubble diagrams assume that the two values discussed in scatter diagrams also have a third value (relative size of the bubble) that relates to the frequency or strength of the point located on two dimensions—e.g. a study that tracks combinations of mortgage rate and mortgage points that must be paid by borrowers. In this case, the size of the bubble is the frequency of the occurrence of specific combinations.

General Issues

a. Is the magnitude of a data value important relative to other data values occurring in the same category or at the same time? (This was the case in Exhibit 2.4.) If so, then consider Stacked and 100% Stacked graph. The Stacked graph preserves the opportunity to compare *across* various time periods or categories—e.g. the revenue contribution of 3 categories of products for 4 quarters provides not only the relative importance of products within a quarter, but also shows how the various quarters compare. Note that this last feature (comparison across quarters) will be lost in a 100% Stacked graph.

b. In general, I find that 3-D graphs can be potentially distracting. The one exception is the display of multiple series of data (usually less than 5 or 6) where the overall pattern of behavior is important to the viewer. Here a 3-D Line graph (ribbon graph) or an Area graph is appropriate, as long as the series do not obscure the view of series with lesser values. If a 3-D graph is still your choice, exercise the 3-D View options that reorient the view of the graph or point and grab a corner of the graph to rotate the axes. This may clarify the visual issues that make a 3-D graph distracting.

c. It may be necessary to use several chart types to fully convey the desired information. Don't be reluctant to organize data into several graphical formats; this is more desirable than creating a single, overly complex graph.

d. Once again, it is wise to invoke a philosophy of simplicity and parsimony.

2.7 Summary

In the next chapter we will concentrate on numerical *analysis* of quantitative data. Chap. 3, and the two chapters that follow, contain techniques and tools that are applicable to the material in this chapter. You may want to return and review what you have learned in this chapter in light of what is to come; this is good advice for all chapters. It is practically impossible to present all the relevant tools for analysis in a single chapter, so I have chosen to "spread the wealth" among the 7 chapters that remain.

Key Terms

Ratio Data	Stacked Column
Interval Data	3-D Column
Categorical/Nominal Data	Line Chart
Ordinal Data	Time Series Data
Data Context	Lags
Records	Linear Trend
Fields	Systematic Behavior
Comment Field	Horizontal (Category) Axis Labels
Relational Database	Select Data, (Format) Chart Area, Chart Type
Charts and Graphs	Chart Title
Chart Wizard	Data Analysis
Ribbons	Collecting
Tabs	Summarizing
Groups	Analyzing
Chart Tools	Eyeballing
Data Range	Presenting
Column or Bar chart	Fishing Expedition
Negative Correlation	Sort Tool
Positive Correlation	Sort Keys

Problems and Exercises

1. Consider the data in Table 2.3 of this chapter.

   a. Replicate the charts that appeared in the chapter and attempt as many other chart types and variations as you like. Use new chart types—pie, doughnut, pyramid, etc.—to see the difference in appearance and appeal of the graphs.
   b. Add another series to the data for a new product, F. What changes in graph characteristics are necessary to display this new series with A-E? (Hint: scale will be an issue in the display).

F
425
560
893
1025
1206
837
451
283

2. Can you find any interesting relationships in Tere's expenditures that Dad has not noticed?
3. Create a graph similar to Exhibits 2.15 and 2.16 that compares *Friday* and *Saturday*.
4. Perform a single sort of the data in Table 2.6 to reflect the following 3 conditions: 1st—credit card expenditures, 2nd—in chronological order, 3rd—if there are multiple entries for a day, sort by quantity in ascending fashion.
5. Create a pie chart reflecting the proportion of all expenditures related to *Food*, *Personal*, and *School* for Dad and Tere's example.
6. Create a scatter diagram of *Day #* and *Amount* for Dad and Tere's example.
7. The data below represent information on bank customers at 4 branch locations, their deposits at the branch, and the percent of the customers over 60 years of age at the branch. Create graphs that show: (1) line graph for the series No. Customers and $ Deposits for the various branches and (2) pie graphs for each quantitative series. Finally, consider how to create a graph that incorporates all the quantitative series (hint: bubble graph).

Branch	No. customers	$ Deposits	Percent of customers over 60 years of age
A	1268	23,452,872	0.34
B	3421	123,876,985	0.57
C	1009	12,452,198	0.23
D	3187	97,923,652	0.41

8. For the following data, provide the summarization and manipulation that will permit you to sort the data by day-of-the-week. Thus, you can sort all Mondays, Tuesdays, etc. (hint: you will need a good day calculator).

Last name, First name	Date of birth	Contribution
Laffercar, Carole	1/24/76	10,000
Lopez, Hector	9/13/64	12,000
Rose, Kaitlin	2/16/84	34,500
LaMumba, Patty	11/15/46	126,000
Roach, Tere	5/7/70	43,000
Guerrero, Lili	10/12/72	23,000
Bradley, James	1/23/48	100,500
Mooradian, Addison	12/25/97	1,000
Brown, Mac	4/17/99	2,000
Gomez, Pepper	8/30/34	250,000
Kikapoo, Rob	7/13/25	340,000

9. *Advanced Problem*—Isla Mujeres is an island paradise located very near Cancun, Mexico. The island government has been run by a prominent family, the Murillos, for most of four generations. During that time, the island has become a

major tourist destination for many foreigners and Mexicans. One evening, while vacationing there, you are dining in a local restaurant. A young man seated at a table next to yours overhears you boasting about your prowess as a quantitative data analyst. He is local politician that is running for the position of Island President, the highest office on Isla Mujeres. He explains how difficult it is to unseat the Murillos, but he believes that he has some evidence that will persuade voters that it is time for a change. He produces a report that documents quantitative data related to the island's administration over 46 years. The data represent 11 four year presidential terms and the initial two years of the current term. Presidents are designated as A–D, for which all are members of the Murillo clan, except for B. President B is the only non-Murillo to be elected and was the uncle of the young politician. Additionally, all quantities have been converted to 2008 USD (US Dollars)

a. The raw data represent important economic development relationships for the Island. How will you use the raw data to provide the young politician information on the various presidents that have served the Island? Hint— Think as an economist might, and consider how the president's investment in the island might lead to improved economic results.

b. Use your ideas in a. to prepare a graphical analysis for the young politician. This will require you to use the raw data in different and clever ways.

c. Compare the various presidents, through the use of graphical analysis, for their effectiveness in running the island. How will you describe the young politician's Uncle?

d. How do you explain the changes in Per Capita Income given that it is stated in 2008 dollars? Hint—There appears to be a sizable increase over time. What might be responsible for this improvement?

Years	President	Municipal tax collected	Salary of island president	Island infrastructure investment	Per capita income
1963–1966	A	120,000	15,000	60,000	1900
1967–1970	A	186,000	15,000	100,000	2100
1971–1974	A	250,000	18,000	140,000	2500
1975–1978	B	150,000	31,000	60,000	1300
1979–1982	B	130,000	39,000	54,000	1000
1983–1986	C	230,000	24,000	180,000	1800
1987–1990	C	310,000	26,000	230,000	2300
1991–1994	C	350,000	34,000	225,000	3400
1995–1998	C	450,000	43,000	320,000	4100
1999–2002	D	830,000	68,000	500,000	4900
2003–2006	D	1,200,000	70,000	790,000	5300
2007–2008*	D	890,000	72,000	530,000	6100

* represents a reminder that in Ex 2.3 numbers are in terms of millions of dollars (U.S.).

Chapter 3
Analysis of Quantitative Data

Contents

3.1 Introduction

In this chapter we continue our study of data analysis, particularly the analysis of quantitative data. In Chap. 2 we explored types and uses of data, and we performed data analysis on quantitative data with graphical techniques. This chapter will delve more deeply into the topic of quantitative data analysis, providing us with a strong foundation and a preliminary understanding of the results of a data collection effort. Some statistical tools will be introduced, but more powerful tools will follow in later chapters.

H. Guerrero, *Excel Data Analysis*, DOI 10.1007/978-3-642-10835-8_3,
© Springer-Verlag Berlin Heidelberg 2010

3.2 What is Data Analysis?

If you perform an internet search on the term *Data Analysis*, it will take years for you to visit every site that is returned, not to mention encountering a myriad of different types of sites, each claiming the title data analysis. Data analysis means many things to many people, but the goal of data analysis is universal. It is to answer one very important question—*what does the data reveal about the underlying system or process from which the data is collected?* For example, suppose you gather data on customers that shop in your retail operation, data that consists of detailed records of purchases and demographic information on each customer transaction. As a data analyst, you may be interested in investigating the buying behavior of different age groups. The data might reveal that the dollar value of purchases by young men is significantly smaller than those of young women. You might also find that one product is often purchased in tandem with another. These findings can lead to important decisions on how to advertise or promote products. If we consider the findings above, we may devise sales promotions targeted at young men to increase the value of their purchases, or we may consider the co-location of products on shelves that makes tandem purchases more convenient. In each case, the decision maker is examining the data for clues of the underlying behavior of the consumer.

Although Excel provides you with numerous internal tools designed explicitly for data analysis, some of which we have seen already, the user is also capable of employing his own ingenuity to perform many types of analytical procedures by using Excel's basic mathematical functions. Thus, if you are able to understand the basic mathematical principles associated with an analytical technique, there are few limits on the type of techniques that you can apply. This is often how an *add-in* is born; an individual creates a clever analytical application and makes it available to others.

An **add-in** is a program designed to work within the framework of Excel. They use the basic capabilities of Excel; for example, its ability to use either Visual Basic for Applications (VBA) or Visual Basic (VB) programming languages to perform Excel tasks. These programming tools are used to automate and expand Excel's reach into areas that are not readily available. In fact, there are many free and commercially available statistical, business, and engineering add-ins that provide capability in user friendly formats.

Now, let us consider what we have ahead of us. In this chapter, we are going to focus on the *built-in* data analysis functionality of Excel and apply it to quantitative data. We will carefully demonstrate how we apply these internal tools to a variety of data, but throughout our discussions, it will be assumed that the reader has a rudimentary understanding of statistics. Further, recall that the purpose of this chapter, and this book for that matter, is not to make you into a statistician, but rather to give you some powerful tools for gaining insight about the behavior of data. I urge you to experiment with your own data, even if you just make it up, to practice the techniques we will study.

3.3 Data Analysis Tools

There are a number of approaches to perform data analysis on a data set stored in an Excel workbook. In the course of data analysis, it is likely that all approaches will be useful, although some are more accessible than others. Let us take a look at the three principle approaches available:

1) Excel provides resident add-in utilities that are extremely useful in basic statistical analysis. The *Data* ribbon contains an *Analysis* group with about twenty statistical *Data Analysis* tools. Exhibit 3.1 shows the location of the *Data Analysis* add-in tool and 3.2 shows some of the contents of the *Data Analysis* menu. These tools allow the user to perform relatively sophisticated analyses without having to create the mathematical procedures from basic cell functions; thus, they usually require interaction through a dialogue box as shown in Exhibit 3.2. Dialogue boxes are the means by which the user makes choices and provides instructions, such as entering parameter values and specifying ranges

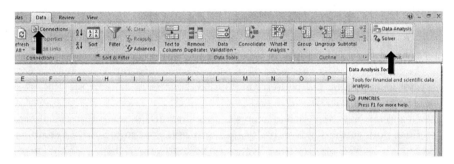

Exhibit 3.1 Data analysis add-in tool

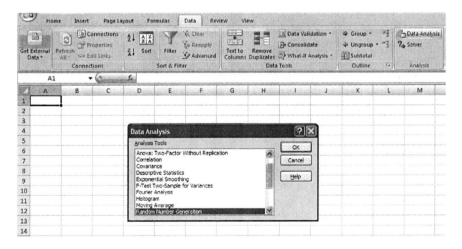

Exhibit 3.2 Data analysis dialogue box

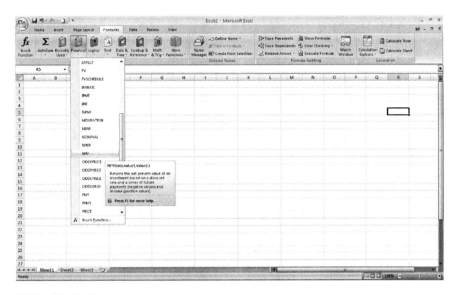

Exhibit 3.3 The insert function

containing data of interest. In Exhibit 3.2 we see a fraction of the analysis tools
available, including *Descriptive Statistics, Correlation,* etc. You simply select a
tool and click the *OK* button. More on this process in the next section—*Data
Analysis for Two Data Sets.*

2) In a more direct approach to analysis, Excel provides dozens of statistical functions through the function utility (*f$_x$* Insert Function) in the *Formulas* ribbon.
Simply choose the *Statistical* category of functions in the *Function Library*
group, select the function you desire, and insert the function in a cell. In the
statistical category there are almost one hundred functions that relate to important theoretical data distributions and statistical analysis tools. In Exhibit 3.3
you can see that the *Financial* function category has been selected, *NPV* (net
present value) in particular. Once the function is selected, Excel takes you to a
dialogue box for insertion of the NPV data as shown Exhibit 3.4. The dialogue
box requests several inputs—e.g. discount rate (*Rate*) and values (Value1, etc.)
to be discounted to the present. The types of functions that can be inserted vary
from *Math & Trig, Date and Time, Statistical, Logical,* to even *Engineering,*
just to name a few. By selecting the *f$_x$* Insert Function at the far left of the
Function Library group, you can also select specific functions. Exhibit 3.5 shows
the dialogue box where these choices are made from the *Or select a category:*
pull-down menu. As you become familiar with a function, you need only begin
the process of keying in the function in a cell preceded by an equal sign; thus,
the process of selection is simplified. You can also see from Exhibit 3.6 that by
placing the cursor in cell C4 and typing =NPV(. Then a small box opens that
guides you through the data entry required by the function. The process also
provides error checking to insure that your data entry is correct.

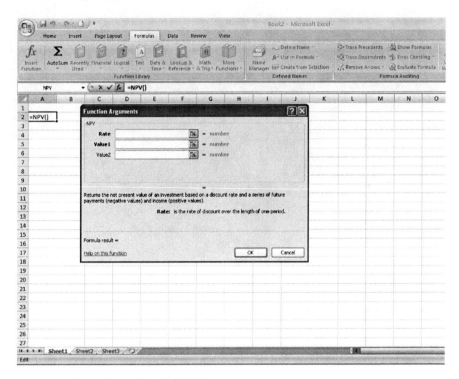

Exhibit 3.4 Example of a financial NPV function

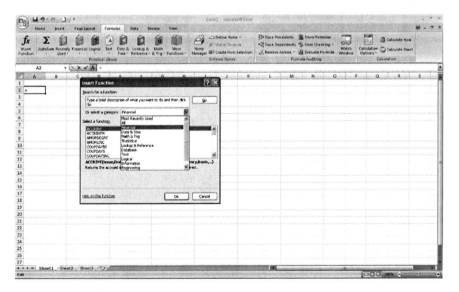

Exhibit 3.5 Function categories

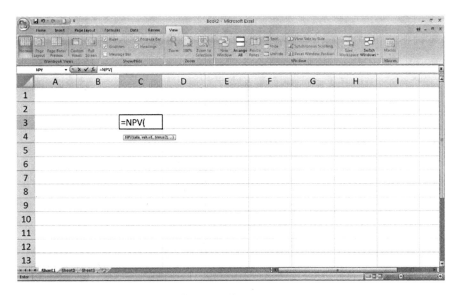

Exhibit 3.6 Keying-in the NPV function in a cell

3) Finally, there are numerous commercially available add-ins: functional programs that can be loaded into Excel that permit many forms of sophisticated analysis. For example, Solver, an add-in that is used in constrained optimization.

Although it is impossible to cover every available aspect of data analysis that is contained in Excel in this chapter, we will focus on techniques that are useful to the average entry-level user, particularly those discussed in (1) above. Once you have mastered these techniques, you will find yourself quite capable of exploring many others on you own. The advanced techniques will require that you have access to a good advanced statistics and/or data analysis reference.

3.4 Data Analysis for Two Data Sets

Let us begin by examining the *Data Analysis* tool in the *Analysis* group. These tools (regression, correlation, descriptive statistics, etc.) are statistical procedures that answer questions about the relationship between multiple data *series*, or provide techniques for summarizing characteristics of a single data set.

A **series**, as the name implies, is a series of data points that are collected and ordered in a specific manner. The ordering can be chronological or according to some other **treatment**: a characteristic under which the data is collected. For example, a treatment could represent customer exposure to high levels of media advertising. These tools are useful in prediction or for the description of data. To access *Data Analysis*, you must first enable the *Analysis ToolPak* box by opening the Excel Options found in the Office Button (round button on extreme left). Exhibit 3.7 shows this operation. In Exhibit 3.8, the arrow indicates where the menu

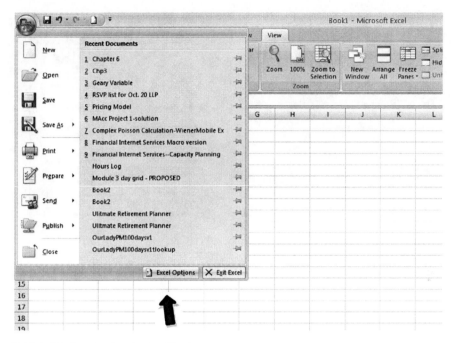

Exhibit 3.7 Excel options in the office button

for selecting the *Analysis ToolPak* can be found. Once enabled, a user has access to the *Analysis Tool Pak*.

We will apply these tools on two types of data: **time series** and **cross-sectional**. The first data set, time series, is data that was introduced in Chap. 2, although the data set has been expanded to provide a more complex example. Table 3.1 presents sales data for five products (A–E) over 24 quarters (six years) in thousands of dollars. In Exhibit 3.9 we use some of the graphing skills we learned in Chap. 2 to display the data graphically. Of course, this type of visual analysis is a preliminary step that can guide our efforts for understanding the behavior of the data, and suggest further analysis. A trained analyst can find many interesting leads to the data's behavior by creating a graph of the data; thus, it is always a good idea to begin the data analysis process by graphing the data.

3.4.1 Time Series Data—Visual Analysis

Time series data is data that is chronologically ordered, and it is one of the most frequently encountered types of data in business. Cross-sectional data is data that is taken at a single point in time or under circumstances where time, as a dimension, is irrelevant. Given the fundamental differences in these two types of data, our approach for analyzing each will be different. Now, let us consider a preliminary approach for time series data analysis.

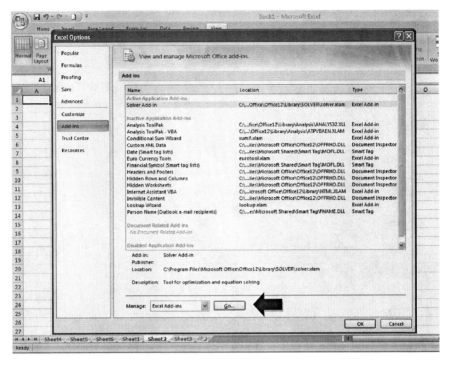

Exhibit 3.8 Enabling the analysis ToolPak add-in

With time series data, we are particularly interested in how the data varies over time and in identifying patterns that occur systematically over time. A graph of the data, as in Exhibit 3.9, is our first step in the analysis. As the British Anthropologist, John Lubbock, wrote: *What we see depends mainly on what we look for,* and herein we see the power of Excel's charting capabilities. We can carefully scrutinize—*look for*—patterns of behavior before we commit to more technical analysis. Behavior like seasonality, co-relationship of one series to another, or one series displaying leading or lagging time behavior with respect to another are relatively easy to observe.

Now, let us investigate the graphical representation of data in Exhibit 3.9. Note that if many series are displayed simultaneously, the resulting graph can be very confusing. Thus, we display each series separately. The following are some of the interesting findings for our sales data:

1. It appears that all of the product sales have some **cyclicality** except for D; that is, the data tends to repeat patterns of behavior over some relatively fixed time length (a cycle). Product D may have a very slight cyclical behavior, but is it not evident by graphical observation.
2. It appears that A and E behave relatively similarly for the first three years, although their cyclicality is out of phase by a single quarter. Cyclicality that

Table 3.1 Sales* data for products A–E

Quarter	A	B	C	D	E
1	98	45	64	21	23
2	58	21	45	23	14
3	23	36	21	31	56
4	43	21	14	30	78
1	89	49	27	35	27
2	52	20	40	40	20
3	24	43	58	37	67
4	34	21	76	40	89
1	81	53	81	42	34
2	49	27	93	39	30
3	16	49	84	42	73
4	29	30	70	46	83
1	74	60	57	42	43
2	36	28	45	34	32
3	17	52	43	45	85
4	26	34	34	54	98
1	67	68	29	53	50
2	34	34	36	37	36
3	18	64	51	49	101
4	25	41	65	60	123
1	68	73	72	67	63
2	29	42	81	40	46
3	20	73	93	57	125
4	24	53	98	74	146

*thousands of dollars

is based on a yearly time frame is referred to as **seasonality**, due to the data's variation with the seasons of the year.
3. The one quarter difference between A and E (phase difference) can be explained as E **leading** A by a period. For example, E peaks in quarter 4 of the first year and A peaks in quarter 1 of the second year, thus the peak in E leads A by one quarter. The quarterly lead appears to be exactly one period for the entire six year horizon.
4. Product E seems to behave differently in the last three years of the series by displaying a general tendency to increase. We call this pattern **trend**, and in this case, a *positive* trend over time. We will, for simplicity's sake, assume that this is **linear trend**; that is, it increases or decreases at a constant rate. For example, a linear trend might increase at a rate of 4,000 dollars per quarter.
5. There are numerous other features of the data that can and will be identified later.

We must be careful not to extend the findings of our visual analysis too far. Presuming we know all there is to know about the underlying behavior reflected in the data without a more formal analysis can lead to serious problems. That is precisely why we will apply more sophisticated data analysis once we have visually inspected the data.

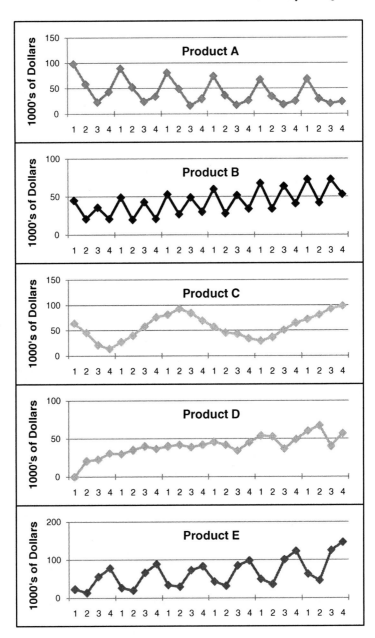

Exhibit 3.9 Graph of sales data for products A–E

3.4.2 Cross-Sectional Data—Visual Analysis

Now, let us consider another set of data that is collected by a web-based **e-tailer** (retailers that market products via the internet) that specializes in marketing to teenagers. The e-tailer is concerned that their website is not generating the number of **page-views** (website pages viewed per visit) that they desire. They suspect that the website is just not attractive to teens. To remedy the situation they hire a web designer to redesign the site with teen's preferences and interests in mind. An experiment is devised that randomly selects 100 teens that have not previously visited the site and exposes them to the old and new website designs. They are told to interact with the site until they lose interest. Data is collected on the number of web-pages each teen views on the old site and on the new site.

In Table 3.2 we organize page-views for individual teens in columns. We can see that teen number 1 (top of the 1st column) viewed 5 pages on the old website and 14 on the new website. Teen number 15 (the bottom of the 3rd column) viewed 10 pages on the old website and 20 on the new website. The old website and the new website represent *treatments* in statistical analysis.

Our first attempt at analysis of this data is a simple visual display—a graph. In Exhibit 3.10 we see a frequency distribution for our pages viewed by 100 teens, before and after the website update. A **frequency distribution** is simply a count of the number of occurrences of a particular quantity. For example, if in Table 3.2 we count the occurrence of 2 page views on the old website, we find that there are 3 occurrences—teen 11, 34, and 76. Thus, the frequency of 2 page views is 3 and can be seen as a bar 3 units high in Exhibit 3.10. Note that Exhibit 3.10 counts all possible values of page views for old and new websites to develop the distribution. The range (low to high) of values for old is 1-15. It is also possible to create *categories* of values for the old, for example 1–5, 6–10 and 11–15 page views. This distribution would have all observations in only 3 possible outcomes and appear quite different from Exhibit 3.10.

Table 3.2   Old web site pages visited

							Old Website												
5	6	2	4	11	4	8	12	10	4	6	15	8	7	5	2	3	9	5	6
4	7	11	7	6	5	9	10	6	8	10	6	4	11	8	8	15	8	4	11
7	6	12	8	1	5	6	10	14	11	4	6	11	6	8	6	11	8	6	6
5	12	5	5	7	7	2	5	10	6	7	5	12	8	9	7	5	8	6	6
7	7	10	10	6	10	6	10	8	9	14	6	13	11	12	9	7	4	11	5
							New Website												
14	5	18	19	10	11	11	12	15	10	9	9	11	9	10	11	8	5	21	8
10	10	16	10	14	15	9	12	16	14	20	5	10	12	21	12	16	14	17	15
12	12	17	7	9	8	11	12	12	12	8	12	11	14	10	16	8	5	6	10
5	16	9	9	14	9	12	11	13	6	15	11	14	14	16	9	7	17	10	15
9	13	20	12	11	10	18	9	13	12	19	6	9	11	14	10	18	9	11	11

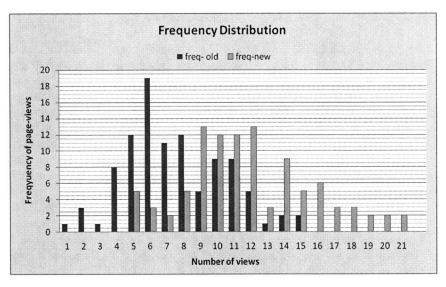

Exhibit 3.10 Frequency distribution of pages viewed

We can see from Exhibit 3.10 that the old website is generally located to the left (lower values of page views) of the new website. Both distributions appear to have a **central tendency**; that is, there is a central area that has more frequent values of page views than the extreme values, either lower or higher. Without precise calculation, it is likely that the average of the pages viewed will be near the center of the distributions. It is also obvious that the average, or mean, pages viewed for the old web site will be less than the average pages viewed for the new web site. Additionally, the **variation**, or spread, of the distribution for the new website is larger than that of the old website: the range of the new values extends from 5 to 21, whereas the range of the old values is 1 to 15.

In preparation for our next form of analysis, descriptive statistics, we need to define a number terms:

1. The average, or **mean**, of a set of data is the sum of the observations divided by the number of observations.
2. A frequency distribution organizes data observations into particular categories based on the number of observations in a particular category.
3. A frequency distribution with a central tendency is characterized by the grouping of observations near or about the center of a distribution.
4. A **standard deviation** is the statistical measure of the degree of variation of observations relative to the mean of all the observations. The calculation of the standard deviation is the square root of the sum of the squared deviations for each value in the data from the mean of the distribution, which is then divided by the number of observations. If we consider the observations collected to

be a sample, then the division is by the number of observations minus 1. The standard deviation formula in Excel for observations that are assumed to be a sample (*unbiased version*) is *STDEV(number1, number2...)*. In the case where we assume our observations represent a **population** (all possible observations), the formula is *STDEVP(number1, number2,...)* .

5. A **range** is a simple, but useful, measure of variation which is calculated as the high observation value minus the low.
6. A population is the set of all possible observations of interest.
7. The **median** is the data point in the middle of the distribution of all data points. There are as many values below as above the median.
8. The **mode** is the most often occurring value in the data observations.
9. The **standard error** is the sample standard deviation divided by the square root of the number of data observations.
10. **Sample variance** is the square of the sample standard deviation of the data observations.
11. **Kurtosis** (peakedness) and **skewness** (asymmetry) are measures related to the shape of a data organized into a frequency distribution.

In most cases, it is likely we are *not* interested in viewing our time series data as a distribution of points, since frequency distributions generally ignore the time element of a data point. We might expect variation and be interested in examining it, but usually with a specific association to time. A frequency distribution does not provide this time association for data observations.

Let us examine the data sets by employing *descriptive statistics* for each type of data: time series and cross-sectional. We will see in the next section that some of Excel's descriptive statistics are more appropriate for some types of data than for others.

3.4.3 Analysis of Time Series Data—Descriptive Statistics

Consider the time series data for our Sales example. We will perform a very simple type of analysis that generally *describes* the sales data for each product— *Descriptive Statistics*. First, we locate our data in columnar form on a worksheet. To perform the analysis, we select the *Data Analysis* tool from the *Analysis* group in the Data ribbon. Next we select the *Descriptive Statistics* tool as shown in Exhibit 3.11. A dialogue box will appear that asks you to identify the input range containing the data. You must also provide some choices regarding the output location of the analysis and the types of output you desire (check the summary statistics box). In our example, we select data for product A. See Exhibit 3.12. We can also select all of the products (A–E) and perform the same analysis. Excel will automatically assume that each column represents data for a different product. The output of the analysis for product A is shown in Exhibit 3.13.

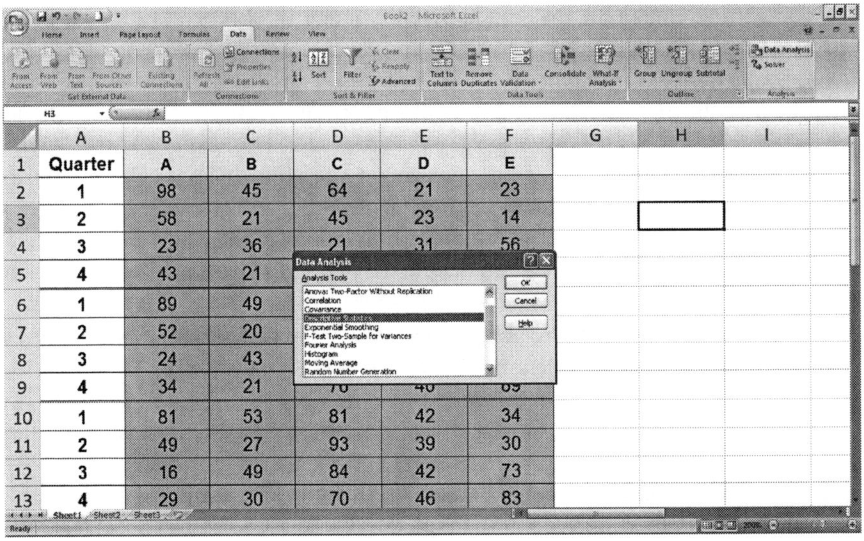

Exhibit 3.11 Descriptive statistics in data analysis

Note that the mean of sales for product A is approximately 43 (thousand). As suggested earlier, this value, although of moderate interest, does not provide much useful information. It is the six year average. Of more interest might be a comparison of each year's average. This would be useful if we were attempting to identify a trend, either up or down. More on the summary statistics for product A later.

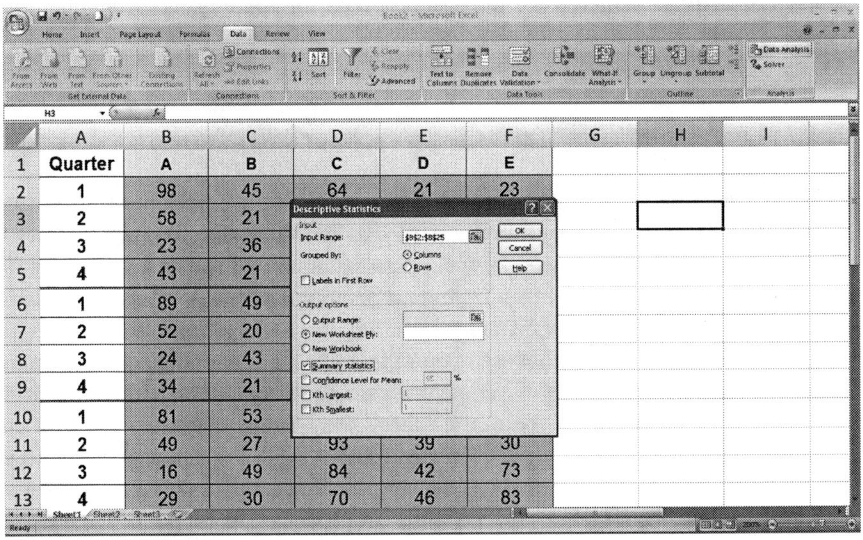

Exhibit 3.12 Dialogue box for descriptive statistics

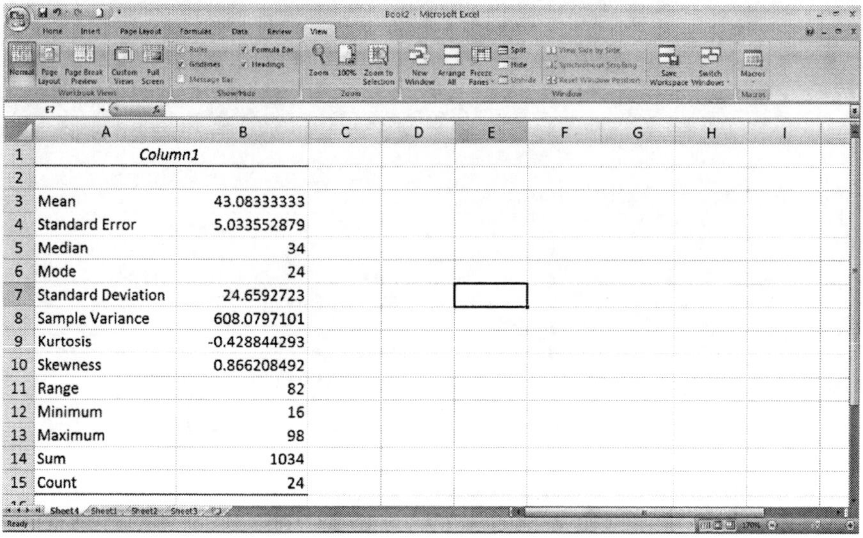

	A	B	C	D	E	F	G	H	I
1		*Column1*							
2									
3	Mean	43.08333333							
4	Standard Error	5.033552879							
5	Median	34							
6	Mode	24							
7	Standard Deviation	24.6592723							
8	Sample Variance	608.0797101							
9	Kurtosis	-0.428844293							
10	Skewness	0.866208492							
11	Range	82							
12	Minimum	16							
13	Maximum	98							
14	Sum	1034							
15	Count	24							

Exhibit 3.13 Product A descriptive statistics

3.4.4 Analysis of Cross-Sectional Data—Descriptive Statistics

Our website data is cross-sectional, thus, the time context is not an important dimension of the data. The descriptive statistics for the old website are shown in Exhibit 3.14. It is a quantitative summary of the old website data graphed in Exhibit 3.10. To perform the analysis, it is necessary to rearrange the data shown in Table 3.2 into a single column of 100 data points since the *Data Analysis* tool assumes that data is organized in either rows *or* columns. Table 3.2 contains data in rows *and* columns; thus, we need to stretch-out the data into either a row *or* a column. This could be a tedious task if we are rearranging a large quantity of data points, but the *Cut* and *Paste* tools in the *Home* ribbon and *Clipboard* group will make quick work of the changes. It is important to keep track of the 100 teens as we rearrange the data, since the old website will be compared to the new and tracking the change in specific teens will be important. Thus, whatever cutting and pasting is done for the new data must be done similarly for the old data.

Now let us consider the measures shown in the *Descriptive Statistics*. As the graph in Exhibit 3.10 suggested, the mean or average for the old website appears to be between 6 and 8, probably on the higher end given the positive skew of the graph—the frequency distribution tails off in the direction of higher or *positive* values. In fact, the mean is 7.54. The skewness is positive, 0.385765, indicating the right tail of the distribution is longer than the left, as we can see from Exhibit 3.10. The measure of kurtosis (peaked or flatness of the distribution relative to the normal

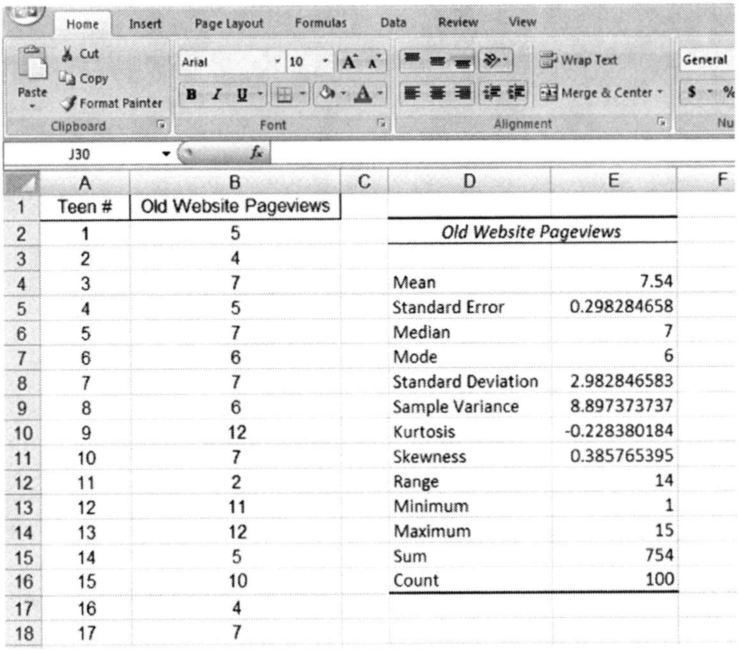

	A	B	C	D	E
1	Teen #	Old Website Pageviews			
2	1	5		*Old Website Pageviews*	
3	2	4			
4	3	7		Mean	7.54
5	4	5		Standard Error	0.298284658
6	5	7		Median	7
7	6	6		Mode	6
8	7	7		Standard Deviation	2.982846583
9	8	6		Sample Variance	8.897373737
10	9	12		Kurtosis	-0.228380184
11	10	7		Skewness	0.385765395
12	11	2		Range	14
13	12	11		Minimum	1
14	13	12		Maximum	15
15	14	5		Sum	754
16	15	10		Count	100
17	16	4			
18	17	7			

Exhibit 3.14 Descriptive statistics of old website data

distribution), –0.22838, is slightly negative, indicating mild relative flatness. The other measures are self-explanatory, including the measures related to samples: standard error and sample variance. We can see that these measures are more relevant to cross-sectional data than to our time series data since the 100 teens are a randomly selected *sample* of the entire population of visitors to the old website for a particular period of time.

There are several other tools that are related to descriptive statistics—Rank and Percentile and Histogram—that can be very useful. Rank and Percentile generates a table that contains an ordinal and percentage rank of each data point in a data set (see Exhibit 3.15). Thus, one can conveniently state that of the 100 viewers of the old website, individuals number 56 and 82 rank highest (number 1 in the table shown in Exhibit 3.15) and hold the percentile position 98.9%, which is the percent of teens that are *at* or *below* their level of views (15). Percentiles are often used to create thresholds; for example, a score on an exam below the 30th percentile is a failing grade.

The *Histogram* tool in the *Data Analysis* group creates a table of the frequency of the values relative to your selection of *bin* values. The results could be used to create the graphs in Exhibit 3.10. Exhibit 3.16 shows the dialogue box entries necessary to create the histogram. Just as the bin values used to generate Exhibit 3.10 are values from the lowest observed value to the largest in increments of one, these are the entry values in the dialogue box in Exhibit 3.16—D2:D17. (Note the

Teen #	Old Website Pageviews		Point	Old Website Pageviews	Rank	Percent
1	5					
2	4		56	15	1	98.90%
3	7		82	15	1	98.90%
4	5		43	14	3	96.90%
5	7		55	14	3	96.90%
6	6		65	13	5	95.90%
7	7		9	12	6	90.90%
8	6		13	12	6	90.90%
9	12		36	12	6	90.90%
10	7		64	12	6	90.90%
11	2		75	12	6	90.90%
12	11		12	11	11	81.80%
13	12		21	11	11	81.80%
14	5		48	11	11	81.80%
15	10		63	11	11	81.80%
16	4		67	11	11	81.80%
17	7		70	11	11	81.80%
18	8		83	11	11	81.80%
19	5		95	11	11	81.80%
20	10		97	11	11	81.80%
21	11		15	10	20	72.70%

Exhibit 3.15 The rank and percentile of old website data

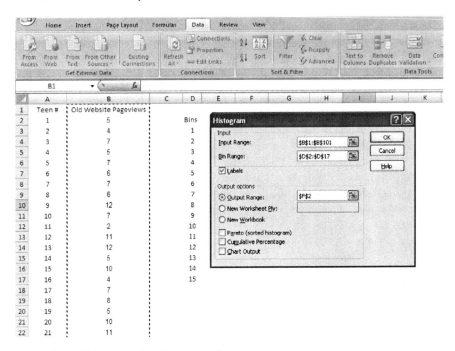

Exhibit 3.16 Dialogue box for histogram analysis

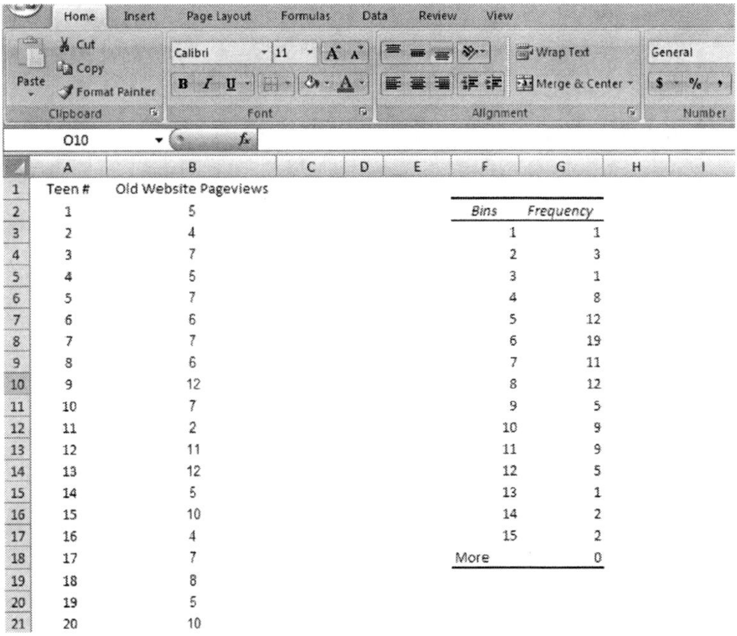

Teen #	Old Website Pageviews	Bins	Frequency
1	5	1	1
2	4	2	3
3	7	3	1
4	5	4	8
5	7	5	12
6	6	6	19
7	7	7	11
8	6	8	12
9	12	9	5
10	7	10	9
11	2	11	9
12	11	12	5
13	12	13	1
14	5	14	2
15	10	15	2
16	4	More	0
17	7		
18	8		
19	5		
20	10		

Exhibit 3.17 Results of histogram analysis for old website views

Labels box is checked to include the title *Bins*). The results of the analysis are shown in Exhibit 3.17. It is now convenient to graph the histogram by selecting the *Insert* ribbon and the *Charts* group. This is equivalent to the previously discussed frequency distribution in Exhibit 3.10.

3.5 Analysis of Time Series Data—Forecasting/Data Relationship Tools

We perform data analysis to answer questions and gain insight. So what are the central questions we would like to ask about our time series data? Put yourself in the position of a data analyst. Here is a list of possible questions you might want to answer:

1. Do the data for a particular series display a repeating and systematic pattern over time?
2. Does one series move with another in a predictable fashion?
3. Can we identify behavior in a series that can predict systematic behavior in another series?
4. Can the behavior of one series be incorporated into a forecasting model that will permit accurate prediction of the future behavior of another series?

Although there are many questions that can be asked, these four are important and will allow us to investigate numerous analytical tools in *Data Analysis*. As a note of caution, let us keep in mind that this example is based on a very small amount of data; thus, we must be careful to not overextend our perceived insight. The greater the amount of data, the more secure one can be in his observations. Let us begin by addressing the first question.

3.5.1 Graphical Analysis

Our graphical analysis of the sales data has already revealed the possibility of **systematic behavior** in the series; that is, there is an underlying *system* that influences the behavior of the data. As we noted earlier, all series, except for product D, display some form of cyclical behavior. How might we determine if systematic behavior exists? Let us select product E for further analysis, although we could have chosen any of the products.

In Exhibit 3.18 we see that the product time series does in fact display repetitive behavior; in fact, it is *quite* evident. Since we are interested in the behavior of both the yearly demand and quarterly demand, we need to rearrange our time series data to permit a different type of graphical analysis. Table 3.3 shows the data from Table 3.1 in a modified format: each row represents a year (1–6) and each column a quarter (1–4); thus, the value 101 represents quarter 3 in year 5. Additionally, the right most vertical column of the table represents yearly totals. This new data configuration will allow us to perform some interesting graphical analysis.

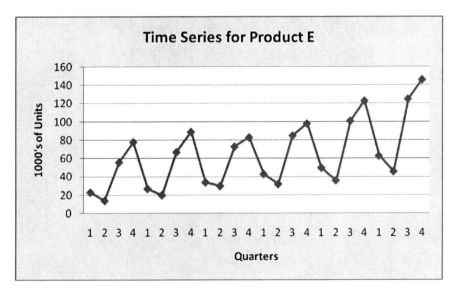

Exhibit 3.18   Product E time series data

Table 3.3 Modified quarterly data* for product E

	Qtr 1	Qtr 2	Qtr 3	Qtr 4	Yearly Total
Yr1	23	14	56	78	171
Yr2	27	20	67	89	203
Yr3	34	30	73	83	220
Yr4	43	32	85	98	258
Yr5	50	36	101	123	310
Yr6	63	46	125	146	380

* sales in thousands of dollars

Now let us proceed with the analysis. First, we will apply the *Histogram* tool to explore the quarterly data behavior in greater depth. There is no guarantee that the tool will provide insight that is useful, but that's the challenge of data analysis—it can be as much an art as a science. In fact, we will find the *Histogram* tool will be of little use. Why? It is because the tool does not distinguish between the various quarters. As far as the *Histogram* tool is concerned, a data point is a data point, without regard to its related quarter; thus we see the importance of the *context* of data points. Had the data points for each quarter been clustered in distinct value groups (e.g. all quarter 3 values clustered together) the tool would have been much more useful. See Exhibit 3.19 for the results of the histogram with bin values in increments of 10 units starting at a low value of 5 and a high of 155. There are

	A	B	C	D	E	F	G
1		Qtr 1	Qtr 2	Qtr 3	Qtr 4	Total	
2	Yr1	23	14	56	78	171	
3	Yr2	27	20	67	89	203	
4	Yr3	34	30	73	83	220	
5	Yr4	43	32	85	98	258	
6	Yr5	50	36	101	123	310	
7	Yr6	63	46	125	146	380	
8							
9		Bin	Frequency				
10		5	0				
11		15	1				
12		25	2				
13		35	4				
14		45	2				
15		55	2				
16		65	2				
17		75	2				
18		85	3				
19		95	1				
20		105	2				
21		115	0				
22		125	2				
23		135	0				
24		145	0				
25		155	1				
26		More	0				

Exhibit 3.19 Histogram results for all product E adjusted data

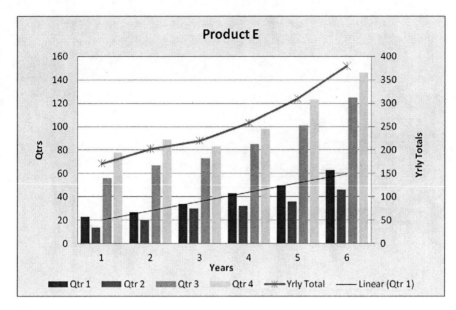

Exhibit 3.20   Product E quarterly and yearly total data

clearly no clusters of data representing distinct quarters that are easily identifiable. For example, there is only 1 value that falls into the category (bin) of *values between 5 and 15*, and that is the 2nd quarter of year 1. Similarly, there are 3 data values that fall into the 75 to 85 bin: quarters 4 of year 1, quarter 4 of year 3, and quarter 3 of year 4. It may be possible to adjust the bins to capture clusters more effectively, but that is not the case for these data values. But don't despair; we still have other graphical tools that will prove useful.

Exhibit 3.20 is a graph that explicitly considers the quarterly position of data by dividing the time series into 4 quarterly sub-series for product E. See Exhibit 3.21 for the data selected to create the graph. It is the same as Table 3.3. From Exhibit 3.20, it is evident that all the product E time series over six years display important data behavior: the 4th quarter in all years is the largest sales value, followed by quarters 3, 1, and 2. Note that the *Yearly Total* is increasing consistently over time (measured on the vertical scale on the right-Yrly Totals), as are all other series except for quarter 4, which has a minor reduction in year 3. This suggests that there is a seasonal effect related to our data, as well as a consistent trend for all series. It may be wise to reserve judgment on quarterly sales behavior in the future, but clearly these are interesting questions to pursue with more advanced techniques.

Before we proceed, let us take stock of what the graphical data analysis has revealed about product E:

1) We have assumed that it is convenient to think in terms of these data having three components—a base level, seasonality effects, and a linear trend.

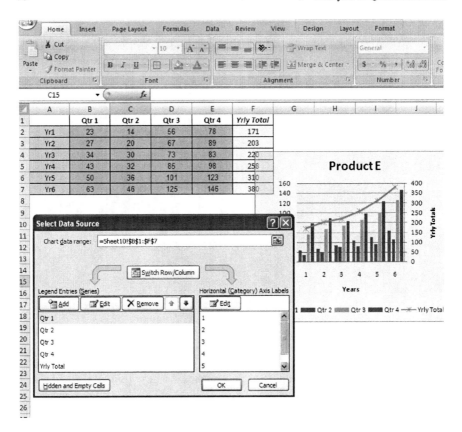

Exhibit 3.21 Selected data for quarters and yearly total

2) The base relates to the value of a specific quarter, and when combined with a quarterly trend for the series, results in a new base in the following year. Trends for the various quarters may be different, but all our series appear to have a positive linear trend, including the total.

3) We have dealt with seasonality by focusing on specific quarters in the yearly cycle of sales. By noting that there is a consistent pattern or relationship within a yearly cycle (quarter 4 is always the highest value), we observe seasonal behavior.

4) Visual analysis suggests that we can build a model of the data behavior that might provide future estimates of quarterly and yearly total values. This is because we understand the three elements that make up the behavior of each quarterly series.

One last comment on the graph 3.21 is appropriate. Note that the graph has two vertical scales. This is necessary due to the large difference in the magnitude of values for the individual quarters and the Yrly Totals. To use a single vertical axis would make viewing the movement of the series difficult. By selecting any data observation associated with the Yrly total with a right-click, a menu appears that permits us to format the data series. One of the options available is to plot the series on a secondary axis. This feature can be quite useful when viewing data that vary in magnitude.

3.5.2 Linear Regression

Now let us introduce a tool that is useful in the prediction of future values of the series. The tool is the forecasting technique *linear regression*, and although it is not appropriate for all forecasting situations, it is very commonly used. There are many sophisticated forecasting techniques that can be used to forecast business and economic data that may be more appropriate depending on the data that is to be analyzed. I introduce linear regression because of its common use and instructive character—understanding the ideas of a linear model can be quite useful in understanding other more complex models. Just as in our graphical analysis, the choice of a model should be an intensive and methodical process.

Linear Regression builds a model that predicts future behavior for a *dependent* variable based on the assumed linear influence of one or more *independent* variables. The **dependent variable** is what we attempt to predict or forecast, in this case sales values for quarters, and the **independent variable** is what we base our forecast on, in this case, the year into the future. The concept of a regression formula is relatively simple: for particular values of an independent variable, we can construct a linear relationship that permits the prediction of a dependent variable. For our product E sales data, we will create a regression model for each quarter. So, we will construct 4 regressions. We do this to avoid the need to explicitly consider seasonality in the linear regression. Our assumption is that there is a linear relationship between the independent variable, *year*, and the dependent variable, *quarterly sales*.

Simple linear regression, which is the approach we will use, can be visualized on an X–Y coordinate system—a single X represents the independent variable and Y the dependent variable. Multiple linear regression uses more than one X to predict Y. Simple regression finds the linear relationship that best fits the data by choosing a slope of the regression line, known as the **beta** (β), and a Y intercept (where the line crosses the Y axis) known as the **alpha** (α). If we examine the individual series in Exhibit 3.20, it appears that all quarters, except for 4, are a good linear fit with years. Notice the dip for quarter 4 in year 3. To more closely understand the issue of a linear *fit*, I have drawn a linear trend line for the quarter 1 series in Exhibit 3.20—marked Linear (Qtr1) in the legend. As you can see, the fit of the line nicely tracks the changes in the quarter 1 series. By selecting a series and right clicking, an option to *Add Trendline* appears.

Before we move on with the analysis, let me caution that creating a regression model from only 6 data points is quite dangerous. Yet, data limitations are often a

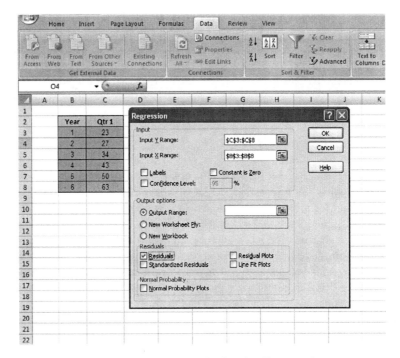

Exhibit 3.22 Dialogue box for regression analysis of product E, quarter 1

fact of life and must be dealt with, even if it means basing predictions on very little data, and assuredly, 6 data points are an extremely small number of data observations. In this case, it is also a matter of using what I would refer to as a *baby problem* to demonstrate the concept. So, how do we perform the regression analysis?

As with the other tools in *Data Analysis*, a dialogue box, shown in Exhibit 3.22 will appear and query you as to the data ranges that you wish to use for the analysis: the dependent variable will be the *Input Y Range* and the independent variable will be the *Input X Range*. The data range for Y is the set of 6 values (C3:C8) of observed quarterly sales data. The X values are the numbers 1–6 (B3:B8) representing the years for the quarterly data. Thus, regression will determine an alpha and beta that when incorporated into a predictive formula (Y = b X + α) will result in the best *model* available for some criteria. This does not mean that you are guaranteed a regression that is a good fit—it could be good, bad, or anything in between. Once alpha and beta have been determined, they can then be used to create a predictive model. The resulting regression statistics and regression details are shown in Exhibit 3.23.

The regression statistics that are returned judge the fit, good or bad, of a regression line to the dependent variable values of quarterly sales. The **R-square** (coefficient of determination) shown in the *Regression Statistics* of Exhibit 3.23

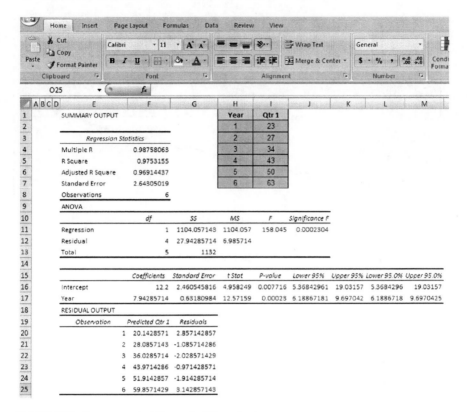

Exhibit 3.23   Summary output for product E quarter 1

measures how well the estimated values of the regression line correspond to the actual quarterly sales data; it is a guide to the *goodness of fit* of the regression model. R-square values can vary from 0 to 1, with 1 indicating perfect correspondence between the estimated value and the data, and with 0 indicating no systematic correspondence whatsoever. In this model the R Square is approximately 97.53%. This is a very high R-square, implying a very good fit.

The analysis can also provide some very revealing graphs: the fit of the regression to the actual data and the **residuals** (the difference between the actual and the predicted values). To produce a residuals plot, check the residuals box in the dialogue box shown in Exhibit 3.22. This allows you to see the accuracy of the regression model. In Exhibit 3.23 you can see the *Residuals Output* at the bottom of the output. The residual for the first observation (23) is 2.857... since the predicted value produced by the regression is 20.143... (23–20.143... = 2.857...).

Finally, the coefficients of the regression are also specified in Exhibit 3.23. The Y intercept or α, 12.2 for the sales data, is where the regression line crosses the Y axis. The coefficient of the independent variable β, approximately 7.94, is the slope of the

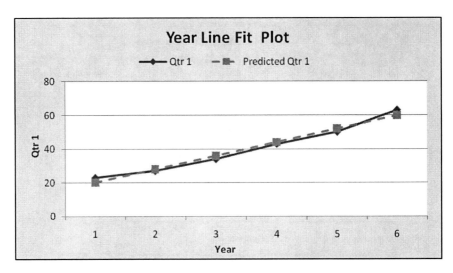

Exhibit 3.24 Plot of fit for product E quarter 1

linear regression for the independent variable. These coefficients specify the model and can be used for prediction. For example, the analyst may want to predict an estimate of the 1st quarterly value for the 7th year. Thus the prediction calculation results in the following:

$$\text{Estimated Y for Year 7} = \alpha + \beta \,(\text{Year}) = 12.1 + 7.94(7) = 67.8$$

Exhibit 3.24 shows the resulting relationship between the actual and predicted values for quarter 1. The fit is almost perfect. Note that regression can be applied to any data set, but it is only when we examine the results that we can determine if regression is a good predictive tool. When the R-square is low and residuals are not a good fit, it is time to look elsewhere for a predictive model. Of course, R-square is a relative measure and should be considered along with other factors. In some applications, analysts might be quite happy with an R-Square of 0.4, in others it is of no value.

Now let us determine the fit of a regression line for quarter 4. As mentioned earlier, a visual observation of Exhibit 3.20 indicates that quarter 4 appears to be the least suitable among quarters for a linear regression model and Exhibit 3.25 indicates a less impressive R-square of approximately 85.37%. Yet, this is still a relatively high value. Exhibit 3.26 shows the predicted and actual plot for quarter 4.

There are other important measures of fit that should be considered for regression. Although we have not discussed this measure yet, the **Significance F** for quarter 1 regression is quite small (0.0002304), indicating that we should conclude that there is *significant* association between the independent and dependant

	Home	Insert	Page Layout	Formulas	Data	Review	View

	A	B	C	D	E	F	G	H	I
3	*Regression Statistics*								
4	Multiple R	0.92396165							
5	R Square	0.85370513							
6	Adjusted R Square	0.81713141							
7	Standard Error	11.3057086							
8	Observations	6							
9									
10	ANOVA								
11		*df*	*SS*	*MS*	*F*	*Significance F*			
12	Regression	1	2983.557148	2983.557	23.34204	0.00845293			
13	Residual	4	511.2761905	127.819					
14	Total	5	3494.883333						
15									
16		*Coefficients*	*Standard Error*	*t Stat*	*P-value*	*Lower 95%*	*Upper 95%*	*Lower 95.0%*	*Upper 95.0%*
17	Intercept	57.1333333	10.52504194	5.428324	0.005586	27.9111321	86.35553	27.911182	86.35553452
18	Year	13.0571429	2.702581281	4.83136	0.008453	5.55357429	20.56071	5.5535743	20.56071142
19									
20	RESIDUAL OUTPUT								
21									
22	*Observation*	*Predicted Qtr 4*	*Residuals*						
23		1	70.1904762	7.80952381					
24		2	83.247619	5.752380952					
25		3	96.3047619	-13.3047619					
26		4	109.361905	-11.3619048					
27		5	122.419048	0.580952381					
28		6	135.47619	10.52380952					

Exhibit 3.25 Summary output for product E quarter 4

variables. For the quarter 4 regression model in Exhibit 3.25, the value is larger (0.00845293), yet there is likely to be a significant association between X and Y. The smaller the Significance F the better the fit.

There are many other important measures of regression fit that we have not discussed for time series errors or residuals—e.g. independence or serial correlation, homoscedasticity, and normality. These are equally important measures to those we have discussed and deserve attention in a serious regression modeling effort, but are beyond the scope of this chapter.

Thus far, we have used data analysis to explore and examine our data, taking what we can from each form of analysis and adding whatever is contributed to our overall insight. Simply because a model, such as regression, does not fit our data does not mean that our efforts have been wasted. It is still likely that we *have* gained insight: this is not an appropriate model and there may be indicators of an alternative to explore. It may sound odd, but often we may be as well informed by what doesn't work, as by what does.

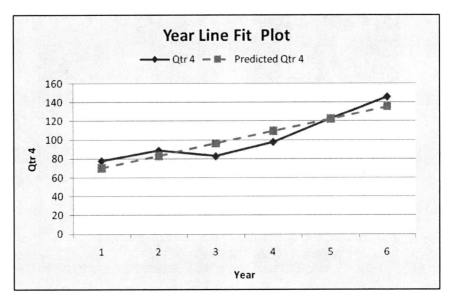

Exhibit 3.26   Plot of fit for product E quarter 4

3.5.3 Covariance and Correlation

Recall the original questions posed about the product sales time series data, and in particular the second question which asked: "Does one series move with another in a predictable fashion?" The **Covariance** tool helps answer this question by determining how the series *co-vary*. We return to the original data in Table 3.1 to determine the movement of one series with another. The *Covariance* tool, which is found in the *Data Analysis* tool, returns a matrix of values for a set of data series that you select. For the product sales data, it performs an exhaustive pairwise comparison of all 6 times series. As is the case with other *Data Analysis* tools, the dialogue box asks for the data ranges of interest and we provide the data in Table 3.1. Each value in the matrix represents either the variance of one time series or the covariance of one time series compared to another. For example, in Exhibit 3.27 we see the covariance of product A to itself (its variance) is 582.7431 and the covariance of product A and C is –74.4896. Large positive values of covariance indicate that large values of data observations in one series correspond to large values in the other series. Large negative values indicate the inverse: small values in one series indicate large values the other.

Exhibit 3.27 is relatively easy to read. The covariance of product D and E is relatively strong at 323.649, while the same is true for product A and E at –559.77. These values suggest that we can expect D and E moving together, or in the same direction; while A and E also move together, but in opposite directions, due to the negative sign of the covariance. Again, we need only refer to Exhibit 3.9 to see that the numerical covariance values bear out the graphical evidence. Small values of

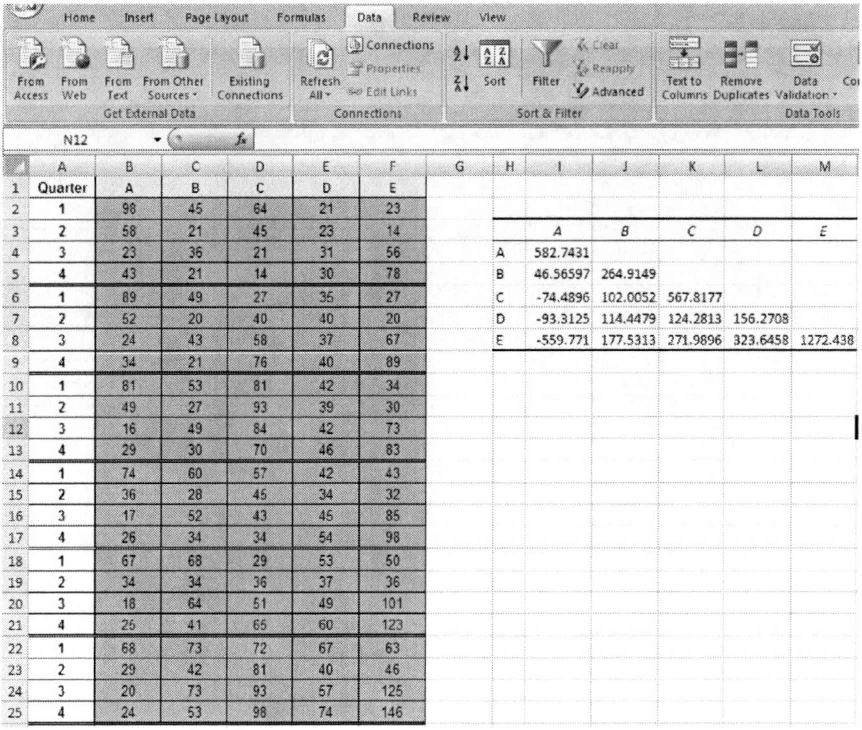

Quarter	A	B	C	D	E
1	98	45	64	21	23
2	58	21	45	23	14
3	23	36	21	31	56
4	43	21	14	30	78
1	89	49	27	35	27
2	52	20	40	40	20
3	24	43	58	37	67
4	34	21	76	40	89
1	81	53	81	42	34
2	49	27	93	39	30
3	16	49	84	42	73
4	29	30	70	46	83
1	74	60	57	42	43
2	36	28	45	34	32
3	17	52	43	45	85
4	26	34	34	54	98
1	67	68	29	53	50
2	34	34	36	37	36
3	18	64	51	49	101
4	25	41	65	60	123
1	68	73	72	67	63
2	29	42	81	40	46
3	20	73	93	57	125
4	24	53	98	74	146

	A	B	C	D	E
A	582.7431				
B	46.56597	264.9149			
C	-74.4896	102.0052	567.8177		
D	-93.3125	114.4479	124.2813	156.2708	
E	-559.771	177.5313	271.9896	323.6458	1272.438

Exhibit 3.27 Covariance matrix for product A–E

covariance like those for product A and B (and C also) indicate little co-variation. The problem with this analysis is that it is not a simple matter to know what we mean by large or small values—large or small relative to what.

Fortunately, statisticians have a solution for this problem—*Correlation* analysis. **Correlation** analysis will make understanding the linear co-variation or co-relation between two variables much easier, because it is measured in values that are standardized between the range of −1 and 1. A correlation coefficient of 1 for two data series indicates that the two series are **perfectly positively correlated**: as one variable increases so does the other. If correlation coefficient of −1 is found, then the series are **perfectly negatively correlated**: as one variable increases the other decreases. Two series are said to be independent if their correlation is 0. The calculation of correlation coefficients involves the covariance of two data series.

In Exhibit 3.28 we see a correlation matrix which is very similar the covariance matrix. You can see that the strongest positive correlation in the matrix is between products D and E, 0.725793, and the strongest negative correlation is between A and E, where the coefficient of correlation is −0.65006. There are also some values that indicate near linear independence; for example, products A and B with a coefficient of 0.118516. Clearly this is a more direct method of determining the linear correlation of one data series with another than the covariance matrix.

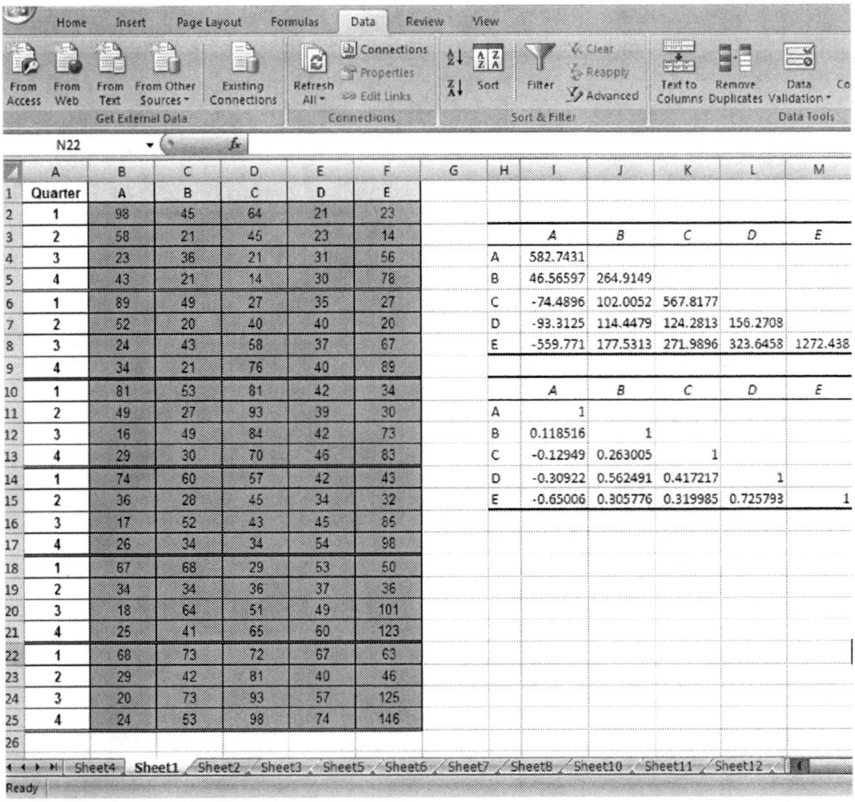

Home   Insert   Page Layout   Formulas   **Data**   Review   View

	From Access	From Web	From Text	From Other Sources ▾	Existing Connections		Refresh All ▾	Connections / Properties / Edit Links		Sort	Filter	Clear / Reapply / Advanced	Text to Columns	Remove Duplicates	Data Validation ▾	Co
			Get External Data					Connections			Sort & Filter				Data Tools	

N22

	A	B	C	D	E	F	G	H	I	J	K	L	M
1	Quarter	A	B	C	D	E							
2	1	98	45	64	21	23							
3	2	58	21	45	23	14			A	B	C	D	E
4	3	23	36	21	31	56		A	582.7431				
5	4	43	21	14	30	78		B	46.56597	264.9149			
6	1	89	49	27	35	27		C	-74.4896	102.0052	567.8177		
7	2	52	20	40	40	20		D	-93.3125	114.4479	124.2813	156.2708	
8	3	24	43	58	37	67		E	-559.771	177.5313	271.9896	323.6458	1272.438
9	4	34	21	76	40	89							
10	1	81	53	81	42	34			A	B	C	D	E
11	2	49	27	93	39	30		A	1				
12	3	16	49	84	42	73		B	0.118516	1			
13	4	29	30	70	46	83		C	-0.12949	0.263005	1		
14	1	74	60	57	42	43		D	-0.30922	0.562491	0.417217	1	
15	2	36	28	45	34	32		E	-0.65006	0.305776	0.319985	0.725793	1
16	3	17	52	43	45	85							
17	4	26	34	34	54	98							
18	1	67	68	29	53	60							
19	2	34	34	36	37	36							
20	3	18	64	51	49	101							
21	4	25	41	65	60	123							
22	1	68	73	72	67	63							
23	2	29	42	81	40	46							
24	3	20	73	93	57	125							
25	4	24	53	98	74	146							
26													

◄ ◄ ► ►| Sheet4 | **Sheet1** | Sheet2 | Sheet3 | Sheet5 | Sheet6 | Sheet7 | Sheet8 | Sheet10 | Sheet11 | Sheet12

Ready

Exhibit 3.28 Correlation matrix for Product A–E

3.5.4 Other Forecasting Models

In a more in-depth investigation of the data, we would include a search for *other* appropriate models to describe the data behavior. These models could then be used to predict future quarterly periods. Forecasting models and techniques abound and require very careful and studied analysis, but a good candidate model for this data is one that is known as **Winters' 3-factor exponential smoothing**. The conceptual fit appears to be excellent. Winters' model assumes 3 components in the structure of a forecast model—a base or level, a linear trend, and some form of cyclicality. All these elements appear to be present in most of the data series for product sales and are also part of our previous analytical assumptions. The Winters' model also incorporates the differences between the actual and predicted values (errors) into its future calculations: that is, it incorporates a self-corrective capability to account for errors made in forecasting. This self-corrective property permits the model to adjust to changes that may be occurring in underlying behavior. A much simpler version of Winters' model is found in *Data Analysis* as **Exponential Smoothing**, which only assumes a base or level component of sales.

3.5.5 Findings

So what have we learned about our product sales data? A great deal has been revealed about the underlying behavior of the data. Some of the major findings are summarized in the list below:

1. The products display varying levels of trend, seasonality, and cyclicality. This can be seen in Exhibit 3.9. Not all products were examined in depth, but the period of the cyclicality varied from seasonal for product A and E, to multi-year for product C. Product D appeared to have no cyclicality, while product B appears to have a cycle length of 2 quarters. These are reasonable observations, although we should be careful given the small number of data points.
2. Our descriptive statistics are not of much value for time series data, but the mean and the range could be of interest. Why? Because descriptive statistics generally ignore the time dimension of the data, and this is problematic for our time series data.
3. There are both positive (products D and E) and negative linear (products A and E) co-relations among a number of the time series. For some (products A and B), there is little to no linear co-relation. This variation may be valuable information for predicting behavior of one series from the behavior of another.
4. Repeating systematic behavior is evident in varying degrees in our series. For example, product D exhibits a small positive trend in early years. In later years the trend appears to increase. Products B, D, and E appear to be growing in sales. Product C might also be included, but it is not as apparent as in B, D, and E. The opposite statement can be made for product A, although its periodic lows seem to be very consistent. All these observations are derived from Exhibit 3.9.
5. Finally, we were able to examine an example of quarterly behavior for the series over six years, as seen in Exhibit 3.20. In the case of product E, we fitted a regression to the quarterly data and determined a predictive model that could be used to forecast future Product E quarterly sales. The results were a relatively good model fit, yet again based on a very, very small amount of data.

3.6 Analysis of Cross-Sectional Data—Forecasting/Data Relationship Tools

Now let us return to our cross-sectional data and apply some of the *Data Analysis* tools to the website data. Which tools shall we apply? We have learned a considerable amount about what works and why, so let us use our new found knowledge and apply techniques that make sense.

First, recall that this is cross-sectional data; thus, the time dimension of the data is not a factor to consider in our analysis. Let us consider the questions that we might ask about our data:

1. Is the average number of pages higher or lower for the new website?
2. How does the frequency distribution of *new* versus *old* pages compare?

3. Can the results for our sample of one hundred teen subjects be generalized to the population of all possible teen visitors to our website?
4. How *secure* are we in our generalization of the sample results to the population of all possible teen visitors to our website?

As with our time series data, there are many other questions we could ask, but these four questions are certainly important to our understanding of the effectiveness of the *new* website design. Additionally, as we engage in the analysis, other questions of interest may arise. Let us begin with a simple examination of the data. Exhibit 3.29 presents the descriptive statistics for the *new* and *old* website data.

Notice that the mean of the *old* website is 7.54 and the *new* website mean is 11.83. This appears to be a significant difference, an increase of 4.29 pages visited. But the difference could also be a matter of the sample of 100 individuals we have chosen for our experiment; that is, the 100 observations may not be representative of the universe of potential website visitors. Yet, in the world of statistics, a random sample of 100 is often a relatively substantial number of observations. The website change in views represents an approximately 57% increase from the *old* page views. Can we be sure that a 4.29 page change is indicative of what will be seen in the universe of all potential teen website visitors? Fortunately, there are statistical tools available for examining the question of our confidence in the outcome of the 100 teens experiment. We will return to this question momentarily, but in the interim, let us examine the changes in the data a bit more carefully.

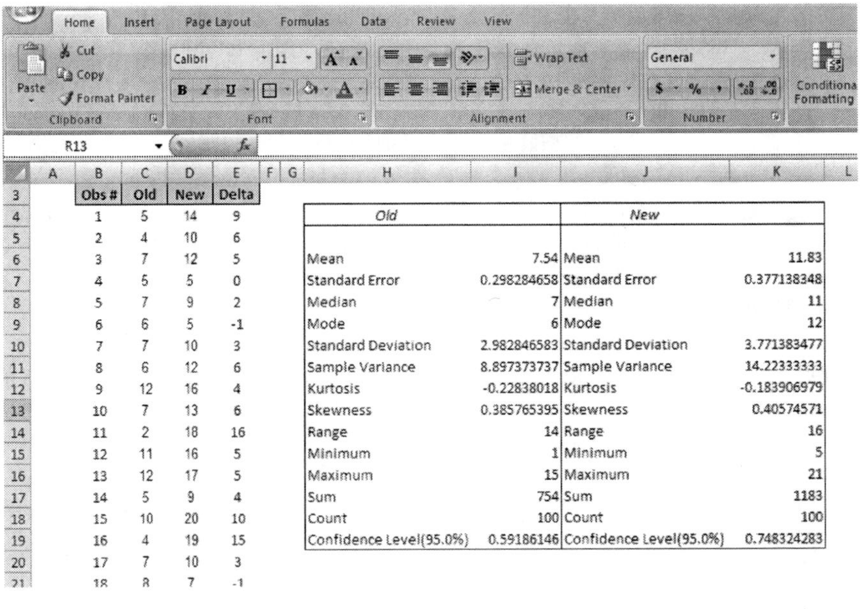

Obs #	Old	New	Delta
1	5	14	9
2	4	10	6
3	7	12	5
4	5	5	0
5	7	9	2
6	6	5	-1
7	7	10	3
8	6	12	6
9	12	16	4
10	7	13	6
11	2	18	16
12	11	16	5
13	12	17	5
14	5	9	4
15	10	20	10
16	4	19	15
17	7	10	3
18	8	7	-1

Old		*New*	
Mean	7.54	Mean	11.83
Standard Error	0.298284658	Standard Error	0.377138348
Median	7	Median	11
Mode	6	Mode	12
Standard Deviation	2.982846583	Standard Deviation	3.771383477
Sample Variance	8.897373737	Sample Variance	14.22333333
Kurtosis	-0.22838018	Kurtosis	-0.183906979
Skewness	0.385765395	Skewness	0.40574571
Range	14	Range	16
Minimum	1	Minimum	5
Maximum	15	Maximum	21
Sum	754	Sum	1183
Count	100	Count	100
Confidence Level(95.0%)	0.59186146	Confidence Level(95.0%)	0.748324283

Exhibit 3.29 New and old website descriptive statistics

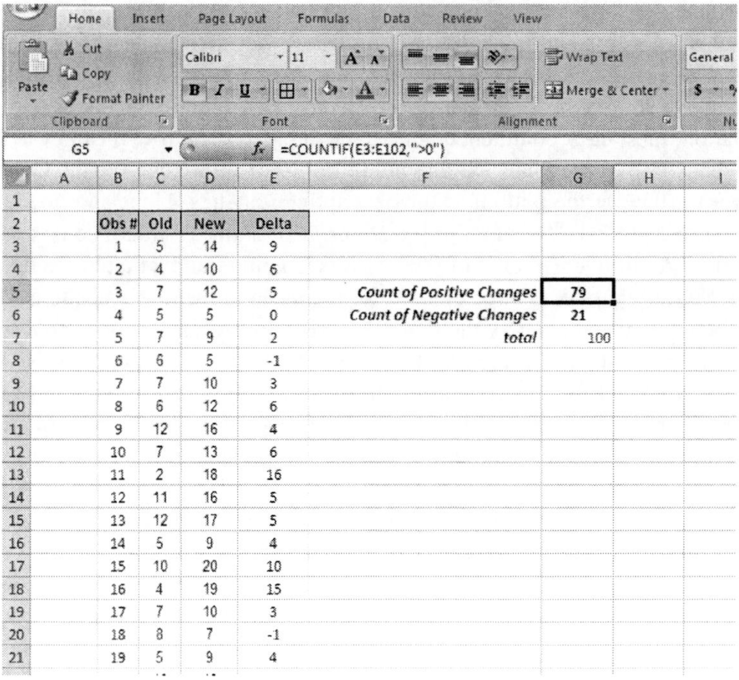

Obs #	Old	New	Delta			
1	5	14	9			
2	4	10	6			
3	7	12	5	*Count of Positive Changes*	79	
4	5	5	0	*Count of Negative Changes*	21	
5	7	9	2	*total*	100	
6	6	5	-1			
7	7	10	3			
8	6	12	6			
9	12	16	4			
10	7	13	6			
11	2	18	16			
12	11	16	5			
13	12	17	5			
14	5	9	4			
15	10	20	10			
16	4	19	15			
17	7	10	3			
18	8	7	-1			
19	5	9	4			

Formula bar: G5 f_x =COUNTIF(E3:E102,">0")

Exhibit 3.30   Change in each teen's page views

Each of the randomly selected teens has two data points associated with the data set: *old* and *new* website views. We begin with a very fundamental analysis: a calculation of the difference between the *old* and *new* web-page views. Specifically, we count the number of teens that increase their number of web-page views and conversely the number that reduce or remain at their current number of views. Exhibit 3.30 provides this analysis for these two categories of results. For the 100 teens in the study, 21 viewed fewer or the same number of web-pages for the *new* design, while 79 viewed more. The column labeled *Delta*, column E, is the difference between the *new* and *old* website views, and the logical criteria used to determine if a cell will be counted is *> 0* placed in quotes. It is shown in the formula bar as *Countif (E3:E102, ">0")*.

Again, this appears to be relatively convincing evidence that the website change has had an effect, but the strength and the certainty of the effect may still be in question. This is the problem with sampling—we can never be absolutely certain that the sample is representative of the population from which it is taken.

Sampling is a fact of life, and living with its shortcomings is unavoidable. We are often forced to sample because of convenience and the cost limitations associated with performing a census, and samples can lead to unrepresentative results for our population. This is one of the reasons why the mathematical science of statistics was invented: to help us quantify our level of comfort with the results from samples.

Fortunately, we have an important tool available in our *Descriptive Statistics* that helps us with sampling results—*Confidence Level*. We can choose a particular **level of confidence**, 95% in our case, and create an interval about the sample mean, above and below. If we sample 100 teens many times from our potential teen population, these new confidence intervals will capture the true mean of new webpage visits 95% of the times we sample. In Exhibit 3.29 we can see the Confidence Interval for 95% at the bottom of the descriptive statistics. Make sure to check the *Confidence Level for Mean* box in the *Descriptive Statistics* dialogue box to return this value. A confidence level of 95% is very common and suitable for our application. So our 95% confidence interval for the mean of the *new* website is 11.83 ± 0.74832..., or approximately the range 11.08168 to 12.57832. For the *old* website, the confidence interval for the mean is 7.54 ± 0.59186..., or the range 6.94814 to 8.13186. Note that the low end of the mean for the *new* website views (11.08168) is larger than the high end of the mean for the *old* views (8.13186). This strongly suggests with statistical confidence, that there is indeed a difference in the page views.

Next, we can expand on the analysis by not only considering the two categories, positive and non-positive differences, but also the magnitude of the differences. This is an opportunity to use the *Histogram* tool in *Data Analysis*. We will use bins values from –6 to 16 in one unit intervals. These are the minimum and maximum observed values, respectively. Exhibit 3.31 shows the graphed histogram results of the column E (*Delta*). The histogram appears to have a central tendency around the range 2 to 6 web-pages, which leads to the calculated mean of 4.29. It also has a very minor positive skew. For perfectly symmetrical distributions, the mean, the median, and the mode of the distribution are the same and there is no positive or negative skew. Finally, if we are relatively confident about our sample of 100 teens being representative of all potential teens, we are ready to make a number of important statements about our data, given our current analysis:

1. If our sample of 100 teens is representative, we can expect an average improvement of about 4.29 pages after the change of the *new* web-site design.
2. There is considerable variation in the difference between *new* and *old* (Delta) evidenced by the range, −6 to 16. There is a central tendency in the graph that places many of the Delta values between 2 and 6.
3. We can also make statements such as: (1) I believe that approximately 21% of teens will respond negatively, or not at all, to the web-site changes; (2) approximately 51% of teens will increase their page views by 2 to 6 pages; (3) approximately 24% of teens will increase page views by 7 or more pages. These statements are based on the 100 teen samples we have taken and will likely vary somewhat if another sample is taken. If these numbers are important to us, then we may want to take a much larger sample to improve of chances of stability in these percentages.
4. Our 95% confidence interval in the *new* website mean can be stated as 11.83 ± 0.74832.... This is a relatively tight interval. If a larger number of observations

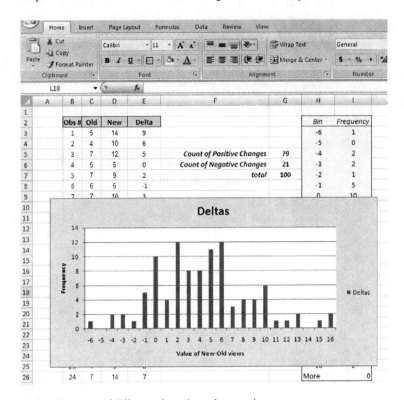

Exhibit 3.31 Histogram of difference in each teen's page views

is taken in our sample, the interval will be even tighter ($< 0.74832\ldots$). The larger the sample, the smaller the interval for a particular confidence interval.

Let us now move to a more sophisticated form of analysis, which answers questions related to our ability to generalize the sample result to the entire teen population. In the *Data Analysis* tool, there is an analysis called a **t-Test**. A t-test examines whether the means from two samples are equal or different; that is, whether they come from population distributions with the same mean or not. Of special interest for our data is the **t-Test: Paired Two Sample for Mean**. It is used when *before* and *after* data is collected from the same sample group, in our case the same 100 teens being exposed to both the *new* web-site and the *old* .

By selecting *t-Test: Paired Two Sample for Means* from the *Data Analysis* menu, the two relevant data ranges can be input, along with a hypothesized mean difference, 0 in our case, because we will assume *no* difference. Finally an *alpha* value is requested. The value of alpha must be in the range 0 to 1. Alpha is the significance level related to the probability of making a **type 1 error** (rejecting a true hypothesis); the more certain you want to be about *not* making a type 1 error, the smaller the

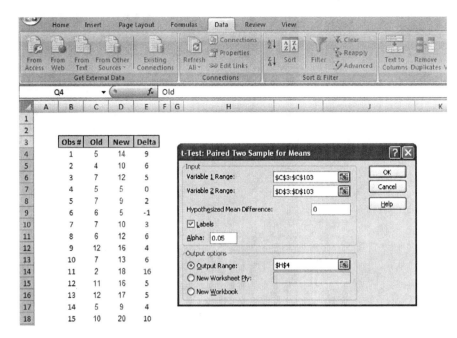

Exhibit 3.32   t-test: Paired two sample for means dialogue box

value of alpha that is selected. Often, an alpha of 0.05, 0.01, or 0.001 is appropriate, and we will choose 0.05. Once the data is provided in the dialogue box, a table with the resulting analysis appears. See Exhibit 3.32 for the dialogue box inputs and Exhibit 3.33 for the results.

The resulting **t-Stat**, 9.843008, is compared with a **critical value** of 1.660392 and 1.984217 for the one-tail and two-tail tests, respectively. This comparison amounts to what is known as a **test of hypothesis**. In hypothesis testing, a **null hypothesis** is established: the means of the underlying populations are the same and therefore their difference is equal to 0. If the calculated t-stat value is larger than the critical values, then the hypothesis that the difference in means is equal to 0 is *rejected* in favor of the alternative that the difference is *not* equal to 0. For a **one-tail test,** we assume that the result of the rejection implies an alternative in one direction. In our case, we might compare the one-tail critical value (1.660392) to the resulting *t-Stat* (9.843008), where we assume that if we reject the hypothesis that the means are equal, we then favor that the *new* website mean is in fact *greater* than the *old*. The preliminary analysis that gave us a 4.29 page increase would strongly suggest this alternative. The one-tail test does in fact reject the null hypothesis since 9.843008 is greater than (>) 1.660392. So the implication is that the difference in means is *not* zero.

If we decide not to impose a direction for the alternative hypothesis, a **two-tail test** of hypothesis is assumed. We might be interested in results in both directions: a possible higher mean suggesting that the *new* website improves page views or a

A	B	C	D	E	F G	H	I	J
3	Obs #	Old	New	Delta				
4	1	5	14	9		t-Test: Paired Two Sample for Means		
5	2	4	10	6				
6	3	7	12	5			New	Old
7	4	5	5	0		Mean	11.83	7.54
8	5	7	9	2		Variance	14.22333333	8.897373737
9	6	6	5	-1		Observations	100	100
10	7	7	10	3		Pearson Correlation	0.183335427	
11	8	6	12	6		Hypothesized Mean Difference	0	
12	9	12	16	4		df	99	
13	10	7	13	6		t Stat	9.843007778	
14	11	2	18	16		P(T<=t) one-tail	1.2036E-16	
15	12	11	16	5		t Critical one-tail	1.660391157	
16	13	12	17	5		P(T<=t) two-tail	2.40721E-16	
17	14	5	9	4		t Critical two-tail	1.9842169	
18	15	10	20	10				
19	16	4	19	15				

Exhibit 3.33   t-test: Paired two sample for means results

lower mean suggesting that the number of *new* site views is lower than before. The critical value (1.9842…) in this case is also much smaller than the *t-Stat* (9.843…). This indicates that we can *reject* the notion that the means for the new and old page views are equal. Thus, outcomes for both the one-tail and two-tail tests suggest that we should believe that the web-site has indeed improved page views.

Although this is not the case in our data, in situations where we consider more than 2 means, and more than a single factor in the sample (currently we consider a visitor's status as a *teen* as a single factor), we can use **ANOVA** (Analysis of Variance) to do similar analysis as we did in the t-tests. For example, what if we determine that gender of the teens might be important and we have an additional new website option? In that case, there are two alternatives. We might randomly select 3 groups of 100 teens each (50 men and 50 women) to view three websites—the old website, a new one from web designer X, and a new one for web designer Y. This is a very different and more complex problem than our paired t-test data analysis, and certainly more interesting. ANOVA is more sophisticated and powerful statistical test than t-tests and they require a basic understanding of inferential statistics. We'll see more of these tests in later chapters.

Finally, we might wonder if most teens are *equally* affected by the new website— Is there a predictable number of additional web-pages that most teens will visit while viewing the new site? Our initial guess would suggest *no* because of the wide distribution of the histogram in Exhibit 3.31. If every teen had been influenced to view exactly 4 more web-pages after viewing the *new* site, then the histogram would indicate a value of 4 for all 100 observations. This is certainly not the results that we

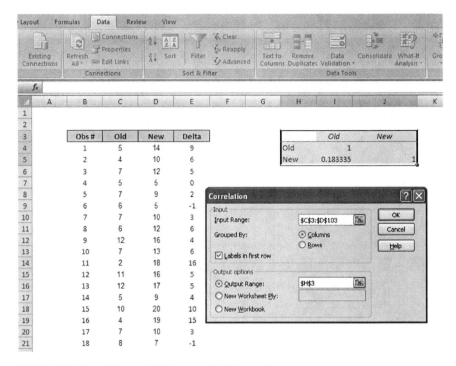

Exhibit 3.34 Correlation matrix for new and old page views

see. One way to test this question is to examine the correlation of the 2 series. Just as we did for the product sales data, we can perform this analysis on the *new* and *old* web-page views. Exhibit 3.34 shows a correlation matrix for the analysis. The result is a relatively low positive correlation (0.183335), indicating a slightly linear movement of the series in the same direction. So, although there is an increase in page views, the increase is quite different for different individuals: some are greatly affected by the *new* site, others are not.

3.6.1 Findings

We have completed a thorough preliminary analysis of the cross-sectional data and we have done so using the *Data Analysis* tools in the *Analysis* group. So what have we learned? The answer is similar to the analysis of the time series data—a great deal. Some of the major findings are presented below:

1. It appears that the change in the website has had an effect on the number of web pages the teens will view when they visit the site. The average increase for the sample is 4.29.

2. There is a broad range in the difference of data (Delta), with 51% occurring from 2 to 6 pages and only 21% of the teens not responding positively to the new website.
3. The 95% confidence interval for our sample of 100 is approximately 0.75 units about (±) the sample mean of 11.83. In a sense, the interval gives us a measure of how uncertain we are about the population mean: larger intervals suggest greater uncertainty.
4. A *t-Test: Paired Two Sample for Means* has shown that it is *highly unlikely* that the means for the *old* and *new* views are equal. This reinforces our growing evidence that the website changes have indeed made a positive difference in page views among teens.
5. To further examine the extent of the change in views for individual teens, we find that our *Correlation* tool in *Data Analysis* suggests a relatively low value of positive correlation. This suggests that although we can expect a positive change with the *new* website, the *magnitude* of change for individuals is not a predictable quantity.

3.7 Summary

Data analysis can be performed at many levels of sophistication, ranging from simple graphical examination of the data to far more complex statistical methods. This chapter has introduced the process of thorough examination of data. The tools we have used are those that are often employed in an initial or preliminary examination of data. They provide an essential basis for a more critical examination of data, in that they guide our future analyses by suggesting new analytical paths that we may want to pursue. In some cases, the analysis preformed in this chapter may be sufficient for an understanding of the data's behavior; in other cases, the techniques introduced in this chapter are simply a beginning point for further analysis.

There are a number of issues that we need to keep in mind as we embark on the path to data analysis:

1. Think carefully about the type of data you are dealing with and ask critical questions to clarify where the data comes from, the conditions under which it was collected, and the measures represented by the data.
2. Keep in mind that not all data analysis techniques are appropriate for all types of data: for example, sampling data versus population data, cross-sectional versus time series, and multi-attribute data versus single attribute data.
3. Consider the possibility of data transformation that may be useful. For example, our cross-sectional data for the new and old website was combined to create a *difference* or Delta data set. In the case of the time series data, we can adjust data to account for outliers (data that are unrepresentative) or one-time events, like promotions.

4. Use data analysis to generate further questions of interest. In the case of the teen's data, we made no distinction between male and female teens, or the actual ages of the teens. It is logical to believe that a 13 year old female web visitor may behave quite differently than a 19 year old male. This data may be available for analysis and it may be of critical importance for understanding behavior.

Often our data is in qualitative form rather than quantitative, or is a combination of both. In the next chapter, we perform similar analyses on qualitative data. It is important to value both data types equally, because they can both serve our goal of gaining insight. In some cases, we will see similar techniques applied to both types of data, but in others, the techniques will be quite different. Developing good skills for both types of analyses is important for anyone performing data analysis.

Key Terms

Add-in

Series

Treatment

Time Series Data

Cross-sectional Data

Cyclicality

Seasonality

Leading

Trend

Linear Trend

E-tailer

Page-views

Frequency Distribution

Central Tendency

Variation

Descriptive Statistics

Mean

Standard Deviation

Population

Range

Median

Mode

Standard Error

Sample Variance

Kurtosis

Skewness

Systematic Behavior

Linear Regression

Dependent Variable

Independent Variable

Simple Linear Regression

Beta

Alpha

R-square

Residuals

Significance F

Covariance

Correlation

Perfectly Positively Correlated

Perfectly Negatively Correlated

Winters' 3-factor Exponential
      Smoothing

Exponential Smoothing

Level of Confidence

t-Test

t-Test: Paired Two Sample for Mean

Type 1 Error

t-Stat

Critical Value

Test of Hypothesis

Null Hypothesis

One-tail Test

Two-tail Test

ANOVA

Problems and Exercises

1. What is the difference between time series and cross-sectional data? Give examples of both?
2. What are the three principle approaches we discussed for performing data analysis in Excel?
3. What is a frequency distribution?
4. Frequency distributions are often of little use with time series data. Why?
5. What are three statistics that provide location information of a frequency distribution?
6. What are two statistics describing the dispersion or variation of frequency distributions?
7. What does a measure of positive skewness suggest about a frequency distribution?
8. If a distribution is perfectly symmetrical, what can be said about its mean, median, and mode?
9. How are histograms and frequency distributions related?
10. What is the difference between a sample and a population?
11. Why do we construct confidence intervals?
12. Are we more of less confident that a sampling process will capture the true population mean if the level confidence is 95 or 99%?
13. What happens to the overall length of a confidence interval as we are required to be more certain about capturing the true population mean?
14. What is the difference between an independent variable and dependent variable in regression analysis?
15. You read in a newspaper article that a Russian scientist has announced that he can predict the fall enrollment of students at Inner Mongolia University (IMU) by tracking last spring's wheat harvest in metric tons in Montana, USA.

    a. What are the scientist's independent and dependent variables?
    b. You are dean of students at IMU, so this announcement is of importance for your planning. But you are skeptical, so you call the scientist in Moscow to ask him about the accuracy of the model. What measures of fit or accuracy will you ask the scientist to provide?

16. The Russian scientist provides you with an alpha (1040) and a beta (38.8) for the regression. If the spring wheat harvest in Montana is 230 metric tons, what is your prediction for enrollment?
17. The Russian scientist claims the sum of all residuals for his model is zero and therefore it is a perfect fit. Is he right? Why or why not?
18. What *Significance F* would you rather have if you are interested in having a model with a significant association between the independent and dependent variables—0.000213 or 0.0213?
19. In the covariance matrix below, answer the following questions:

a. What is the variance of C?
b. What is the covariance of B and D?

	A	B	C	D
A	432.1			
B	−345.1	1033.1		
C	19.23	−543.1	762.4	
D	123.81	−176.4	261.3	283.0

20. What is the correlation between amount of alcohol consumption and the ability to operate an automobile safely—Negative or positive?
21. Consider the sample data in the table below.

Obs. #	Early	Late
1	3	14
2	4	10
3	7	12
4	5	7
5	7	9
6	6	9
7	7	10
8	6	12
9	2	16
10	1	13
11	2	18
12	4	16
13	3	17
14	5	9
15	2	20

a. Perform an analysis of the descriptive statistics for each data category (Early and Late).
b. Graph the two series and predict the correlation between Early and Late—positive or negative?
c. Find the correlation between the two series.
d. Create a histogram for the two series and graph the results.
e. Determine the 99% confidence interval for the Early data.

22. Assume the Early and Late data in problem 21 represent the number of clerical tasks correctly performed by college students, who are asked to perform the tasks Early in the morning and then Late in the morning. Thus, student 4 performs 5 clerical tasks correctly Early in the morning and 7 correctly Late in the morning.

a. Perform a test to determine if the means of the two data categories come from population distributions with the same mean. What do you conclude about the one-tail test and the two-tail test?
b. Create a histogram of the differences between the two series—Late minus Early. Are there any insights that are evident?

23. *Advanced Problem*—Assume the Early and Late data in problem 21 is data relating to energy drinks sold in a college coffee shop on individual days—on day 1 the Early sales of energy drinks were 3 units and Late sales were 14 units, etc. The manager of the coffee shop has just completed a course in data analysis and believes she can put her new found skills to work. In particular, she believes she can use one of the series to predict future demand for the other.

a. Create a regression model that might help the manager of the coffee shop to predict the Late purchases of energy drinks. Perform the analysis and specify the predictive formula.
b. Do you find anything interesting about the relationships between Early and Late?
c. Is the model a good fit? Why?
d. Assume you would like to use the Late of a particular day to predict the Early of the next day—on day 1 use Late to predict Early on day 2. How will the regression model change?
e. Perform the analysis and specify the predictive formula.

Chapter 4
Presentation of Qualitative Data

Contents

4.1 Introduction—What is Qualitative Data?

In Chaps. 2 and 3 we concentrated on approaches for collecting, presenting, and analyzing quantitative data. Here, and in Chap. 5, we turn our attention to qualitative data. Quantitative data is simple to identify; for example, sales revenue in dollars, number of new customers purchasing a product, and units of a SKU (Stock Keeping Unit) sold in a quarter. Similarly, **qualitative data** is easily identifiable. It can be in the form of such variables as date of birth, country of origin, and revenue status among a sales force (1st, 2nd, etc.).

H. Guerrero, *Excel Data Analysis*, DOI 10.1007/978-3-642-10835-8_4,
© Springer-Verlag Berlin Heidelberg 2010

Do quantitative and qualitative data exist in isolation? The answer is a resounding *No*! Qualitative data is very often linked to quantitative data. Recall the Payment example in Chap. 2 (see Table 2.2). Each record in the table represented an invoice and the data fields for each transaction contained a combination of quantitative and qualitative data; for example, *$ Amount* and *Account*, respectively. The *Account* data is associated with the *$ Amount* to provide a set of circumstances and conditions (the context) under which the quantitative value is observed. Of course, there are many other fields in the invoice records that add context to the observation: *Date Received, Deposit, Days to Pay,* and *Comment*.

The distinction between qualitative and quantitative is often subtle. The *Comment* field will clearly contain data that is non-quantitative, yet in some cases we can apply simple criteria to convert non-quantitative data into a quantitative value. Suppose the *Comment* field contained customer comments that could be categorized as either positive of negative. By counting the number in each category we have made such a conversion, from qualitative to quantitative. We could also, for example, categorize the number of invoices in the ranges of *$1-$200* and *>$200* to convert quantitative data into qualitative, or categorical, data. This is how qualitative data is dealt with in statistics—by counting or categorizing.

The focus of Chap. 4 will be to prepare data for eventual analysis. We will do so by utilizing the built-in data presentation and manipulation functionality of Excel. We also will demonstrate how we apply these tools to a variety of data. Some of these tools are available in the *Data* ribbon—*Sort, Filter,* and *Validation*. Others will be found in the cell functions that Excel makes available or in non-displayed functionality, like *Forms*. As in Chap. 3, it will be assumed that the reader has a rudimentary understanding of data analysis, but every attempt will be made to progress through the examples slowly and methodically, just in case those skills are dormant.

4.2 Essentials of Effective Qualitative Data Presentation

There are numerous ways to present qualitative data stored in an Excel worksheet. Although, for the purposes of this book, I make a distinction between the *presentation* and the *analysis* of data, this distinction is often subtle. Arguably, there is little difference between the two, since well conceived presentation often can provide as much insight as mathematical analysis. I prefer to think of presentation as a **soft** form of data analysis, but do not let the term soft imply a form of analysis that is *less* valuable. These types of analyses are often just as useful as sophisticated mathematical analyses. Additionally, soft analysis is often an initial step toward the formal analytical tools (**hard** analysis) that we will encounter in Chap. 5.

4.2.1 Planning for Data Presentation and Preparation

Before we begin our data presentation and preparation, it is essential that we plan and organize our data collection effort. Without thoughtful planning, it is possible to waste enormous amounts of time and energy, and to create frustration. In Chap. 2

we offered some general advice on the collection and presentation of quantitative data. It is worth repeating that advice at this point, but now from the perspective of qualitative data presentation.

1. *Not all data are created equal*—Spend some time and effort considering the type of data that you will collect and how you will use it. Do you have a choice in the type of data? For example, it may be possible to collect ratio data relating to individual's annual income ($63,548), but it may be easier and more convenient to collect the annual income as categorical data (in the category $50,000 to $75,000). Thus, it is important to know prior to collection, how we will use the data for analysis and presentation.

2. *More is better*—If you are uncertain of the specific dimensions of the observation that you will require for analysis, err on the side of recording a greater number of dimensions. For example, if an invoice in our payment data (see Table 4.1) also has an individual responsible for the transaction's origination, then it might be advisable to also include this data as a field for each observation. Additionally, we need to consider the granularity of the categorical data that is collected. For example, in the collection of annual income data from above, it may be wise to make the categories narrower rather than broader: categories of $50,000–$75,000 and $75,001–$100,000 rather than a single category of $50,000–$100,000. Combining more granular categories later is much easier than returning to the original source data to collect data in narrower categories.

3. *More is **not** better*—If you can communicate what you need to communicate with less data, by all means do so. Bloated databases and presentations can lead to misunderstanding and distraction. The ease of data collection may be important here. It may be much easier to obtain information about an individual's income if we provide categories, rather than asking them for an exact number that they may not remember or want to share.

4. *Keep it simple and columnar*— Select a simple, descriptive, and unique title for each data dimension (e.g. *Revenue, Branch Office,* etc.), and enter the data in a column, with each row representing a **record** or *observation* of recorded data. Different variables of the data should be placed in different columns. Each variable in an observation will be referred to as a **field** or *dimension* of the observation. Thus, rows represent records and columns represent fields. See Table 4.1 for an example of columnar formatted data entry.

5. *Comments are useful*—It may be wise to include a miscellaneous dimension reserved for general comments—a comment field. Be careful, because of the variable nature of comments, they are often difficult, if not impossible, to query. If a comment field contains a relatively limited variety of entries, then, it may not be a *general* comment field. In the case of our payment data, the comment field provides further specificity to the account information. It identifies the project or activity that led to the invoice. For example, we can see in Table 4.1 that the record for *Item 1* was *Office Supply* for *Project X.* Since there is a limited number of these project categories, we might consider using this field differently. The title *Project* might be an appropriate field to record for each observation. The *Comment* field could then be preserved for more free form data entry.

Table 4.1   Payment example

Item	Account	$ Amount	Date rcvd.	Deposit	Days to pay	Comment
1	Office Supply	$123.45	1/2/2004	$10.00	0	Project X
2	Office Supply	$54.40	1/5/2004	$0.00	0	Project Y
3	Printing	$2,543.21	1/5/2004	$350.00	45	Feb. Brochure
4	Cleaning Service	$78.83	1/8/2004	$0.00	15	Monthly
5	Coffee Service	$56.92	1/9/2004	$0.00	15	Monthly
6	Office Supply	$914.22	1/12/2004	$100.00	30	Project X
7	Printing	$755.00	1/13/2004	$50.00	30	Hand Bills
8	Office Supply	$478.88	1/16/2004	$50.00	30	Computer
9	Office Rent	$1,632.00	1/19/2004	$0.00	15	Monthly
10	Fire Insurance	$1,254.73	1/22/2004	$0.00	60	Quarterly
11	Cleaning Service	$135.64	1/22/2004	$0.00	15	Water Damage
12	Orphan's Fund	$300.00	1/27/2004	$0.00	0	Charity
13	Office Supply	$343.78	1/30/2004	$100.00	15	Laser Printer
14	Printing	$2,211.82	2/4/2004	$350.00	45	Mar. Brochure
15	Coffee Service	$56.92	2/5/2004	$0.00	15	Monthly
16	Cleaning Service	$78.83	2/10/2004	$0.00	15	Monthly
17	Printing	$254.17	2/12/2004	$50.00	15	Hand Bills
18	Office Supply	$412.19	2/12/2004	$50.00	30	Project Y
19	Office Supply	$1,467.44	2/13/2004	$150.00	30	Project W
20	Office Supply	$221.52	2/16/2004	$50.00	15	Project X
21	Office Rent	$1,632.00	2/18/2004	$0.00	15	Monthly
22	Police Fund	$250.00	2/19/2004	$0.00	15	Charity
23	Printing	$87.34	2/23/2004	$25.00	0	Posters
24	Printing	$94.12	2/23/2004	$25.00	0	Posters
25	Entertaining	$298.32	2/26/2004	$0.00	0	Project Y
26	Orphan's Fund	$300.00	2/27/2004	$0.00	0	Charity
27	Office Supply	$1,669.76	3/1/2004	$150.00	45	Project Z
28	Office Supply	$1,111.02	3/2/2004	$150.00	30	Project W
29	Office Supply	$76.21	3/4/2004	$25.00	0	Project W
30	Coffee Service	$56.92	3/5/2004	$0.00	15	Monthly
31	Office Supply	$914.22	3/8/2004	$100.00	30	Project X
32	Cleaning Service	$78.83	3/9/2004	$0.00	15	Monthly
33	Printing	$455.10	3/12/2004	$100.00	15	Hand Bills
34	Office Supply	$1,572.31	3/15/2004	$150.00	45	Project Y
35	Office Rent	$1,632.00	3/17/2004	$0.00	15	Monthly
36	Police Fund	$250.00	3/23/2004	$0.00	15	Charity
37	Office Supply	$642.11	3/26/2004	$100.00	30	Project W
38	Office Supply	$712.16	3/29/2004	$100.00	30	Project Z
39	Orphan's Fund	$300.00	3/29/2004	$0.00	0	Charity

6. *Consistency in category titles*—Although upon casual viewing, you may not consider that there is a significant difference between the category entries *Office Supply* and *Office Supplies* for the *Account* field, Excel will view them as completely distinct entries. As intelligent as Excel is, it requires you to exercise very precise and consistent use of entries. Even a hyphen makes a difference; the term *H-G* is different that *H G* in the mind of Excel.

Now, let us reacquaint ourselves with data we first presented in Chap. 2. Table 4.1 contains this quantitative and qualitative data and it will serve as a basis for some of our examples and explanations. We will concentrate our efforts on the three qualitative records in the data table: *Account* (account type), *Date Rcvd.* (date received), and *Comment.* Recall that the data is structured as 39 records with 7 data fields. One of the 7 data fields is an identifying number (*Item*) associated with each record that can also be considered categorical data, since it merely identifies that record's chronological position in the data table. Note that there is also a date, *Date Rcvd.*, that provides similar information, but in different form. We will see later that *both* these fields serve useful purposes.

Next, we will consider how we can reorganize data into formats that enhance understanding and facilitate preparation for analysis. The creators of Excel have provided a number of extremely practical tools: *Sort, Filter, Form, Validation,* and *PivotTable/Chart.* (We will discuss PivotTable/Chart in Chap. 5). Besides these tools, we will also use cell functions to *prepare* data for graphical presentation. In the forthcoming section, we concentrate of data entry and manipulation. Later sections will demonstrate the sorting and filtering capabilities of Excel, which are some of the most powerful utilities in Excel's suite of tools.

4.3 Data Entry and Manipulation

Just as was demonstrated with quantitative data, it is wise to begin your analytical journey with a thorough visual examination of qualitative data before you begin the process of formal analysis. It also may be necessary to manipulate the data to permit clearer understanding. This section describes how clarity can be achieved with Excel's data manipulation tools. Additionally, we examine a number of techniques for secure and reliable data entry and acquisition. After all, data that is incorrectly recorded will most likely lead to analytical results that are incorrect; or to repeat an old saw—garbage in, garbage out!

4.3.1 Tools for Data Entry and Accuracy

We begin with the process of acquiring data; that is, the process of taking data from some outside source and transferring it to an Excel worksheet. Excel has two very useful tools, *Form* and *Validation*, that help the user enter data accurately. Data entry is often tedious and uninteresting, and as such, it can lead to entry errors. If we are going to enter a relatively large amount of data, then these tools can be of great benefit. An alternative to data entry is to *import* data that may have been stored in software other than Excel—a database, a text file, etc. Of course, this does not eliminate the need to thoroughly examine the data for errors, specifically, someone else's recording errors. Let us begin by examining the *Form* tool. This tool permits a highly structured and error proof method of data entry. The *Form* tool is one that is not shown in a ribbon and therefore must be added to the Quick Access toolbar

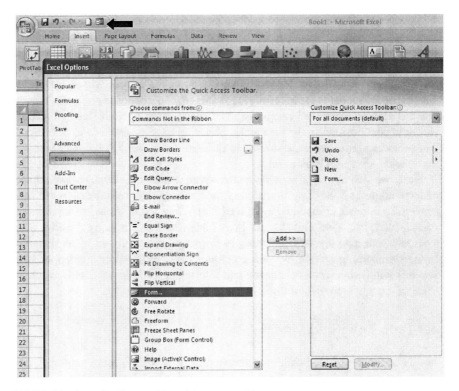

Exhibit 4.1   Accessing form tool in quick access tool bar

shown at the top left of a workbook near the Office button. The Excel Options at the bottom of the Office button menu permits you to *Customize* the *Quick Access Toolbar*. This process is shown in Exhibit 4.1 and the result is an icon in the Quick Access that looks like a form (see the arrow).

Form, as the name implies, allows you to create a convenient form for data entry. We begin the process by creating titles in our familiar columnar data format. As before, each column represents a field and each row is a record. The *Form* tool assumes these titles will be used to guide you in the data entry process (see Exhibit 4.2). Begin by capturing the range containing the titles, B2:C2, with your cursor and then find the *Form* tool in the *Quick Access Toolbar*. As you can see from Exhibit 4.2, the tool will prompt you to enter new data for each field of the two fields identified—*Name* and *Age*. Each time you depress the button entitled *New*, the data is transcribed to the data table, just below the last data entry. In this case the name Maximo and age 22 will be entered below Sako and 43. This process can be repeated as many times as necessary and results in the creation of a simple worksheet database.

The *Form* tool also permits convenient search of the data entered into the database. Begin by selecting the entire range containing the database. Then using the *Form* tool, select the *Criteria* button to specify a search criterion for the search:

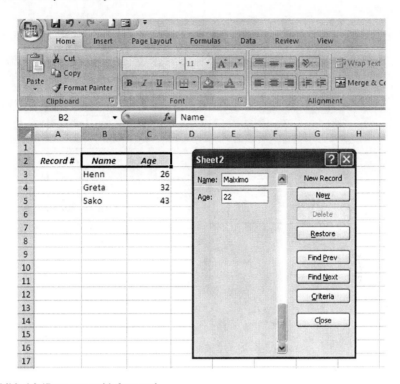

Exhibit 4.2 Data entry with form tool

a particular *Name* or *Age*. Next, select the *Find Next* or *Find Prev* option, to search your database. This permits a search of records containing the specific search criteria, for example, the name Greta. By placing Greta in the *Name* field of the form and then depressing *Find Next*, the form will return the relative position of the field name Greta above the New button—2 of 4: the second record out of four total records. This is shown in Exhibit 4.3, as well as all of the other fields related to that record. Note that we could have searched the age field for a specific age, for example 26. Later in this chapter we will use *Filter* and *Advance Filter* to achieve the same end, with far greater search power.

Validation is another tool in the *Data* menu that can be quite useful for promoting accurate data entry. It permits you to set a simple condition on values placed into cells and returns an error message if that condition is not met. For example, if our database in Exhibit 4.2 is intended for individuals with names between 3 and 10 characters, you are able to set this condition with *Validation* and return a message if the condition is not met. Exhibit 4.4 shows how the condition is set: (1) capture the data range for validation, (2) find the *Validation* tool in the *Data* ribbon and *Data Tools* group, (3) set the criterion, in this case *Text length*, but many others are available, (4) create a message to be displayed when a data input error occurs (a default message is available), and (5) proceed to enter data.

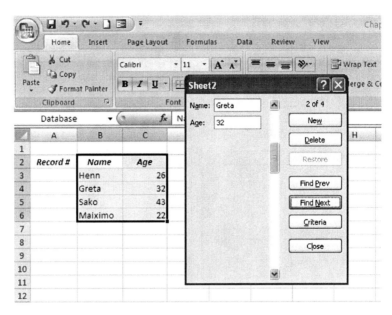

Exhibit 4.3 Search of database with the form tool

Together these tools can make the process of data entry less tedious and far more accurate. Additionally, they permit maintenance and repair capability by allowing search of the database for records that you have entered. This is important for two reasons. First, databases acquire an enduring nature because of their high costs and the extensive effort required to create them; they tend to become sacrosanct. Any tools that can be made available for maintaining them are, therefore, welcomed. Secondly, because data entry is simply not a pleasant task, tools that lessen the burden are also welcomed.

4.3.2 Data Transposition to Fit Excel

Occasionally, there is the need to manipulate data to make it useful. Let's consider a few not uncommon examples where manipulation is necessary:

1. We have data located in a worksheet, but the rows and columns are interchanged. Thus, rather than each row representing a record, each column represents a record.
2. We have a field in a set of records that is not in the form needed. Consider a situation where ages for individuals are found in a database, but what is needed is an alphabetic or numeric character that indicates membership in a category. For example, an individual of 45 years of age should belong to the category 40–50 years which is designated by the letter "D".

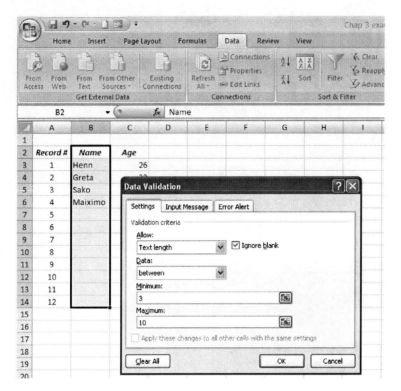

Exhibit 4.4   Data entry with data validation tool

3. We have data located in an MS Word document, either as a table or in the form
    of structured text, that we would like to import and duplicate in a worksheet.

There are many other situations that could require manipulation of data, but
these cover some of the most commonly encountered. Conversion of data is a very
common activity in data preparation.

Let's begin with data that is not physically oriented as we would like; that is, the
inversion of records and fields. Among the hundreds of cell functions in Excel is the
Transpose function. It is used to transpose rows and columns of a table. The use
of this cell formula is relatively simple, but does require one small difficulty—entry
of the transposed data as an **Array**. Arrays are used by Excel to return multiple
calculations. It is a convenient way to automate many calculations with a single
formula and arrays are used in many situations. We will find other important uses
of arrays in future chapters. The difference in an array formula and standard cell
formulas is that the entry of the formula in a cell requires the keystrokes *Ctrl-Shift-
Enter* (simultaneous key strokes), as opposed to simply keying of the *Enter* key.

The steps in the transposition process are quite simple and are shown in
Exhibit 4.5:

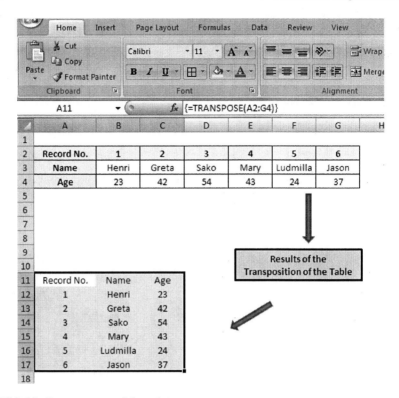

Exhibit 4.5 Data transpose cell formula

1. Identify the source data to be transposed (A2:G4): simply know where it is located and the number of columns and rows it contains.
2. Select and capture a target range where the data transposition will take place— A11:C17. The target range for transposition must have the same number of *columns* as the source has *rows* and the same number of *rows* as the source has *columns*—A2:G4 has 3 rows and 7 columns which will be transposed to the target range A11:C17 which has 7 rows and 3 columns.
3. While the entire target range is captured, enter the *Transpose* formula. It is imperative that the entire target range remain captured throughout this process.
4. The last step is very important, in that it creates the array format for the target range. Rather than depressing the *Enter* key to complete the formula entry, simultaneously depress *Ctrl-Shift-Enter*, in that order of key strokes.
5. Interestingly, the formula in the target range will be the same for all cells in the range: {=TRANSPOSE(A2:G4)}
6. The brackets ({}) surrounding the formula, sometimes called curly brackets, designate the range as an array. The only way to create an array is to use the *Ctrl-Shift-Enter* sequence of key strokes (the brackets are automatically produced). Note that physically typing the brackets will not create an array in the range.

4.3.3 Data Conversion with the Logical IF

Next, we deal with the conversion of a field value from one form of an alpha-numeric value to another. Why convert? Often data is entered in a particular form that appears to be useful, but later the data must be changed or modified to suit new circumstances. Thus, this makes data conversion necessary. For example, we often collect and enter data in the greatest detail possible (although this may seem excessive), anticipating we might need less detail later. How data will be used later is uncertain, so generally we err on the side of collecting data in the greatest detail. This could be the case for the quantitative data in the payment example in Table 4.1. This data is needed for accounting purposes, but we may need far less detail for other purposes. We could categorize the payment transactions into various ranges, for example $0–$250. Later these categories could be used to provide specific personnel the authority to make payments in specific ranges. For example, Mary is allowed to make payments up to $1000 and Naomi is allowed to make payments up to $5000. Setting rules for payment authority of this type is quite common.

To help us with this conversion, I introduce one of Excel's truly useful cell functions, the **logical IF**. I guarantee that you will find hundreds of applications for the logical *IF* cell function. As the name implies, a logical *IF* asks a question or examines a condition. If the question is answered positively (the condition is *true*) then a particular action is taken; otherwise (the condition is *false*), an alternative action is taken. Thus, an *IF* function has a dichotomous outcome: either the condition *is* met and action A is taken, or it *is not* met and action B is taken. For example, what if we would like to know the number of observations in the payment data in Table 4.1 that correspond to four categorical ranges which include: $0–$250; $251–$1000; $1001–$2000; $2001–$5000. As we suggested above, this information might be important to assign individuals with authority to execute payment for particular payment categories. Thus, payment authority in the $2000–$5000 range may be limited to a relatively few individuals, while authority for the range $0–$250 might be totally unrestricted.

The basic structure of a logical IF cell function is: *IF (logical test, value if true, value if false)*. This structure will permit us to identify two categories of authority only; for example, if a cell's value is between 0 and 500 return *Authority 1*, otherwise return *Authority 2*. So how do we use a logical *IF* to distinguish between more than two categorical ranges? The answer to this question is to insert another *IF* for the *value if false* argument. Each *IF* inserted in this manner results in identifying an additional condition. This procedure is known as **nested IF's**. Unfortunately, there is a limit of 7 *IF* functions that can be nested, which will provide 8 conditions that can be tested.

Let us consider an example of a nested *IF* for the payment ranges above. Since there are 5 distinct ranges (including the out of range values greater than or equal to $5001), we will need 4 (one less than the number of conditions) *IF*'s to test for the values of the 5 categorical ranges. The logic we will use will test if a cell value, *$ Amount*, is below a value specified in the *IF* function. Thus, we will successively, and in ascending order, compare the cell value to the upper limit of each range.

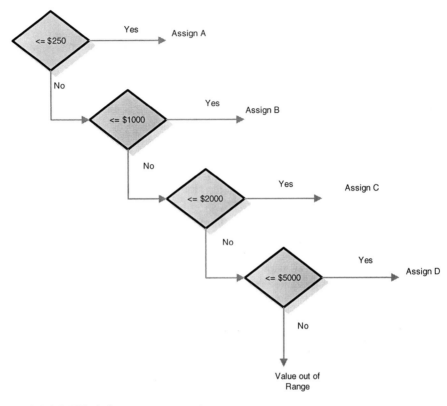

Exhibit 4.6 IF logic for payments categories

Exhibit 4.6 shows the logic necessary to designate each range with an alphabetic
value, *A–E*. Note that if we do not nest the *IF's* in successively increasing value,
we run the risk of prematurely assigning an alphabetic value that is indeed true,
but not as accurate as it could be in assigning categories. For example, if we make
the second comparison <= $2000 and the third <=$1000, a value of $856 would
be assigned a value at the second condition rather than the third. Clearly the third
condition is preferred, but unavailable due to the structure of the conditions. Note
in Exhibit 4.6 that the last test can lead to a *Value out of Range* condition. This
condition is equivalent to *E*.

So how do we convert the logic in Exhibit 4.6 to a logical *IF*? We do so by nesting
each condition in the *value if false* argument of the *IF's*. Assuming the cell value we
are testing is in a cell location such as D2, the cell formula will appear as follows:

$$= \text{IF (D2} <= 250, \text{"A", IF (D2} <= 1000, \text{"B", IF (D2} <= 2000, \text{"C", IF}$$
$$\text{(D2} <= 5000, \text{"D" "E"}))))$$

The cell H2 in Exhibit 4.7 shows the *IF* function that performs the comparisons,
and returns an appropriate alphabetic value—A, B, etc. By reading the *IF* function

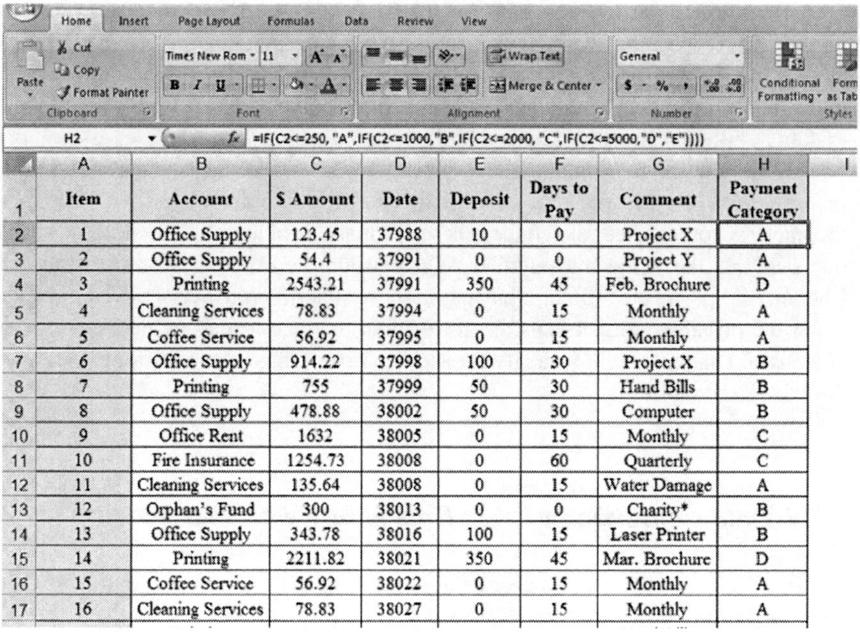

	A	B	C	D	E	F	G	H	I
1	Item	Account	$ Amount	Date	Deposit	Days to Pay	Comment	Payment Category	
2	1	Office Supply	123.45	37988	10	0	Project X	A	
3	2	Office Supply	54.4	37991	0	0	Project Y	A	
4	3	Printing	2543.21	37991	350	45	Feb. Brochure	D	
5	4	Cleaning Services	78.83	37994	0	15	Monthly	A	
6	5	Coffee Service	56.92	37995	0	15	Monthly	A	
7	6	Office Supply	914.22	37998	100	30	Project X	B	
8	7	Printing	755	37999	50	30	Hand Bills	B	
9	8	Office Supply	478.88	38002	50	30	Computer	B	
10	9	Office Rent	1632	38005	0	15	Monthly	C	
11	10	Fire Insurance	1254.73	38008	0	60	Quarterly	C	
12	11	Cleaning Services	135.64	38008	0	15	Water Damage	A	
13	12	Orphan's Fund	300	38013	0	0	Charity*	B	
14	13	Office Supply	343.78	38016	100	15	Laser Printer	B	
15	14	Printing	2211.82	38021	350	45	Mar. Brochure	D	
16	15	Coffee Service	56.92	38022	0	15	Monthly	A	
17	16	Cleaning Services	78.83	38027	0	15	Monthly	A	

Cell H2 formula: `=IF(C2<=250, "A",IF(C2<=1000,"B",IF(C2<=2000, "C",IF(C2<=5000,"D","E"))))`

Exhibit 4.7   IF function conversion to payments categories

left to right, the function first tests the condition—is the C2 cell content *less than or equal* to 250? If the answer is *yes* (true) the function returns the text value A in the cell where the function is located; If the answer is *no* (false), then the function tests the next condition— are the contents of C2 less than or equal to 1000? The process is repeated until a condition is met. If the last condition, C2<=5000, is not met, a default value of E terminates the *IF*, implying that the value in C2 is greater than 5000. Note that the category to be returned is placed in quotes ("A") to designate it as text and whatever is placed in the quotes will be returned in the cell as text. Of course there are many other comparisons, like *less than and equal*, that can be made. These include *greater than and equal*, *greater*, and *equal to*, just to name a few. Additionally, there are numerous logic functions that can be used in conjunction with *IF* to perform other logical comparisons: **AND, OR, NOT, FALSE** and **TRUE**.

Note that our example has violated one of our best spreadsheet Feng Shui practices—a formula used in numerous locations should not commonly contain numeric values since a change in values will necessitate a change in every cell location. It would be wise to create a table area where these numerical values can be stored and referenced by the *IF*, and thus, a single change to the table will be reflected in all cells using the value. Be sure to use an **absolute address** (e.g. $A$1), one that has $'s in front of the row and column, if you plan to copy the function to other locations, otherwise the reference to the table will change and result in errors.

As mentioned earlier, you can use the nested *IF* function structure to accommodate up to 8 comparisons. If your comparisons exceed 8, I recommend vertical (**VLOOKUP**) and horizontal (**HLOOKUP**) lookups. We will learn more about these functions in later chapters. These functions make comparisons of a cell value to data in either vertically or horizontally oriented tables. For example, assume that you are calculating marginal tax rates. A **LOOKUP** will compare a particular gross income, to values located in a table, and return the rate associated with the gross income. As you can see, the concept is to compare a value to information in a table that *converts* the value into another related value, and then returns it to the cell. Lookups are performed often in business. These functions are very convenient, even when the possible comparisons are fewer than 8, since it may be necessary occasionally to change a table value used for comparison; thus, it can be done directly in the LOOKUP table.

4.3.4 Data Conversion of Text from Non-Excel Sources

Finally, it is not unusual to receive data in a text file. This data may be in tables that we request from other sources, or data in a non-table format in a text file. Consider the following example. Suppose we request data from a sales force (three sales agents) regarding sales of a particular product for their three largest accounts. In particular, you are interested in the agents providing the names, age, and quantity of year-to-date sales for their top three accounts. Depending on their understanding of Excel, you are likely to receive data in numerous forms, and quite likely in a text file, either as a table or simply as word-processed text. Our job is to insert this data into a worksheet, where we can examine it closely and perform analysis.

Your initial reaction upon receipt of the data might be frustration—Why can't these sales people simply learn to use a spreadsheet? In fact, you wish they would use a spreadsheet that allows entry of their data in a standard format, using the *Forms* and *Validation* capability of Excel we have already studied. Let's assume we solicit data from the 3 sales agents. One sends an MSWord file containing a table, another sends an MSWord file with data delimited with tabs (data elements that are separated by tabs), while the last (and the laziest) sends a Word file with no data delimiters other than spaces. Exhibit 4.8 shows the three documents received. Note that the data contains three records with three fields for each sales associate.

Transcribing the table, section *a* of Exhibit 4.8, is quite easy. Simply copy the entire table in the Word document by capturing all cells and applying the copy command, and then find the location in your Excel sheet where you want to locate the data. The paste command will then place each item contained in each cell of the table into a corresponding cell of the worksheet. Thus, the mapping of the Excel worksheet data will be exactly that of the original text data. The paste also transfers the format of the table. For example, the shaded titles will be transferred to the worksheet causing those cells to also be shaded. This type of transfer is a preferred

Name	Age	Quantity
Morris	56	734
Lopez	76	1237
Hamadi	43	873

Name, ages, and quantity are:		
Chen	46	544
Gandi	43	837
Oh	63	785

Names, ages, and quantities are:		
McGee	49	649
Geller	52	647
Aquino	77	930

(a)   (b)   (c)

Exhibit 4.8   Various text formats for entry into excel

approach to transfer text since it leads to very little effort. The next example is also quite simple to transcribe.

The process for transcription of the tabbed list is essentially the same as that of the table. Just as before, the entire text table, section *b* in Exhibit 4.8, is captured and copied, then it is pasted to the worksheet. This fills the cells of the worksheet with the **tab delimited** text (text that is separated by tabs). Each tab on a line represents a movement to an adjacent column in the worksheet, so take care to insure that the text is tabbed precisely as you want it to appear on the worksheet.

Finally, the last type of data, with no tabs and no table structure, is the most problematic. Yet, there is a solution for the transcription of the unformatted text data to a worksheet. The Excel software designers seem to have thought of everything, well, almost everything. The process, as before, begins by copying the text data from section *c* in Exhibit 4.8. Next the text is pasted into a cell in the worksheet. This places all data elements into a single cell. Yet, as we know, our goal is to place each data item into an individual cell. The distribution of the data elements to cells is achieved by locating the **Text to Column Command** in the *Data* menu and *Data Tools* group, as shown in the wizard steps in Exhibits 4.9, 4.10, 4.11, and 4.12. This command assesses the type of data that you would like to distribute to columns and step one of the wizard, shown in Exhibit 4.10, identifies the data as delimited. Our data is not delimited by tabs or commas, but it is delimited by spaces, as shown in step 2 and Exhibit 4.11. You are even provided a preview of the distribution of the data. The final results are shown in Exhibit 4.12.

Although this process is relatively simple to complete, it can become cumbersome. Therefore, I would strongly suggest to the sales associate that they enter the data in table or in a tabbed list. Regardless, these tools can turn what appears to be a monumental headache into a relatively simple procedure.

We have spent a considerable amount of time discussing the collection, entry, and modification of data, yet there remains an inherent problem with the preparation of data—we can't be totally certain how the data will be used. Thus, it is important to invest substantial effort at this stage to anticipate how the data might be used. A few hours of planning and preparation at this point can save hundreds later. In the next section we begin the process of asking some basic questions of our data by using the *Sort, Filter* and *Advanced Filter* functions.

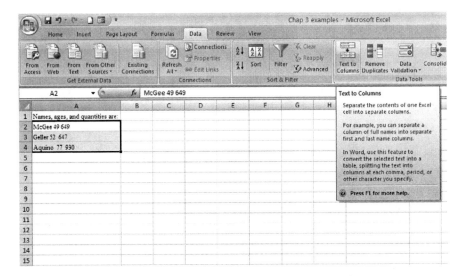

Exhibit 4.9   Text to columns tool in data menu

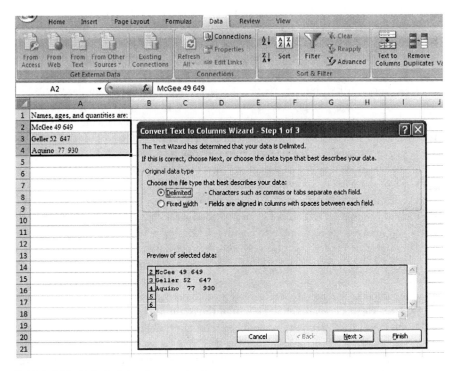

Exhibit 4.10   Delimited data format selection

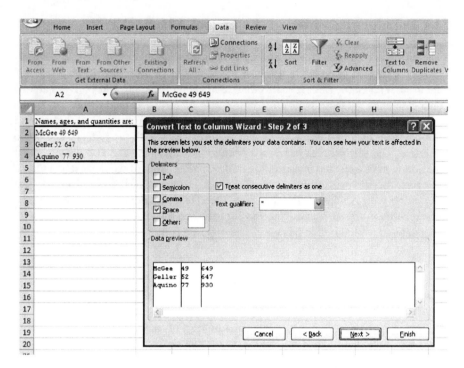

Exhibit 4.11 Space delimited text conversion

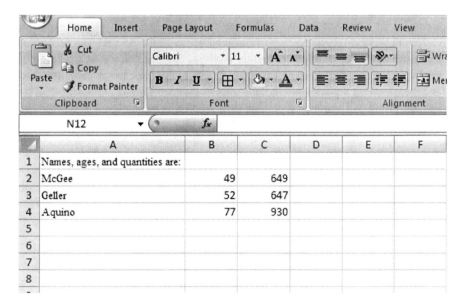

Exhibit 4.12 Results of text to column conversion

4.4 Data queries with Sort, Filter, and Advanced Filter

The difference between a **Sort** and a **Filter** is quite simple: sorts are used to *rearrange* the records of an entire database based on one or more **sort keys,** and filters are used to *eliminate* (hide) records that do not conform to **filter keys. Keys** are the data elements, database field *titles* in the case of sorts and field *conditions* in the case of filters, which permit execution of the function. For example, in the case of our payment data in Table 4.1, we can perform a sort with the primary key set to *Account.* The sort process requests a primary key (*Sort by*) and also asks whether the sort is to be executed in ascending or descending order. If we request an ascending sort, the sort will return *Account* records in alphabetical order with any text beginning with A's listed first, B's next, etc. Thus, *Cleaning Services* will be the first records to appear because the sort is conducted on the first alphabetic character encountered, which is C in this case. The opposite is true for descending sorts.

4.4.1 Sorting Data

A single key is valuable in a sort, but a hierarchy of several keys is even more useful. Two additional keys (*Then by*) are permitted that will sort the result of each prior sort in consecutive order. These sorts will *re-sort* the database after each higher level sort in executed. Thus, a *Then by* sort by *$ Amount* will maintain the initial sort and perform another sort within each alphabetic cluster of records. This is convenient when an initial sort results in large clusters of first key records that need further sorting.

Exhibit 4.13 displays the *Sort* of our database with the primary key as *Account* and a secondary key as *$Amount* and the dialogue box that permits search key entry. We begin the *Sort* execution by capturing the entire database range, including titles, and then locating the *Sort* tool in the *Data* ribbon (also located in the Home ribbon Editing groups). The range identifies the data that is to be sorted. Upon entry of primary (Account) and secondary ($ Amount) key, execution of the sort results in the accounts being arranged in ascending alphabetical order, and within a particular *Account* (Cleaning Services), the *$ Amount* is also sorted in ascending numerical value. Thus, item number 11 is the last *Cleaning Service* to appear because it has the highest *$ Amount* ($135.64).

As you can see, the Sort tool is a convenient method for quickly organizing records for presentation. To return to the original data organization, select the *Undo* arrow in the *Quick Access Toolbar.* Alternatively, the *Item* number is always available to use as a sort key. In fact, this is another good reason why we use an item number or record number to identify each record. It permits us to reconstruct the original data by using a field that identifies the original order of records.

As useful as sorts are, there are many situations when we are interested in viewing only particular records. In the world of database management, this process is referred to as a *query.* As we noted earlier, the *Data* ribbon contains a group

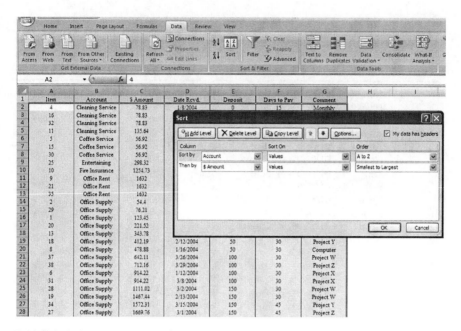

Exhibit 4.13   Sort by account and $ amount

entitled *Sort* and *Filter*, within which we find two tools for querying data—*Filter* and *Advanced Filter*. These tools enable you to perform both simple and complex queries of the database. For example, suppose you want to see all records that occur between January 1, 2004 (01/01/04) and February 23, 2004 (02/23/04). This requires a relatively simple query since it is based on a single dimension—time. We will see that it is possible to perform more complex queries with both *Filter* and *Advanced Filter*.

If you are dealing with a large database and you have a need for very sophisticated queries and report generation, you should consider using a software package designed explicitly for that purpose, like MS Access. But Excel does permit considerable capability for a reasonably large database, and it does not require expertise in SQL (Structured Query Language), which are languages used to query sophisticated relational databases. As a rule of thumb, I use 500 records and 50 fields as a maximum limit for a database that I will build in Excel. Beyond that size I am inclined to move the data to MS Access. There should also be the consideration of the *relational* nature of data elements for maintenance of the database. These relationships are best handled by a **relational database** like Access. Obviously, in business there may many occasions for creating databases that are large: tens of thousands of records and hundreds of fields. These could include databases for Customer Relationship Management (**CRM**) and Enterprise Resource Planning (**ERP**). Yet, this does not diminish the usefulness and convenience of Excel's database capabilities, and it is often the case that we download portions of large databases to Excel for analysis.

Table 4.2   Auto sales example

Auto Sales Data—01/01/2005—01/31/2005

Rcd No.	Slsprn	Date	Make	Model	Amt paid	Rebates	Sales com
1	Bill	01/02/05	Ford	Wgn	24000	2500	2150
2	Henry	01/02/05	Toyota	Sdn	26500	1000	2550
3	Harriet	01/03/05	Audi	Sdn	34000	0	3400
4	Ahmad	01/06/05	Audi	Cpe	37000	0	5550
5	Ahmad	01/06/05	Ford	Sdn	17500	2000	2325
6	Henry	01/08/05	Toyota	Trk	24500	1500	2300
7	Lupe	01/10/05	Ford	Wgn	23000	2500	2050
8	Piego	01/12/05	Ford	Sdn	14500	500	1400
9	Kenji	01/13/05	Toyota	Trk	27000	1200	2580
10	Ahmad	01/14/05	Audi	Cpe	38000	0	5700
11	Kenji	01/16/05	Toyota	Trk	28500	1500	2700
12	Bill	01/16/05	Toyota	Sdn	23000	2000	2100
13	Kenji	01/18/05	Ford	Wgn	21500	1500	2000
14	Ahmad	01/19/05	Audi	Sdn	38000	0	5700
15	Bill	01/19/05	Ford	Wgn	23000	1000	2200
16	Kenji	01/21/05	Toyota	Trk	26500	1500	2500
17	Lupe	01/24/05	Ford	Sdn	13500	500	1300
18	Piego	01/25/05	Ford	Sdn	12500	500	1200
19	Bill	01/26/05	Toyota	Trk	22000	1000	2100
20	Ahmad	01/29/05	Audi	Cpe	36500	0	5475
21	Bill	01/31/05	Ford	Sdn	12500	500	1200
22	Piego	01/31/05	Ford	Sdn	13000	500	1250

4.4.2 Filtering Data

Now let's take a look at the use of filters—**Filter** and **Advanced Filter**. Recall that filtering differs from sorting in that it filters out records that do not match user provided criteria, while sorts rearrange records according to a key. Consider an Excel database that contains an entire quarter's sales transactions related to an auto dealership's sales force. The database documents the details of individual sales of autos as they occur. Such data will likely include the name of the salesperson, the vehicle sold, the amount paid for the vehicle, the commission earned by the salesperson, any rebates or bonuses the buyer receives on the sale, the amount of time from first contact with the customer until sale, etc. Table 4.2 shows an example of this database. There are 22 records and 8 fields and many queries that we can perform on this database.

4.4.3 Filter

Let us begin by using the *Filter* tool in the *Data* ribbon. As we have done before, we must capture the range in the worksheet containing the database to be used in the queries. In this case, we have a choice. We can either capture the entire range

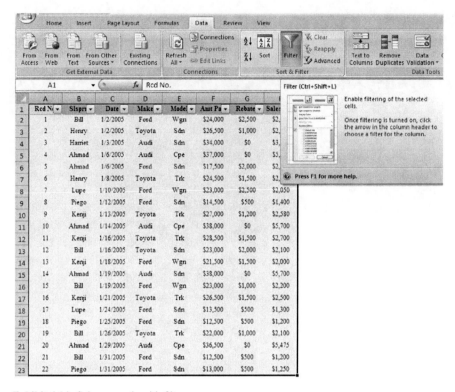

Exhibit 4.14   Sales example with filter

that contains the database (A1:H23), or we can simply capture the row containing the field titles (A1:H1). Excel assumes that if the field title row is captured, the rows containing data that follow directly below represent the records of interest. Exhibit 4.14 shows the steps involved: (1) I prefer to capture the database including titles (A1: H23), (2) select the *Data* ribbon, and (3) select *Filter* in the *Sort and Filter* group. Excel will place a pull-down menu into each title cell and will make available for selection as many unique items as appear in each column containing data. This is shown in Exhibit 4.15. For example, if we select the pull-down menu in column B, *Slsprn* (salesperson), all the salesperson names will be made available—Bill, Henry, Harriet, Ahmad, etc. Selecting an individual name, say Bill, will return all records related to the name Bill and temporarily hide all others that do not pertain. To return to the entire set of records, simply *de-select* the *Filter* by selecting the *Filter* again, or select (*All. . .*) in the pull-down menu of the *Slsprn* field. The former removes the *Filter* pull-down menus and exits, and the latter returns all records without exiting the *Filter* tool. Thus, the database is never permanently changed, although some records may be hidden temporarily. Also, we can progressively select other fields for further filtering of the data. This makes each successive query more restrictive.

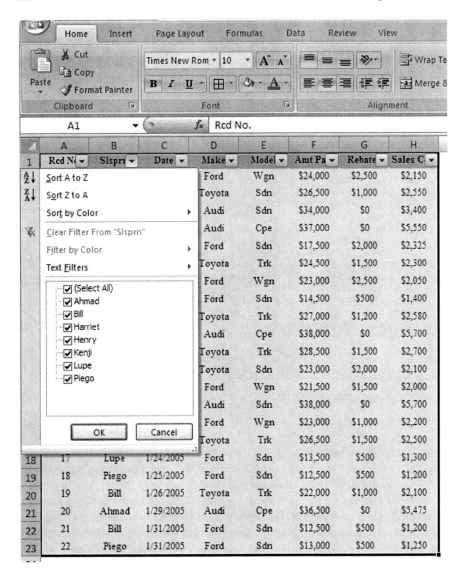

Exhibit 4.15 Filter keys in salesperson (Slsprn)

Now let's query the database to demonstrate the power of *Filter*. We will begin by determining which sales occurred on and between 01/01/05 and 01/014/05. Exhibit 4.16 shows the unique data items in the *Date* pull-down menu as identified by Excel. Note that every unique date in the database is represented. There is also a menu item entitled *Custom*. This selection will allow you to create a query for a range of dates rather than a single unique date. By selecting the *Custom* menu item, you are permitted numerous logical conditions by which a query can be constructed. Exhibit 4.17 shows the dialogue box that contains the logical options needed to

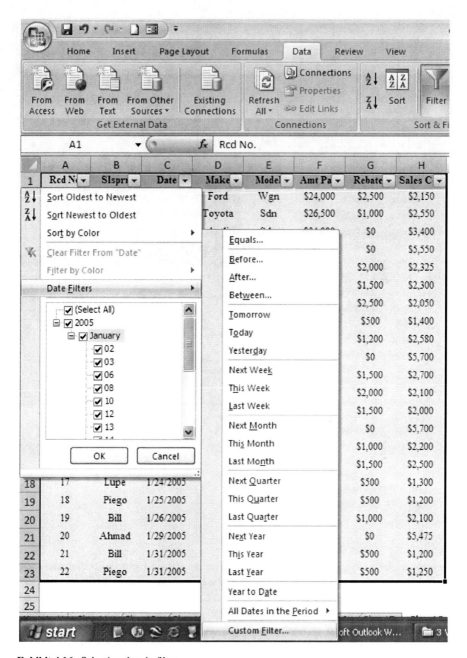

Exhibit 4.16   Selecting date in filter

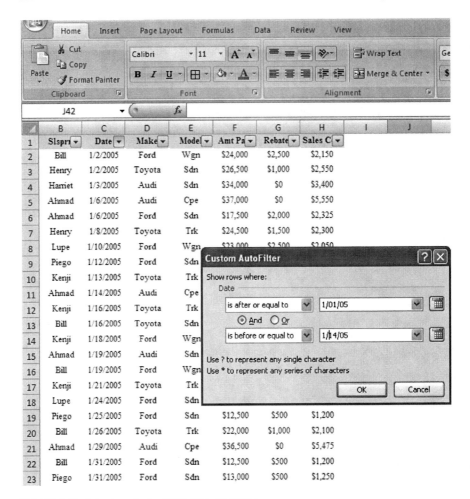

Exhibit 4.17   Selection of sales 01/01/05 to 01/14/05

create a query. We select the *is after or equal to* option from the pull-down menu available for *Date* options, and then enter the date of interest (1/01/05) to establish one end of our range. The process is repeated for the other end of the range (1/14/05) with *is before or equal to*, and the *And* button is also selected to meet both logical conditions. The results are shown in Exhibit 4.18.

Suppose we want to query our database to determine the sales for Audi or Toyota. Exhibit 4.19 demonstrates how we search the *Make* field for either Audi *or* Toyota. A total of 12 records, shown in Exhibit 4.20, are returned. Although a query for two makes of autos is quite valuable, a query for a third auto is not possible, thus *Filter* has limitations. But it is possible to select another field and apply an additional query condition. For example, the dates previously queried could be simultaneously

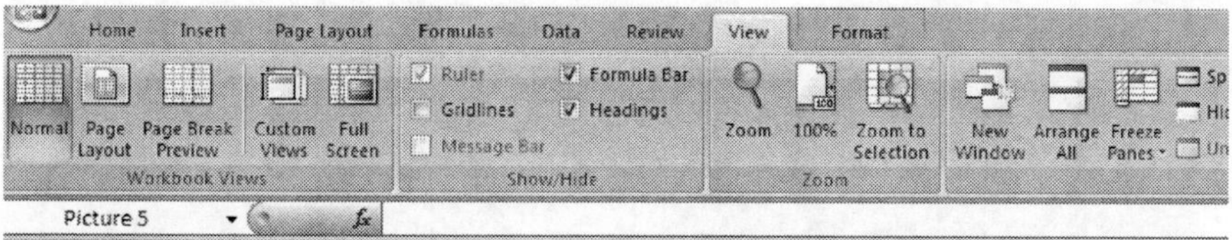

	A	B	C	D	E	F	G	H
1	Rcd No	Slsprn	Date	Make	Model	Amt Pai	Rebate	Sales Co
2	1	Bill	1/2/2005	Ford	Wgn	$24,000	$2,500	$2,150
3	2	Henry	1/2/2005	Toyota	Sdn	$26,500	$1,000	$2,550
4	3	Harriet	1/3/2005	Audi	Sdn	$34,000	$0	$3,400
5	4	Ahmad	1/6/2005	Audi	Cpe	$37,000	$0	$5,550
6	5	Ahmad	1/6/2005	Ford	Sdn	$17,500	$2,000	$2,325
7	6	Henry	1/8/2005	Toyota	Trk	$24,500	$1,500	$2,300
8	7	Lupe	1/10/2005	Ford	Wgn	$23,000	$2,500	$2,050
9	8	Piego	1/12/2005	Ford	Sdn	$14,500	$500	$1,400
10	9	Kenji	1/13/2005	Toyota	Trk	$27,000	$1,200	$2,580
11	10	Ahmad	1/14/2005	Audi	Cpe	$38,000	$0	$5,700

Exhibit 4.18 Results of the query

applied to the current Audi and Ford *Make* condition, and the results are six records which are shown in Exhibit 4.21.

4.4.4 Advanced Filter

Now, let us consider more sophisticated queries that can be performed with the *Advanced Filter* tool. The limitations of *Filter* are easily overcome with this tool. For example, the problem experienced above that limited the *Make* to two entries, Audi or Ford, can be extended to multiple (more than two) entries. Additionally, the *Advanced Filter* tool permits *complex* searches based on **Boolean** logic queries (using *AND* and *OR* conditions) that are not possible, or are difficult, with the *Filter* tool.

The process for creating queries for the *Advanced Filter* is not much more difficult than the process for *Filters*. As before, we must identify the database that will be queried. We must also create a range on the worksheet where we will provide the *criteria* used for the query. As usual, a dialogue box will lead you through the process. We begin by copying the titles of our database and placing them in a range on the worksheet where we will execute the query. All of the database titles, or some subset, can be copied. The copy process insures that the titles appear exactly in the query area as they do in the database. If they are not exactly as they appear in the titles of the database, the query will not be performed correctly.

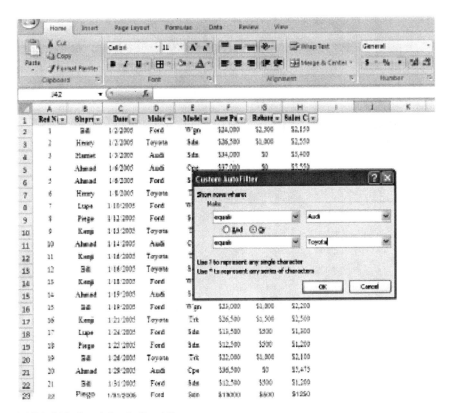

Exhibit 4.19 Search for Audi and Toyota

Let us now consider the use of criteria to create complex queries. Each row that is included below the titles will represent a set of conditions that will be applied to the database. Using a single row is equivalent to using *Filter*. Placing a specific criterion in a cell is similar to the selection of a particular criterion from the pull down menus of the *Filter* function; leaving a cell *blank* permits all values to be returned. Thus, if a row is placed in the criteria range that has no criteria entries in the fields, all records will be returned, regardless of what is contained in the other rows.

Consider the auto sales database in Table 4.2. If we are interested in records containing Audi for sales *on or after* 1/014/05, the *Advanced Filter* tool will return records 10, 14, and 20. The condition *on or after* is set by using the ">=" key characters preceding the date, 1/14/05. Exhibit 4.22 shows the Advanced Filter in action. Note how the query is executed:

1. Titles of the fields are copied and pasted to a range—A27:H27.
2. Go to the *Data* ribbon and select *Advanced Filter*.
3. A dialogue box will appear, and you identify the database—A3:H25.
4. A criteria row (or rows) is selected (A27:H28); each row implies an *Or* condition (look for data records matching the criteria provided in the first row, *Or* criteria provided in second row, *Or*. ...etc.).

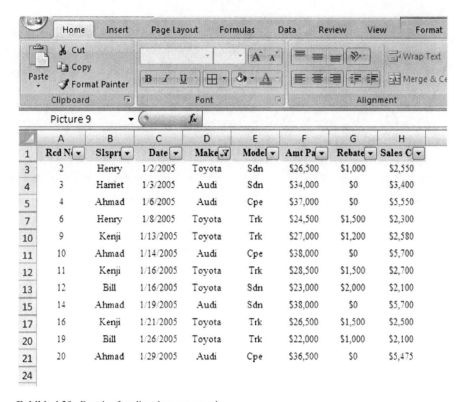

Exhibit 4.20 Result of audi and toyota search

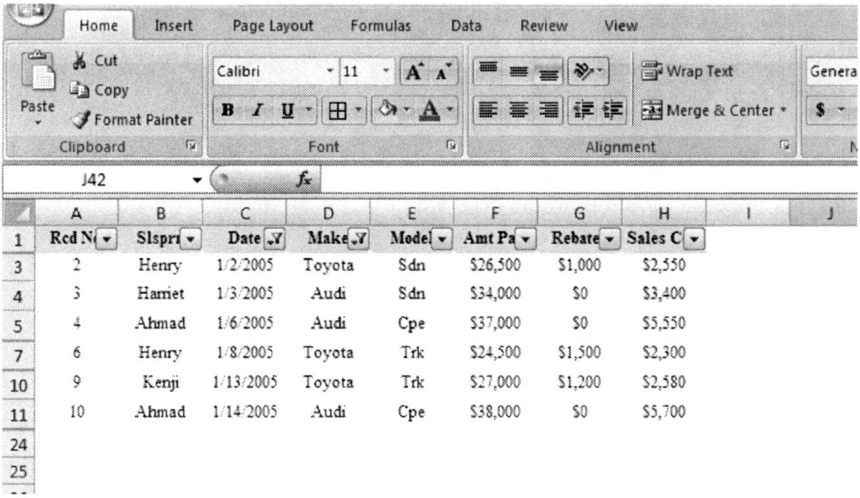

	A	B	C	D	E	F	G	H
1	Rcd N	Slspr	Date	Make	Model	Amt Pa	Rebate	Sales C
3	2	Henry	1/2/2005	Toyota	Sdn	$26,500	$1,000	$2,550
4	3	Harriet	1/3/2005	Audi	Sdn	$34,000	$0	$3,400
5	4	Ahmad	1/6/2005	Audi	Cpe	$37,000	$0	$5,550
7	6	Henry	1/8/2005	Toyota	Trk	$24,500	$1,500	$2,300
10	9	Kenji	1/13/2005	Toyota	Trk	$27,000	$1,200	$2,580
11	10	Ahmad	1/14/2005	Audi	Cpe	$38,000	$0	$5,700

Exhibit 4.21 Date and make filter for database

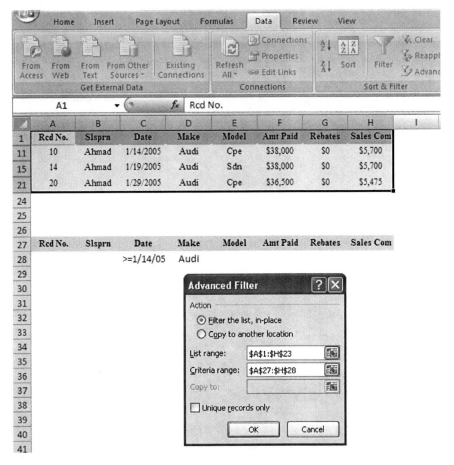

Exhibit 4.22 Advanced filter example

5. It is important to think of the cells within a row as creating *And* conditions and the various rows creating *Or* conditions.
6. This logic can be confusing, so I suggest you perform a number of tests on the database to confirm your understanding of the query results.
7. To return to the full database, click on the *Clear* button in the *Sort and Filter* group.

Let us consider some more complex *Advanced Filter* queries. The query will focus on three fields and use three rows in the *Criteria range* to determine the following:

1. What sales records of Toyota are due to Bill for amounts above $22 k?
2. What sales records of Ford are due to Kenji for amounts below $25 k?
3. What sales records are for Sedans (Sdn)?

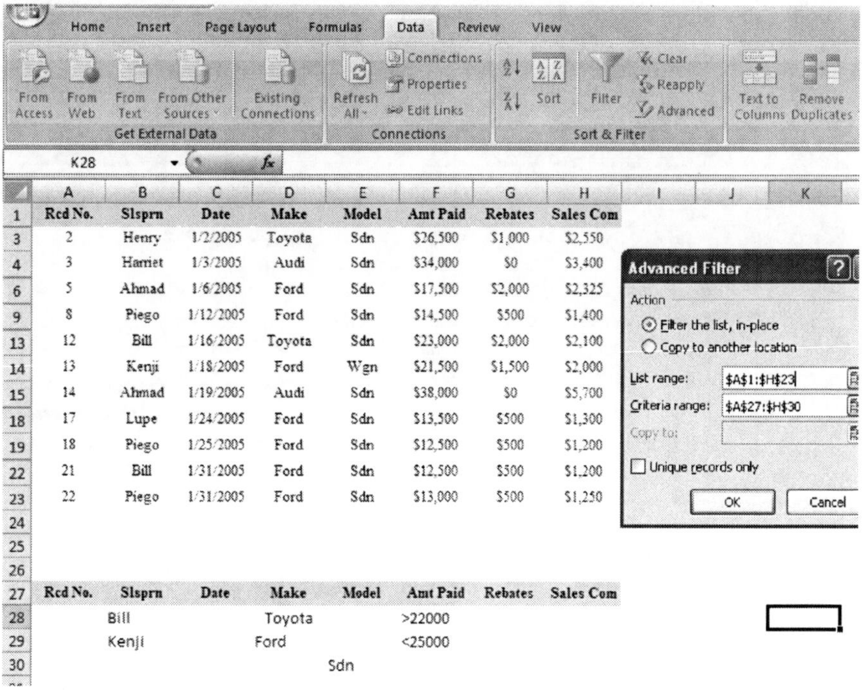

	Rcd No.	Slsprn	Date	Make	Model	Amt Paid	Rebates	Sales Com
3	2	Henry	1/2/2005	Toyota	Sdn	$26,500	$1,000	$2,550
4	3	Harriet	1/3/2005	Audi	Sdn	$34,000	$0	$3,400
6	5	Ahmad	1/6/2005	Ford	Sdn	$17,500	$2,000	$2,325
9	8	Piego	1/12/2005	Ford	Sdn	$14,500	$500	$1,400
13	12	Bill	1/16/2005	Toyota	Sdn	$23,000	$2,000	$2,100
14	13	Kenji	1/18/2005	Ford	Wgn	$21,500	$1,500	$2,000
15	14	Ahmad	1/19/2005	Audi	Sdn	$38,000	$0	$5,700
18	17	Lupe	1/24/2005	Ford	Sdn	$13,500	$500	$1,300
19	18	Piego	1/25/2005	Ford	Sdn	$12,500	$500	$1,200
22	21	Bill	1/31/2005	Ford	Sdn	$12,500	$500	$1,200
23	22	Piego	1/31/2005	Ford	Sdn	$13,000	$500	$1,250
27	Rcd No.	Slsprn	Date	Make	Model	Amt Paid	Rebates	Sales Com
28		Bill		Toyota		>22000		
29		Kenji		Ford		<25000		
30					Sdn			

Exhibit 4.23 Example of more complex search

The results of the queries are shown in Exhibit 4.23. Of the 21 records in the database, 11 total records satisfy the 3 queries. The first query in the list above (1.) returns the record number 12. The second query (2.) returns the record 13. The third query (3.) returns all other records shown (2, 3, 5, 8, 14, 17, 18, 21, 22), as well as record 12 since this record also indicates the sale of a sedan; thus, the first and third queries have record number12 in common.

To return a *split* (non-contiguous) range of dates, or any variable for that matter, place the lower end of the range on a row and the upper end of the range on the same row under a duplicate title, in this case *Date*. This may seem a bit cumbersome, but given the advantages in simplicity of *Advanced Filter* over true database software, it is a minor inconvenience for small scale applications. For example, consider a *Date* query where we were interested in records for sales occurring *on and between* 1/01/05—1/12/05, and *on or after* 1/22/05. The first row in the *Advanced Filter* will contain ">=1/01/05" and "<=1/12/05". The second row will contain ">=1/22/05". Make sure that all other criteria for each row are the same, in this case blank, or you will not be querying the database with a precise focus on the *Date* range intended. Again, this is a relatively complex query that can be executed with a very small amount of effort. Exhibit 4.24 shows the filter criteria, results, and the dialogue box required to execute this query.

As you can see, the power of *Advanced Filter* is substantial. Its ability to handle complex queries makes it an excellent alternative to a dedicated database program

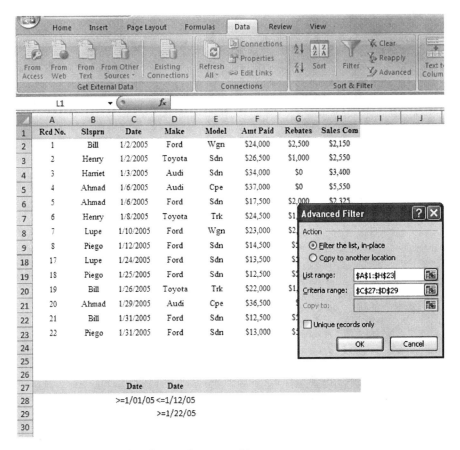

Exhibit 4.24 Advanced filter for complex range of dates

such as MS Access, but only for certain circumstances. The questions that you must answer when considering Excel as a database are:

1. When does the database become so large that it becomes cumbersome to use Excel? Databases have a tendency to grow ever larger. At what point do I make the transition from Excel to a relational database.
2. Are there special features like report writing, relational tables, and complex queries that can best be handled by Access or some other relational database (Sequel Server)?
3. Would I be better off by using a relational database and exporting data occasionally to an Excel workbook for analysis and presentation?

Now, let us query the data in Table 4.1 using the *Filter* and *Advanced Filter* tools in the example below. The data is familiar and will provide some economy of effort.

4.5 An Example

The forensic accounting department of Lafferman Messer and Fuorchette (LMF) has been hired to examine the recent financial activity of a small Eastern European firm, Purvanov Imports (PI), with operations in Houston, Texas. PI is involved in international trade and is suspected of money laundering. In particular, LMF has been asked to examine invoices in a database received from what are believed to be *front* organizations. Thus, PI may be using illegally obtained funds and passing them on to other firms in what appears to be a legitimate business transaction. As **forensic accounting** experts, LMF is skilled in preparing evidence for potential litigation.

The investigation is being conducted by the federal prosecutor, Trini Lopez, in the Houston district. The chief forensic investigator from LMF is McKayla Chan. Also, recall that the data we will use in the analysis is that found in Table 4.1.

Lopez and Chan meet in Houston to discuss the investigation and plan their strategy:

Lopez: Ms. Chan, we need to conduct this investigation quickly. We fear that LMF may soon know of our investigation to uncover money laundering.

Chan: I understand the need for quick action, but I need access to their financial statements.

Lopez: That is no problem. I have the statements here on a portable mass storage device.

Chan: Great. I'll get started tomorrow morning when I return to my office in Chicago.

Lopez: I'm afraid that will be too late. We need to have evidence to make an arrest today. In fact, we need to do so in less than an hour.

Chan: Well, that may be a difficult time limit to satisfy, but if you have the data and you know the records that you want to examine, we can give it a go. I did bring my laptop.

After importing the database into Excel from a Word file, Lopez provides Chan with the specifics of the queries that he is interested in conducting:

1. What is the low-to-high ranking of the *$ Amount* for the entire quarter.
2. What is the monthly *$ Amount* total for each account—*Office Supply*, *Printing*, etc. during the month of February?
3. What is the max, min, and count for *$ Amount* of each account also in the month of February?
4. What is the average, max, sum, and count for *$ Amount* of each Account?
5. We are particularly interested in Projects X and Y. What *$ Amount* charges to these projects occurred between 1/1/04 and 1/13/04 and after 3/1/04?

Lopez believes that there are many more issues of interest, but it is sufficient to begin with these questions. The results of the queries can easily provide sufficient evidence to obtain an arrest warrant and charge PI with fraud.

We will use all we have learned in this chapter to perform these database manipulations, focusing on the *Sort*, *Filter*, and *Advanced Filter* tools. In the process, we

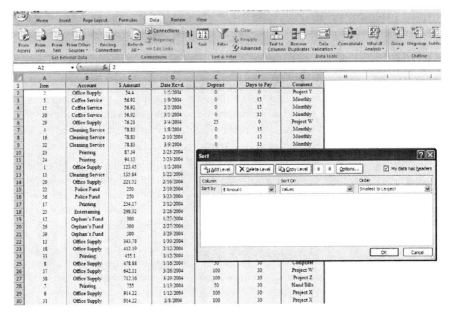

Exhibit 4.25 Sort by $ amount for question 1

will introduce a very valuable and somewhat neglected financial cell function called *Subtotal (function_num, ref1,...).*

Question 1 is relatively simple to answer and requires the use of the *Sort* function with *$ Amount* as the key. Exhibit 4.25 shows the results of an ascending value sort. Record 2 is the first record listed and the smallest *$ Amount* at $54.40 and record 3 is the last record listed and the largest at $2,543.21. Question 2 is more complex because it requires the summation of *monthly $ Amount* totals for all account types. There are two obvious approaches that we can consider. We can sort by *Account*, then we can select a cell to calculate the *$ Amount* totals. This will work nicely, but there is another approach worth considering, and one that introduces a cell function that is extremely useful—**Subtotal (function_num, ref1,...).** *Subtotal* allows you to perform 11 functions on a subtotal—average, min, max, sum, standard deviation, and others. As an added advantage, *Subtotal* functions with the *Filter Tool* to ignore (not count) hidden rows that result from a filtered database. Why is this important? Recall that *Filter* and *Advanced Filter* simply hides rows that are not relevant to our query. Thus, if we use the *Sum* function to aggregate the *$ Amount* for filtered rows, the sum will return the sum of *all* rows, visible *and* hidden. *Subtotal* will only return the values of visible rows.

Exhibit 4.26 shows the initial step to answering question 2. The *Filter* is used to filter for the dates of interest, 2/1/04 and 2/29/04. In Exhibit 4.27 we demonstrate the use of *Subtotal* for all accounts during the month of February. It calculates the sum, max, min, and count for the filter dates. Exhibit 4.27 also shows the arguments codes (designated by the numbers 1–11) of the *Subtotal* function, *function_num,*

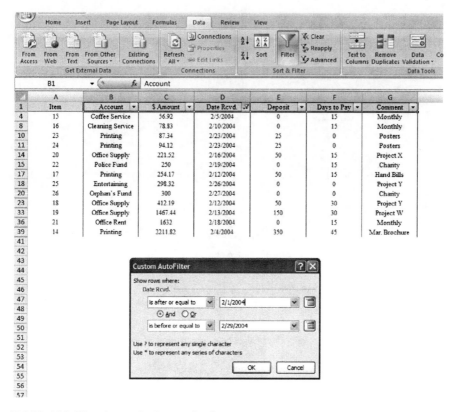

Exhibit 4.26   Filter date entries for question 2

for selecting the operation to be performed—9 is the code that executes the sum of all *visible* rows, 4 is max, 5 is min, and 2 executes the count of all visible rows. See the Excel Help utility to find the appropriate code numbers for each function.

The following are the steps required to perform the filter and analysis discussed above and are shown in Exhibit 4.26:

1. Set the *Filter* tool in the *Data* ribbon for the titles of the fields in the database.
2. Select the field *Date Rcvd.* From the pull down menu, and then select *Custom*.
3. Within *Custom*, set the first condition to "*is after or equal to*" and specify the desired date—02/1/04.
4. Next, set the *And* button.
5. Finally, set the second condition to "*is before or equal to*" and enter the second desired date—02/29/04.

Once the data is filtered, Questions 2 and 3 can be answered by using the *Subtotal* functions - sum, max, min, and count. We see in Exhibit 4.27 that the sum is $7,364.67, the max is $2,211.82, the min is $56.92, and the count is 13.

	A	B	C	D	E	F	G
1	Item	Account	$ Amount	Date Rcvd.	Deposit	Days to Pay	Comment
4	15	Coffee Service	56.92	2/5/2004	0	15	Monthly
8	16	Cleaning Service	78.83	2/10/2004	0	15	Monthly
10	23	Printing	87.34	2/23/2004	25	0	Posters
11	24	Printing	94.12	2/23/2004	25	0	Posters
14	20	Office Supply	221.52	2/16/2004	50	15	Project X
15	22	Police Fund	250	2/19/2004	0	15	Charity
17	17	Printing	254.17	2/12/2004	50	15	Hand Bills
18	25	Entertaining	298.32	2/26/2004	0	0	Project Y
20	26	Orphan's Fund	300	2/27/2004	0	0	Charity
23	18	Office Supply	412.19	2/12/2004	50	30	Project Y
33	19	Office Supply	1467.44	2/13/2004	150	30	Project W
36	21	Office Rent	1632	2/18/2004	0	15	Monthly
39	14	Printing	2211.82	2/4/2004	350	45	Mar. Brochure

Codes	Subtotals	Formulas
subtotal (sum=9)	$ 7,364.67	<==subtotal(9,C1:C39)
subtotal (max=4)	$ 2,211.82	<==subtotal(4,C1:C39)
subtotal (min=5)	$ 56.92	<==subtotal(5,C1:C39)
subtotal (count=2)	13	<==subtotal(2,C1:C39)

Exhibit 4.27 Calculation of subtotals of $ amount for questions 2 and 3

Question 4 is relatively easy to answer with the tools and functions that we have used—*Filter* and *Subtotal*. Exhibit 4.28 shows the outcome of the query for the account *Office Supply* and the specific subtotals of interest—average ($714.24), max ($1,669.76), sum ($10,713.67), and count (15).

Question 5 requires the use of the *Advanced Filter* tool. Exhibit 4.29 shows the outcome of the *Filter* for *Project X* and *Y*. Note that the pull down menu (in box) associated with Comment field permits you to identify the Projects of interest—*Project X* and *Y*. Unfortunately, the *Filter* tool is unable to perform the *Date Rcvd.* filtering. Thus, we are forced to employ the *Advanced Filter* tool. In Exhibit 4.30, the *Advanced Filter* query is shown. It employs some repetitive entries, but it is relatively simple to construct. Let us concentrate on the first two rows in the *Advanced Filter* criteria range (Excel rows 45 and 46), because they are similar in concept to the two that follow. The first two rows can be read as follows:

1. Select records that contain *Date Rcvd* dates *after or equal to* 1/1/04 and records that contain *Date Rcvd* dates *before or equal to* 1/13/04 and records that contain *Comment* entry *Project X*. These queries are found on rows 45 of Exhibit 4.30. Note the importance of the *and* that is implied by the entries in a single row.
2. Recall that each row introduces and *or* condition. Thus, the second row of the criteria can be read as—*or* select records that contain *Date Rcvd* dates *after or equal to* 3/1/04 and that contain *Comment* entry *Project X*. These queries are found on row 46 of Exhibit 4.30.
3. For *Project Y*, rows 47 and 48 repeat the process.

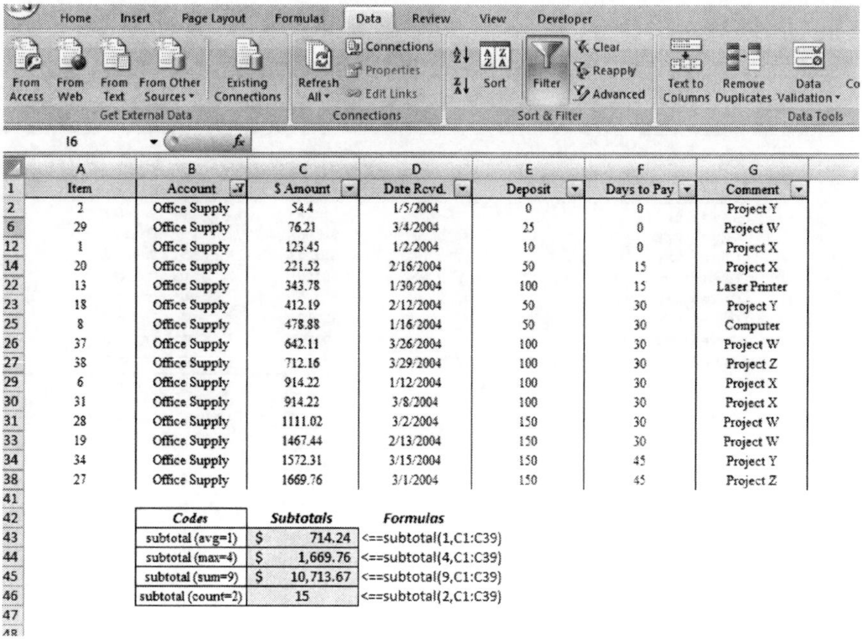

	A	B	C	D	E	F	G
1	Item	Account	$ Amount	Date Rcvd.	Deposit	Days to Pay	Comment
2	2	Office Supply	54.4	1/5/2004	0	0	Project Y
6	29	Office Supply	76.21	3/4/2004	25	0	Project W
12	1	Office Supply	123.45	1/2/2004	10	0	Project X
14	20	Office Supply	221.52	2/16/2004	50	15	Project X
22	13	Office Supply	343.78	1/30/2004	100	15	Laser Printer
23	18	Office Supply	412.19	2/12/2004	50	30	Project Y
25	8	Office Supply	478.88	1/16/2004	50	30	Computer
26	37	Office Supply	642.11	3/26/2004	100	30	Project W
27	38	Office Supply	712.16	3/29/2004	100	30	Project Z
29	6	Office Supply	914.22	1/12/2004	100	30	Project X
30	31	Office Supply	914.22	3/8/2004	100	30	Project X
31	28	Office Supply	1111.02	3/2/2004	150	30	Project W
33	19	Office Supply	1467.44	2/13/2004	150	30	Project W
34	34	Office Supply	1572.31	3/15/2004	150	45	Project Y
38	27	Office Supply	1669.76	3/1/2004	150	45	Project Z
41							
42		Codes	Subtotals	Formulas			
43		subtotal (avg=1)	$ 714.24	<==subtotal(1,C1:C39)			
44		subtotal (max=4)	$ 1,669.76	<==subtotal(4,C1:C39)			
45		subtotal (sum=9)	$ 10,713.67	<==subtotal(9,C1:C39)			
46		subtotal (count=2)	15	<==subtotal(2,C1:C39)			
47							

Exhibit 4.28 Example for question 4

The resulting five items shown in Exhibit 4.30 include records 2, 1, 6, 31, and 34. This is an example of the valuable utility that is provided by the *Advanced Filter* function of Excel. Carefully constructed advanced queries can be somewhat difficult to create, but the power of this tool is undeniable. I suggest that you carefully consider the *and* and *or* conditions implied by the row and column entries by attempting the same analysis. This should help you understand the logic necessary to conduct complex queries.

4.6 Summary

This chapter provides the Excel user with a comprehensive set of tools for the presentation and preparation of qualitative data: data entry and manipulation tools, data querying, filtering, and sorting tools. As important as data presentation is, it is equally important to efficiently enter and manipulate data as we begin to build the database that we will use in our analysis. We often encounter data that has been created and organized in formats other than Excel worksheets. The formidable task of data importation can be easily managed by understanding the format from which data is to converted—tables, tab delimited data, and data with no delimiter. We also examined the use of the *Form* and *Validation* tools to insure the accuracy of data input.

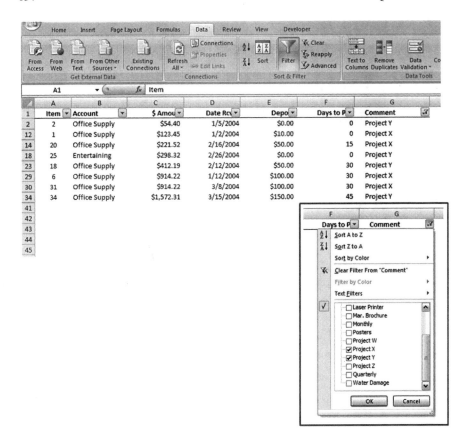

Exhibit 4.29   Filter of projects X and Y for question 5

Once we have constructed our database, we are ready to begin *soft* or non-mathematical analysis of the data, in the form of sorts, filtering, and queries. I like to think of this type of analysis as isolation, summarization, and cross-tabulation (more on this topic in Chap. 5) of the records in our database. Summarization can be as simple as finding totals for categories of variables, while isolation focuses on locating a specific record, or records, in the database. **Cross-tabulation** allows the database user to explore more complex relationships and questions. For example, we may want to investigate if a particular sales associate demonstrates a propensity to sell a particular brand of automobile. Although Excel is not designed to be a *full service* database program, it contains many important tools and functions that will allow relatively sophisticated database analysis on a small scale. More on Cross-tabulation in later in chapters.

Chapter 4 also introduced a number of powerful cell functions that aided our database analysis, although these functions have utility far beyond the analysis of qualitative data. Among them were *Subtotal, Transpose,* and the logical *IF, AND,*

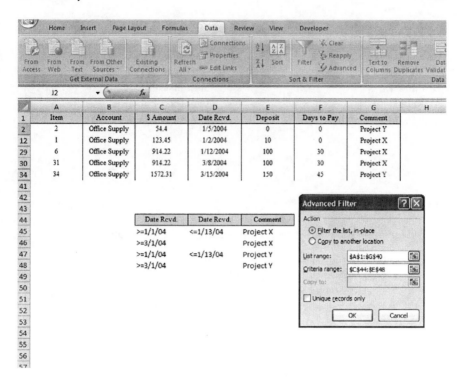

Exhibit 4.30   Advanced filter of projects X and Y for question 5

OR, NOT, FALSE, and *TRUE* functions. There will many opportunities in later chapters to use logical cell functions.

In the Chap. 5 we will examine ways to perform *hard* analysis of qualitative data. Also, we will learn how to generate *PivotChart* and *PivotTable* reports. Beside their analytical value, these tools can also greatly simplify the presentation of data, and although it is somewhat similar to the material in this chapter, because of its importance and widespread use, it deserves a separate exposition.

Key Terms

Qualitative Data	Array
Soft Data Analysis	Logical IF
Hard Data Analysis	Nested IF's
Record	AND, OR, NOT, FALSE, TRUE
Field	Absolute Address
Form	VLOOKUP/HLOOKUP
Validation	LOOKUP
Transpose	Tab Delimited

Text to Column Command   ERP
Sort   Filter
Filter   Advanced Filter
Sort Keys   Boolean
Filter Keys   Forensic Accounting
Keys   Subtotal Function
Relational Database   Cross-tabulation
CRM

Problems and Exercises

1. Identity the following data as either qualitative or quantitative:

   a. Quarterly sales data in dollars for a SKU (stock keeping unit)
   b. The country of birth for a job applicant
   c. Your rank among your fellow high school graduates
   d. Your random assignment to one of 26 rental cars

2. What is the likely information gathered in a credit card transaction. In other words, what are the fields associated with a transaction record?

3. Every customer that visits my restaurant has data recorded on their visit. I use a comment field to enter information on whether customers are returning or new. Is this the proper use of a comment field? Why or why not?

4. Soft data analysis is not as useful as hard data analysis—T or F?

5. Create an Excel form that records the last name, age in years, annual income, and average daily caloric intake of members of your hometown.

6. Create the following validation rules that pertain to anyone in your hometown that is providing information for the fields in problem 5 above:

   a. a last name of between 4 and 12 letters
   b. ages must be older than 14 and less than 67
   c. annual income must be below $500,000
   d. caloric intake cannot include values up to and including 1000 calories

7. Place the Excel Help icon in your Quick Access Toolbar.

8. Copy and paste the following into a workbook:

   a. A table from a Word document (so you will have to create the table below in Word) into appropriate cells of a workbook:

Observation	1	2	3	4	5	6
Type	Bulldog	Chihuahua	Mutt	Poodle	Schnauzer	Airedale
Age	3	14	5	7	9	7

     b. Text from a Word document separated by spaces into appropriate cells of a
       workbook:

       Observation 1 2 3 4 5 6

       Type Bulldog Chihuahua Mutt Poodle Schnauzer Airedale

       Age 3 14 5 7 9 7

9. *Transpose* the table in a. into a worksheet by first copying it into the worksheet
   then applying the transpose function.

10. Write a logical *if* statement that will test for the following conditions:

    a. If the cell value in A1 is larger than the cell value in A2 then write in cell
      A3—"A1 beats A2", otherwise write "A2 beats A1". Test this with values of:
      (1) A1=20 and A2=10; (2) A1=15 and A2=30; (3) A1=15 and A2=15.
      What happens with (3) and can you suggest how you can modify the *if*
      statement to deal with ties?

    b. If the content of cell A1 is between 0 and 50 write "Class A"; If the content
      of A1 is between 51 and 100 write "Class B"; If neither then write "Class
      C". Place the function in cell A2 and test with 23, 56, and 94.

    c. If the content of cell A1 is between 34 and 76 *or* 145 and 453 write "In
      Range", otherwise write "Out of Range". Place the function in A2 and test
      with 12, 36, 87, 243, and 564. Hint: this will require you to use other logical
      functions (*OR, AND*, etc).

    d. If the contents of cell A1>23 *and* the contents of cell A2<56 then write
      "Houston we have a problem" in cell A3. Otherwise write "Onward through
      the fog" in cell A3.

11. Sort the data in Table 4.1 as follows:

    a. by size of deposit—smallest to largest
    b. primarily by account and secondarily by deposit-ascending for both
    c. after a. and b., reconstruct the original table by sorting on the field you think
      is best.

12. Filter Table 4.2 as follows:

    a. by Salesperson Kenji and Lupe and then between and including 1/10/2005
      and 1/21/2005
    b. by Salesperson Kenji and Lupe and then the complement (not included in)
      of the dates in a., above
    c. all sales on the dates 1/06/2005 or 1/31/2005
    d. all Ford sales above and including $13,000.

13. Use the Advanced Filter on Table 4.2 as follows:

    a. any Toyotas sold by Bill on or after 1/15/2005 *or* any Fords sold by Piego
      on or after 1/20/2005- calculate the subtotal sum, average, and count for the
      Amt Paid

    b. any cars sold for more than $25,000 on or after 1/6/2005 and before or on
      1/16/2005 *or* any cars sold for less than $20,000 on or after 1/26/2005 -

calculate the subtotal max, min, standard deviation, and sum for the *Sales Com*

c. for cars sold with a rebate, calculate the sum, max, and min for the *Rebates.*

14. *Advanced Problem*—A wine supplier, Sac Rebleu, deals with exclusive restaurants in New York City. The wine supplier locates hard to find and expensive French wines for the wine stewards of some of the finest restaurants in the world. Below is a table of purchases for the month of January. The supplier is a meticulous man who believes that data reveals *le vérité* (the truth). Answer the following questions for Monsieur Rebleu by using Sort, Filter, and Advanced Filter:

   a. Which customer can be described as most generous based on the proportion of Tip to $Purchase that they provide?
   b. Sort the transactions on and between 1/08/2007 and 1/20/2007 as follows:

      i. Alphabetically.
      ii. By $ Purchase in descending order.
      iii. Primary sort same as (ii) and secondary as Tip (descending).

   c. What is the average $ Purchase for each steward?
   d. What is the sum of Tips for the 3 most frequently transacting stewards?

Obs.	Customer	Date	$ Purchase	Tip
1	Hoffer	1/2/2007	$ 249	20
2	Aschbach	1/2/2007	$ 131	10
3	Jamal	1/3/2007	$ 156	20
4	Johnson	1/6/2007	$ 568	120
5	Johnson	1/6/2007	$ 732	145
6	Aschbach	1/8/2007	$ 134	10
7	Rodriguez	1/10/2007	$ 345	35
8	Polari	1/12/2007	$ 712	125
9	Otto	1/13/2007	$ 219	10
10	Johnson	1/14/2007	$ 658	130
11	Otto	1/16/2007	$ 160	10
12	Hoffer	1/16/2007	$ 254	20
13	Otto	1/18/2007	$ 155	10
14	Johnson	1/19/2007	$ 658	135
15	Hoffer	1/19/2007	$ 312	20
16	Otto	1/21/2007	$ 197	10
17	Rodriguez	1/24/2007	$ 439	40
18	Polari	1/25/2007	$ 967	200
19	Hoffer	1/26/2007	$ 250	20
20	Johnson	1/29/2007	$ 661	130
21	Hoffer	1/31/2007	$ 254	20
22	Polari	1/31/2007	$ 843	160

e. Calculate the subtotal sum, max, min, and average for Johnson and Polari $ Purchase.

f. Use the Advanced Filter to find all records with Tips greater than or equal to $100 *or* Tips less than and equal to $ 25.

g. Use the Advanced Filter to find all records with $ Purchase greater than or equal to $500 *or* dates on or before 1/7/2007 *or* dates on or after 1/25/2007.

h. Use the Advanced Filter to find all records with $ Purchase greater than $250 *and* Tips greater than or equal to 15% *or* Tips greater than $35.

Chapter 5
Analysis of Qualitative Data

Contents

5.1 Introduction

There are many business questions that require the collection and analysis of *qualitative* data. For example, how does a visitor's opinion of a commercial website relate to her purchases at the website? Does a positive opinion of the website, relative to a bad or mediocre opinion, lead to higher sales? This type of information is often gathered in the form of an opinion and measured as a categorical response. Also accompanying these opinions are some quantitative characteristics of the respondent; for example, their age or income. Thus, a data collection effort will include various forms of qualitative and quantitative data elements. Should we be concerned with the type of data we collect? In the prior chapters we have answered this question with a resounding yes. It is the type of the data—categorical, interval, ratio, etc.—that often dictates the form of analysis we can perform.

H. Guerrero, *Excel Data Analysis*, DOI 10.1007/978-3-642-10835-8_5,
© Springer-Verlag Berlin Heidelberg 2010

Table 5.1 Auto sales data example

Record No.	Slsprn	Date	Make	Model	Amt Paid	Rebates	Sales Com
		Auto Sales Data—01/01/2005—01/31/2005					
1	Bill	01/02/05	Ford	Wgn	$24,000	$2,500	$2,150
2	Henry	01/02/05	Toyota	Sdn	26,500	1,000	2,550
3	Harriet	01/03/05	Audi	Sdn	34,000	0	3,400
4	Ahmad	01/06/05	Audi	Cpe	37,000	0	5,550
5	Ahmad	01/06/05	Ford	Sdn	17,500	2,000	2,325
6	Henry	01/08/05	Toyota	Trk	24,500	1,500	2,300
7	Lupe	01/10/05	Ford	Wgn	23,000	2,500	2,050
8	Piego	01/12/05	Ford	Sdn	14,500	500	1,400
9	Kenji	01/13/05	Toyota	Trk	27,000	1,200	2,580
10	Ahmad	01/14/05	Audi	Cpe	38,000	0	5,700
11	Kenji	01/16/05	Toyota	Trk	28,500	1,500	2,700
12	Bill	01/16/05	Toyota	Sdn	23,000	2,000	2,100
13	Kenji	01/18/05	Ford	Wgn	21,500	1,500	2,000
14	Ahmad	01/19/05	Audi	Sdn	38,000	0	5,700
15	Bill	01/19/05	Ford	Wgn	23,000	1,000	2,200
16	Kenji	01/21/05	Toyota	Trk	26,500	1,500	2,500
17	Lupe	01/24/05	Ford	Sdn	13,500	500	1,300
18	Piego	01/25/05	Ford	Sdn	12,500	500	1,200
19	Bill	01/26/05	Toyota	Trk	22,000	1,000	2,100
20	Ahmad	01/29/05	Audi	Cpe	36,500	0	5,475
21	Bill	01/31/05	Ford	Sdn	12,500	500	1,200
22	Piego	01/31/05	Ford	Sdn	13,000	500	1,250

In this chapter, we examine some of the many useful Excel resources available to analyze qualitative data. This includes exploring the uses of *PivotTable* and *PivotChart* reports: built-in Excel capability that permits quick and easy *cross-tabulation* analysis, sometimes referred to as crosstab analysis. Crosstabs permit us to determine how two or more variables in a set of data interact. Consider the auto sales data we introduced in Chap. 4, which now appear in Table 5.1 in this chapter. There are many questions a decision maker might consider in examining these data. For example, is there a relationship between sales associates and the models of automobiles they sell? More specifically, is there a propensity for some of the sales associates to promote the sale of a particular automobile to a particular customer demographic[1]?

Although this type of analysis can be performed with sophisticated statistics, in this chapter, we will use less rigorous numerical techniques to generate valuable insights. The simple, numerical information that results may be all that is necessary for good decision making. Returning to our decision maker's question,

---

[1] In marketing, the term *demographic* implies the grouping or segmentation of customers into groups with similar age, gender, family size, income, professions, education, religious affiliation, race, ethnicity, national origin, etc. The choice of the characteristics to include in a demographic is up to the decision maker.

if we find that associates are concentrating on the sale of higher priced station wagons to a small number of demographics, a decision maker may want to take steps to change this focused selling. It is possible that other demographics will be interested in similar vehicles if we apply appropriate sales incentives.

In Chap. 6 we will focus on *statistical* analysis that can be performed with techniques appropriate for qualitative and quantitative data. Among the techniques that will be examined are Analysis of Variance (ANOVA), tests of hypothesis with t-tests and z-tests, and chi-square tests. These statistical tools will allow us to study the effect of independent variables on dependent variables contained in a data set, and allow us to study the similarity or dissimilarity of data samples. Although these technical terms may sound a bit daunting, I will establish clear rules for their application, certainly clear enough to permit a non-statistician to apply the techniques correctly. Now back to the techniques we will study in this chapter.

5.2   Essentials of Qualitative Data Analysis

In Chap. 4 we discussed the essential steps to prepare, organize, and present qualitative data. The preparation of qualitative data for *presentation* should also lead to preparation for data analysis; thus, most of the work done in presentation will complement the work necessary for the data analysis stage. Yet, there are a number of problems relating to **data errors** that can occur due to problems in data collection or transcription that require special consideration. These errors must be dealt with early in the data analysis process or the analysis will lead to inaccurate and unexplainable results.

5.2.1   Dealing with Data Errors

Data sets, especially large ones, can and usually do, contain errors. Some errors can be uncovered, but others are simply absorbed, never to be detected. Errors can occur due to a variety of reasons: problems with manual keying or electronic transmission of data onto spreadsheets or databases, mistakes in the initial recording of data by a respondent or data collector, and many other sources too numerous to list. Thus, steps insuring the quality of the data entry process need to be taken. As we saw in the previous chapter, where we assumed direct data entry in worksheets, we can devise data entry mechanisms to facilitate entry and to protect against entry errors.

Now let us consider data that has been transcribed onto an Excel worksheet from an outside source. We will focus on the rigorous inspection of the data for unexpected entries. This can include a broad range of data inspection activities, ranging from sophisticated sampling of a subset of data elements to exhaustive (100%) inspection of all data. If a low level of errors can be tolerated, then only a sample of the data need be reviewed for accuracy. This is usually the case when a data set is very large, and the cost of errors is low relative to the cost of verification. If the cost of errors is high, then 100% inspection of the data may be necessary. For example, data collected in the clinical trial of a drug may require 100% inspection due to the gravity of the acceptance or rejection of the trail results.

So what capabilities does Excel provide to detect errors? In this section we examine a number of techniques for verification: (1) do two independent entries of similar data match and, (2) do data entries satisfy some range of characteristics? We will begin with a number of cell functions that permit the comparison of one data entry in a range to an entry in another range. Let us first assume a data collection effort for which data accuracy is of utmost importance. Thus, you employ two individuals to simultaneously key the data into two Excel worksheets. The entry is done independently and the process is identical. Once the data is keyed into separate worksheets of a workbook, a third worksheet is used to compare each data entry. We will assume that if no differences are found in data entries, the data is without error. Of course, it is possible that two entries, though identical, can both be in error, but such a situation is likely to be a rare event.

This is an ideal opportunity to use the *logical IF* cell function to query whether a cell is identical to another cell. The *IF* function will be used to test the *equality* of entries in two cells, and return the cell value *OK* if the test results are equal, or *BAD* if the comparison is not equal. For simplicity's sake, assume that we have three ranges on a single worksheet where data is located—the first is data entry by Otto, the second is the identical data entry by Maribel, and the third is the test area to verify data *equality*. See Exhibit 5.1 for the data entries and resulting test for equality.

Note Otto's data entry in cell B4 and the identical data element for Maribel in E4. The test function will appear in cell H4 as: $= IF (B4=E4, "OK", "BAD")$. (Note that quotation marks must surround all text entries in Excel formulas.) The result of the **error checking** is a value of BAD since the entries are not equal. The cell range

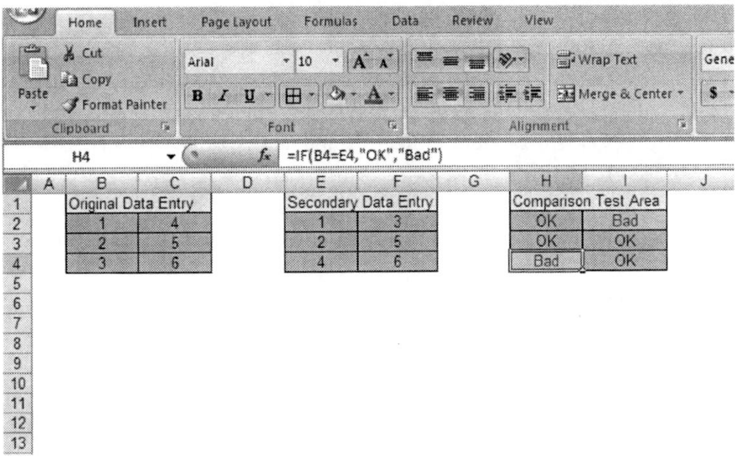

Exhibit 5.1 If function error checking of data

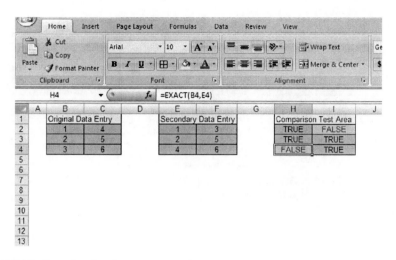

Exhibit 5.2   Exact function data entry comparison

of H2:I4 displays the results of all 9 comparisons. The comparison determines two disagreements in data entry. Of course we are not in a position to suggest which entry is in error, but we are aware that the entry resulting in BAD for both tables must be investigated.

Regardless of the size of the data sets, the *IF* function can be written once and copied to a range equivalent in size and dimensions to the data entry ranges. Thus, if Otto's data entry occurs in range A1:H98, then the data entry range for Maribel could be A101:H198, and the *comparison area* could be in A201:H298. Our only restriction is that the dimension of the ranges containing data must be similar; that is, the number of rows and columns in the entry ranges must be the same.

A more direct approach to a comparison of data elements is the use of the Excel cell function **EXACT(text1,text2)**. As the title implies, the function compares two text data elements, and *if* an exact match is found, it returns **TRUE** in the cell, *else* it returns **FALSE**. Exhibit 5.2 compares the 6 data items and performs error checking similar to that in Exhibit 5.1. Note that the third data element of the first column and first of the second column are different (as before) and return a cell value result of *FALSE*. All other cells result in a value of *TRUE*.

It is also wise to test data for values outside the range of those that are anticipated. This is particularly true of numeric values. In order to perform statistical analysis, we must convert qualitative variables (good, bad, male, female, etc.) to numeric values; thus, if the numeric values are incorrect, the analysis will also be incorrect. For example, it is easy to make a transcription error for data that must be converted from a *text* value (e.g. gender) to a *numeric* value (male=1, female=2).

Consider the data table shown in Exhibit 5.3, consisting of values that are anticipated to be in the range of 1 to 6. We can use the *logical IF* function to test the values occurring in the range of 1 to 6. But rather than testing for each specific value (1, 2, 3,..., 6) by nesting multiple *IF* conditions and testing if the value is 1,

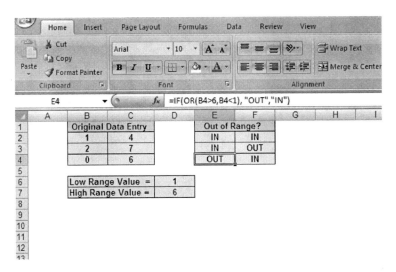

Exhibit 5.3 Out of range data check

2, 3. . .etc., we can employ another logical function, the **OR** function, to capture the entire range and test whether a value is *in* or *out* of the range. The *OR* function is used in Boolean logic, as are **AND**, **NOT**, **FALSE**, and **TRUE**. The combination of these functions can be used for a test of the data entries in an arbitrary cell location, say E4, by using the following logical conditions:

> *IF* a value in cell B4 is less than 1 *OR* is greater than 6
> *then* return the text *"OUT"* to cell E4
> *else* return the text *"IN"* to cell E4

Note that the results in Exhibit 5.3 cells E4 and F3 are *OUT*, since the cell values B4=0 and C3=7 are outside the required range. Assuming the cell location B4 contains the data of interest in Exhibit 5.3, the *IF* function used to perform the comparison in E4 is written as: = IF (OR(B4>6, B4<1), "OUT", "IN"). Of course, we could also replace the values 1 and 6 in the cell formula with cell references D6 and D7, respectively. This permits us to change the range of values in the future, if the need arises, without having to change cell formulas.

What happens when we are anticipating integer values from the data entry process and we instead encounter decimal values? The test above will not indicate that the value 5.6 is an incorrect entry. We can, though, use another Excel Math and Trig cell function, the **MOD (number, divisor)** function, to logically test for a non-integer value. The *MOD* function returns only the remainder (also called the *modulus*) of the division of the *number* by the *divisor*—e.g. if the function argument *number* is 5.6 and the *divisor* is 1, the function will return the value 0.6. We can then include *MOD* as one of the tests in our *IF* function, just as we did with *OR*. It will test for integer values, while the other *OR* conditions test for values in the range of 1 to 6.

The resulting function is now: = IF(OR(MOD(B4,1)>0,B4>6,B4<1), "OUT", "IN"). The first *OR* condition, *MOD(B4,1)>0*, divides B4 by 1, and returns the

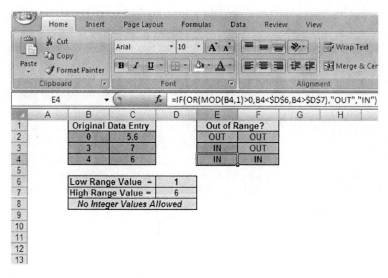

Exhibit 5.4   Out of range and non-integer data error check

remainder. If that remainder is not 0, then the condition is *TRUE* (there is a remainder) and the text message *OUT* is returned. If the condition is *FALSE* (no remainder), then the next condition, B4>6, is examined. If the condition is found to be *TRUE*, then *OUT* is returned, and so on. Note that the *MOD* function simply becomes one of the three arguments for the *OR*. Thus, we have constructed a relatively complex *IF* function for error checking.

Exhibit 5.4 shows the results of both tests. As you can see, this is a convenient way to determine if a non-integer value is in our data, and the values 0 in B2, 5.6 in C2, and 7 in C3 are identified as either out of range or non-integer. The value 0, satisfies both conditions, but is only detected by the out of range condition. This application shows the versatility of the *IF* function, as well as other related logical functions, such as *OR*, *AND*, *NOT*, *TRUE*, and *FALSE*. Any logical test that can be conceived can be handled by some combination of these logical functions.

Once we have verified the accuracy of our data, we are ready to perform several types of descriptive analyses. This includes **cross-tabulation** analysis, through the use of Excel's **PivotChart** and **PivotTable** tools. *PivotChart*s and *PivotTable*s are frequently used tools for analyzing qualitative data—e.g. data contained in customer surveys, operations reports, opinion polls, etc. They are also used as exploratory techniques to guide us to more sophisticated types of analyses.

5.3 PivotChart or PivotTable Reports

Cross-tabulation provides a methodology for observing the interaction of several variables in a set of data. For example, consider an opinion survey that records demographic and financial variables for respondents. Among the variables recorded is age, which is organized into several mutually exclusive age categories (18–25, 26–34, 35–46, and 47 and older). Respondents are also queried for a response or opinion, *good* or *bad*, about some consumer product. Cross-tabulation permits the analyst to determine the number of respondents in the 35–46 age category that report the product to be *good*. The analysis can also determine the number of respondents that fit both our conditions (age and response) as a percentage of the total.

The *PivotTable* and *PivotChart* report functions are found in the *Insert Ribbon*. Both reports are identical, except that the table provides numerical data in table form, while the chart converts the numerical data into a graphical format. The best way to proceed with a discussion of the cross-tabulation capabilities of *PivotTable* and *PivotChart* is to begin with an illustrative problem, one that will allow us to exercise all the capabilities of these powerful functions.

5.3.1 An Example

Now let us consider an example, a consumer survey, to demonstrate the uses of *PivotTables* and *PivotCharts*. The data of interest for the example is shown in Table 5.2. A web-based business, TiendaMía.com,[2] is interested in testing various web designs that customers will use to order products. The owners of TiendaMía.com hire a marketing firm to help them conduct a preliminary survey of 30 randomly selected customers to determine their preferences. Each of the customers is given a gift coupon to participate in the survey and is instructed to visit a website for a measured amount of time. The customers are then introduced to four web-page designs and asked to respond to a series of questions. The data are self-reported by the customers on the website as they experience the four different webpage designs. The marketing firm has attempted to control each step of the survey to eliminate extraneous influences on the respondents. Although this is an example, it is relatively typical of consumer opinion surveys and website tests.

In Table 5.2, the data collected from 30 respondents regarding questions about their gender, age, income, and the region of the country where they live are organized as before. Each respondent, often referred to as a case, has his data recorded in a row. Respondents have provided an *Opinion* on each of the 4 products in one section of the data, and demographic characteristics, *Category*, in another. As is often the case with data, there may be some data elements that are either out of range or simply ridiculous responses; for example, respondent number 13 in Table 5.2 claims to be a 19 year old female that has an income of $55,000,000 and resides in outer space. This is one of the pitfalls of survey data: it is not unusual to receive information that is unreliable. In this case, it is relatively easy to see that our respondent is not providing information that we can accept as true. My position, and that of most

---

[2] *TiendaMía* in Spanish translates to *My Store* in English

Table 5.2 Survey opinions on 4 webpage designs

	Category				Opinion			
Case	Gender	Age	Income	Region	Product 1	Product 2	Product 3	Product 4
1	M	19	2,500	east	good	good	good	bad
2	M	25	21,500	east	good	good	bad	bad
3	F	65	13,000	west	good	good	good	bad
4	M	43	64,500	north	good	good	bad	bad
5	F	20	14,500	east	bad	good	bad	good
6	F	41	35,000	north	bad	good	bad	bad
7	F	77	12,500	south	good	bad	bad	bad
8	M	54	123,000	south	bad	bad	bad	bad
9	F	31	43,500	south	good	good	bad	bad
10	M	37	48,000	east	bad	good	good	bad
11	M	41	51,500	west	good	good	bad	bad
12	F	29	26,500	west	bad	good	bad	bad
13	F	19	55,000,000	outer space	bad	bad	bad	bad
14	F	32	41,000	north	good	bad	good	bad
15	M	45	76,500	east	good	bad	good	good
16	M	49	138,000	east	bad	bad	bad	bad
17	F	36	47,500	west	bad	bad	bad	bad
18	F	64	49,500	south	bad	good	bad	bad
19	M	26	35,000	north	good	good	good	bad
20	M	28	29,000	north	good	bad	good	bad
21	M	27	25,500	north	good	good	good	bad
22	M	54	103,000	south	good	bad	good	good
23	M	59	72,000	west	good	good	good	bad
24	F	30	39,500	west	good	bad	good	good
25	F	62	24,500	east	good	bad	bad	bad
26	M	62	36,000	east	good	bad	bad	good
27	M	37	94,000	north	bad	bad	bad	bad
28	F	71	23,500	south	bad	bad	good	bad
29	F	69	234,500	south	bad	bad	bad	bad
30	F	18	1,500	east	good	good	good	bad

analysts, on this respondent is to remove the record or case completely from the survey. In other cases, it will not be so easy to detect errors or unreliable information, but the validation techniques we developed in Chap. 4 might help catch such cases.

Now let's consider a few questions that might be of business interest to the owners of TiendaMía:

1. Is there a webpage design that dominates others in terms of positive customer response?
2. If we consider the various demographic and financial characteristics of the respondents, how do these characteristics relate to their webpage design preferences; that is, is there a particular demographic group(s) that responded with generally positive, negative, or neutral preferences to the particular webpage designs?

These questions cover a multitude of important issues that TiendaMía will want to explore. Let us assume that we have exercised most of the procedures for ensuring data accuracy discussed earlier, but we still have some **data scrubbing**[3] that needs to be done. We surely will eliminate respondent 13 in Table 5.2 who claims to be from outer space; thus, we now have data for 29 respondents to analyze.

The *PivotTable* and *PivotChart* Report tools can be found in the *Tables Group* of the *Insert Ribbon*. As is the case with other tools, the *PivotTable* and *PivotChart* have a wizard that guides the user in the design of the report. Before we begin to exercise the tool, I will describe the basic elements of a *PivotTable*.

5.3.2 PivotTables

A *PivotTable* organizes large quantities of data in a 2-dimensional format. For a set of data, the combination of 2 dimensions will have an intersection. Recall that we are interested in the respondents that satisfy some set of conditions. For example, in our survey data, the dimension Gender and Product 1 can intersect in how the two categories of gender rate their preference for Product 1, a *count* of either *good* or *bad*. Thus, we can identify all females that choose *good* as an opinion for the webpage design Product 1, or similarly, all males that choose *bad* as an opinion for Product 1. Table 5.3 shows the cross-tabulation of *Gender* and preference for *Product 1*. The table accounts for all 29 respondents, with the 29 respondents distributed into the four mutually exclusive and collectively exhaustive categories—7 in Female/Bad, 7 in Female/Good, 4 in Male/Bad, and 11 in Male/Good. The categories are **mutually exclusive** in that a respondent can only belong to one of the 4 Gender/Opinion categories; they are **collectively exhaustive** in that the four Gender/Opinion categories, taken as a whole, contain all possible respondents. Note we could also construct a similar cross-tabulation for each of the three remaining webpage designs (Product 2, 3, 4) by examining the data and *counting* the respondents that meet the conditions in each cell. This could obviously be a tedious chore, especially if a large data set is involved. That is why we depend on *PivotTables* and *PivotCharts*: they automate the process of creating cross-tabulations.

Table 5.3 Cross-tabulation of gender and product 1 preference in terms of respondent count

Gender	Product 1		
	Bad	Good	Totals
Female	7	7	14
Male	4	11	15
Totals	11	18	29

[3] The term *scrubbing* refers to the process of removing or changing data elements that are contaminated or incorrect, or that are in the wrong format for analysis.

In Excel, the internal area of the cross-tabulation table is referred to as the **data area**, and the data elements captured within the data area represent a **count** of respondents (7, 7, 4, 11). The dimensions are referred to as the *row* and *column*. On the margins of the table we can also see the totals for the various values of each dimension. For example, the *data area* contains 11 total *bad* respondent preferences and 18 *good*, regardless of the gender of the respondent. Also, there are 14 total females and 15 males, regardless of their preferences.

The marginal dimensions are selected by the user, and the sub-totals, for example 11 Bad opinions, can be useful in analysis. In the *data area* we currently display a count of respondents, but there are other values we could include, for example, the respondents average age or the total income for respondents meeting the *row* and *column* criteria. There are also many other values that could be selected. We will provide more detail on this topic later.

We can expand the number of data elements along one of the dimensions, *row* or *column*, to provide a more detailed and complex view of the data. Previously, we had only *Gender* on the *row* dimension. Consider a new combination of *Region* and *Gender*. *Region* has 4 associated categories and *Gender* has 2; thus, we will have 8 (4×2) rows of data, plus a totals row. Table 5.4 shows the expanded and more detailed cross-tabulation of data. This new breakdown provides detail for *Male* and *Female* by region. For example, there are 3 females and 6 males in the East region. There is no reason why we could not continue to add other dimensions, either to the *row* or *column*, but from a practical point of view, adding more dimensions can lead to visual overload of information. Therefore, we are careful to consider the confusion that might result from the indiscriminant addition to dimensions. In general, two characteristics on a row or column are usually a maximum for easily understood presentations.

You can see that adding dimensions to either the row or column results in a data reorganization and different presentation of Table 5.2; that is, rather than organizing based on all respondents (observations), we organize based on the specific categories of the dimensions that are selected. All our original data remains intact. If we add all the counts found in the totals column for females in Table 5.4, we still

Table 5.4 Cross-tabulation of gender/region and product 1 preference in terms of respondent count

| Region | Gender | Product 1 | | |
		Bad	Good	Totals
East	Female	1	2	3
	Male	2	4	6
West	Female	2	2	4
	Male	0	2	2
North	Female	1	1	2
	Male	1	4	5
South	Female	3	2	5
	Male	1	1	2
	Totals	11	18	29

have a count of 14 females (marginal column totals. . .3+4+2+5=14). By not including a dimension, such as Age, we ignore age differences in the data. The same is true for Income. More precisely we are not ignoring Age or Income, but we are simply not concerned with distinguishing between the various categories of these two dimensions.

So what are the preliminary results of the cross-tabulation that is performed in Table 5.3 and 5.4? Overall there appears to be more *good* than *bad* evaluations of Product 1, with 11 *bad* and 18 *good*. This is an indication of the relative strength of the product, but if we dig a bit more deeply into the data and consider the gender preferences in each region, we can see that females are far less enthusiastic about Product 1, with 7 bad and 7 good. Males on the other hand, seem far more enthusiastic, with 4 *bad* and 11 *good*. The information that is available in Table 5.4 also permits us to see the regional differences. If we consider the South region, we see that both males and females have a mixed view of Product 1, although the number of respondents in the South is relatively small.

Thus far, we have only considered count for the data field of the cross-tabulation table; that is, we have counted the respondents that fall into the various intersections of categories—e.g. two Female observations in the West have a *bad* opinion of Product 1. There are many other alternatives for how we can present data, depending on our goal for the analysis. Suppose that rather than using a count of respondents, we decide to present the average income of respondents for each cell of our data area. Other options for the income data could include the sum, min, max, or standard deviation of the respondent's income in each cell. Additionally, we can calculate the percentage represented by a count in a cell relative to the total respondents. There are many interesting and useful possibilities available.

Consider the cross-tabulation in Table 5.3 that was presented for respondent counts. If we replace the respondent count with the average of their Income, the data will change to that shown in Table 5.5. The value is $100,750 for the combination of Male/Bad in the cross-tabulation table. This is the average[4] of the four respondents found in Table 5.2: #8–$123,000, #10–$48,000, #16–$138,000, and #27–$94,000. TiendaMía.com might be very concerned that these males with substantial spending power do not have a good opinion of Product 1.

Table 5.5 Cross-tabulation of gender and product 1 preference in terms of average income

Gender	Product 1		
	Bad	Good	Totals
Female	61,571.4	25,071.4	43,321.4
Male	100,750.0	47,000.0	61,333.3
Totals	75,818.2	38,472.2	52,637.9

---

[4] (123,000 + 48,000 + 138,000 + 94,000) / 4 = $100,750.

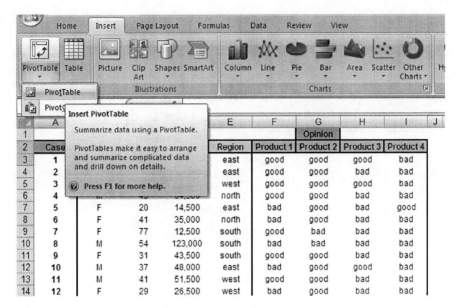

Exhibit 5.5   Insert *PivotTable* command

Now, let us turn to the *PivotTable* and *PivotChart* tool in Excel to produce the results we have seen in Tables 5.3, 5.4, and 5.5. The steps for constructing the tables follow:

1. Exhibit 5.5 shows the *Tables Group* found in the *Insert Ribbon*. In this step, you can choose between a *PivotTable* and a *PivotChart*. We will begin by selecting a *PivotTable*, although a *PivotChart* contains the same information.
2. Step 2, shown in Exhibit 5.6, opens a dialogue box that asks you to identify the data range you will use in the analysis—in our case $A$2:$I$31. Note that I have included the titles (dimension labels such as *Gender, Age*, etc.) of the fields, just as we did in the data sorting and filtering process. This permits a title for each data field—*Case, Gender, Income*, etc. The dialogue box also asks where you would like to locate the *PivotTable*. We choose to locate the table in the same sheet as the data, cell L10, but you can also select a new worksheet.
3. A convenient form of display will enable the drop-and-drag capability for the table. Once the table is established, right click and select Pivot Table Options. Under the Display Tab, select *Classic Pivot Table Layout.* See Exhibit 5.7.
4. Exhibit 5.7 shows the general layout of the *PivotTable*. Note that there are four fields that form the table and that require an input—**Page, Column, Row**, and **Data**. With the exception of the *Page*, the layout of the cross-tabulation table is similar to our previous Tables 5.3, 5.4, and 5.5. On the right, *Pivot Table Field List*, you see nine buttons that represent the dimensions that we identified earlier as the titles of columns in our data table. You can drag-and-drop the buttons into the four fields shown below—*Report Filter (Page Fields), Column Labels,*

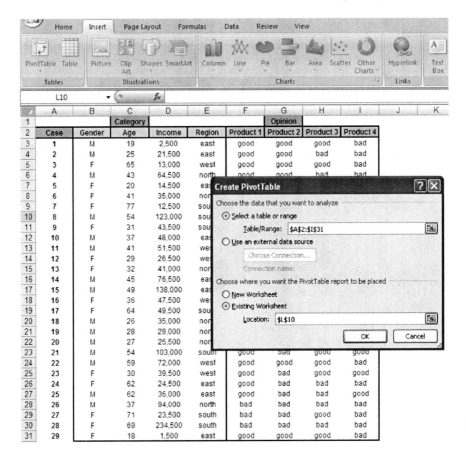

Exhibit 5.6   Create *PivotTable*

Row Labels, and *Values*. Exhibit 5.8 shows the *Row Labels* area populated with *Gender*. Of the four fields, the *Page* field is the only one that is sometimes not populated. Note that the *Page* field provides a third dimension for the *PivotTable*.

5. Exhibit 5.9 shows the constructed table. I have selected *Gender* as the row, *Product 1* as the column, *Region* as the page, and *Case* as the values fields. Additionally, I have selected *count* as the measure for *Case* in the data field. By selecting the pull down menu for the *Values* field (see Exhibit 5.10) or by right clicking a value in the table you can change the measure to one of many possibilities—**Sum, Count, Average, Min, Max**, etc. Even greater flexibility is provided in the *Show Values As* menu tab. For example, the *Value Field Settings* dialogue box allows you to select additional characteristics of how the count will be presented—e.g. as a *% of total Count*. See Exhibit 5.11.

6. In Exhibit 5.12 you can see one of the extremely valuable features of *PivotTables*. A pull-down menu is available for the *Page*, *Row*, and *Column* fields. These correspond to *Region*, *Gender*, and *Product 1*, respectively. These menus will

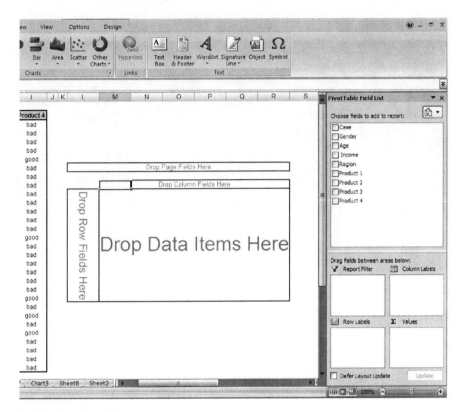

Exhibit 5.7 *PivotTable* field entry

allow you to change the data views by limiting the categories within dimensions, and to do so without reengaging the wizard. For example, currently *all Region* categories (East, West, etc.) are included in the table shown in Exhibit 5.12, but you can use the pull-down menu to limit the regions to *East* only. Exhibit 5.13 shows the results and the valuable nature of this built-in capability. Note that the number of respondents for the *PivotTable* for the *East* region results in only 9 respondents, whereas, the number of respondents for all regions is 29 as seen in Exhibit 5.12. Combining this capability with the changes we can perform for the *Values* field, we can literally view the data in our table from a nearly unlimited number of perspectives.

7. Exhibit 5.14 demonstrates the change proposed in (4) above. We have changed the *Value Field Settings* from count to percent of total, just one of the dozens of views that is available. The choice of views will depend on your goals for the data presentation.

8. We can also extend the chart quite easily to include *Region* in the *Row* field. This is accomplished by pointing to any cell in the table which displays the field list, and dragging and dropping the *Region* button into the *Row Labels*. This converts the table to resemble Table 5.4. This action provides subtotals for the

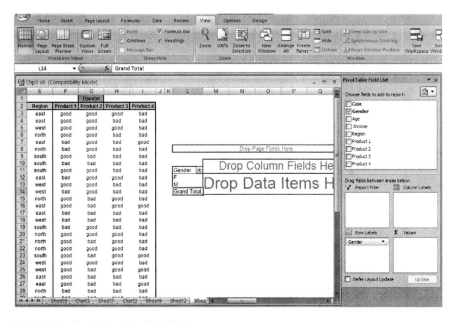

Exhibit 5.8   Populating *PivotTable* fields

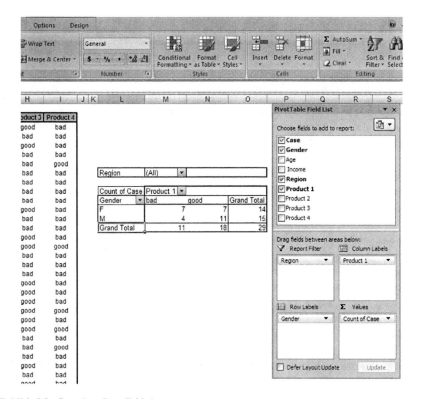

Exhibit 5.9   Complete *PivotTable* layout

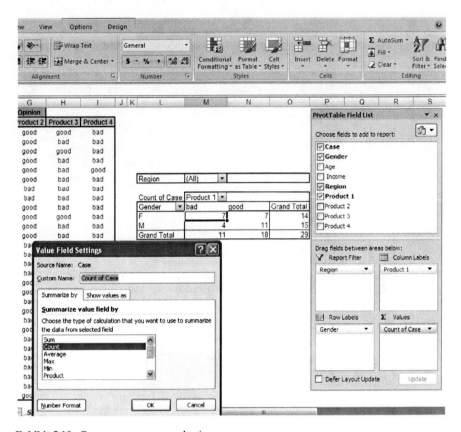

Exhibit 5.10   Case count summary selection

various regions and gender combinations—*East*, *West*, etc. Exhibit 5.15 shows
the results of the extension. Note that I have selected Age as the new *Report
Filter.*

9. One final feature of special note is the capability to identify the specific respon-
    dents' records that are in the cells of the data field. Simply double-click the cell
    of interest in the Pivot Table, and a separate sheet is created with the records of
    interest. This allows you to *drill-down* into the records of the specific respondents
    in the cell count. In Exhibit 5.16, the 11 males that responded *Good* to *Product 1*
    are shown; they are respondents 1, 2, 25, 4, 22, 21, 20, 19, 18, 14, and 11. If you
    return to Exhibit 5.12, you can see the cell of interest to double-click, N13.

5.3.3 PivotCharts

Now, let us repeat the process steps described above to construct a *PivotChart*. There
is little difference in the steps to construct a table versus a chart. In fact, the process

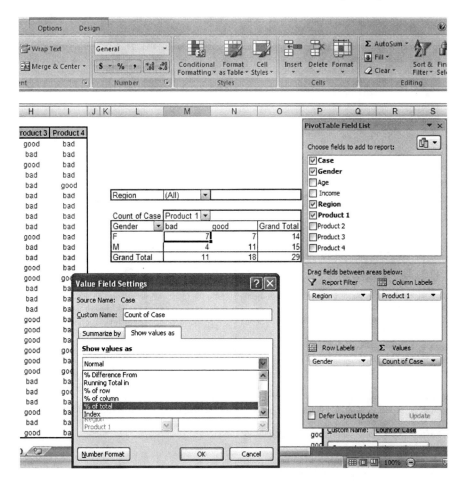

Exhibit 5.11 Case count summary selection as % of total count

of constructing a *PivotChart* leads to the simultaneous construction of a *PivotTable*.
The obvious difference is in step 1: rather than select the *PivotTable*, select the
PivotChart option. The process of creating a *PivotChart* can be difficult; we will not
invest a great deal of effort on the details. It may be wise to always begin by creating
a *PivotTable*. Creating a *PivotChart* then follows more easily. By simply selecting
the *PivotTable*, a tab will appear above the row of ribbons – *PivotTable Tools*. In
this ribbon you will find a tools group that permits conversion of a *PivotTable* to a
PivotChart. See Exhibit 5.17.

The table we constructed in Exhibit 5.9 is presented in Exhibit 5.18, but as a
chart. Note that a *PivotChart Filter Pane* is available in the *Analyze Ribbon* when
the chart is selected. Just as before, it is possible to filter the data that is viewed by
manipulating the *Filter Pane*. Exhibit 5.19 shows the result of changing the *Page*
field from all regions to only the *South*. From the chart, it is apparent that there

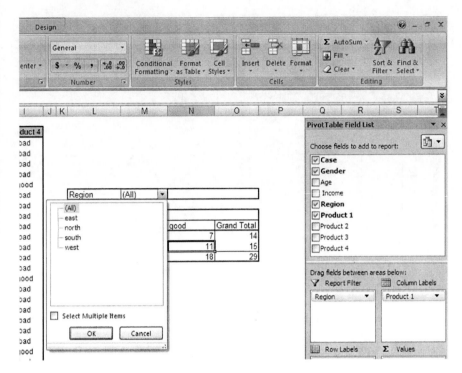

Exhibit 5.12 *PivotTable* drop-down menus

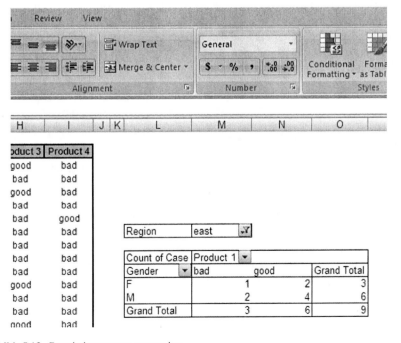

Exhibit 5.13 Restricting page to east region

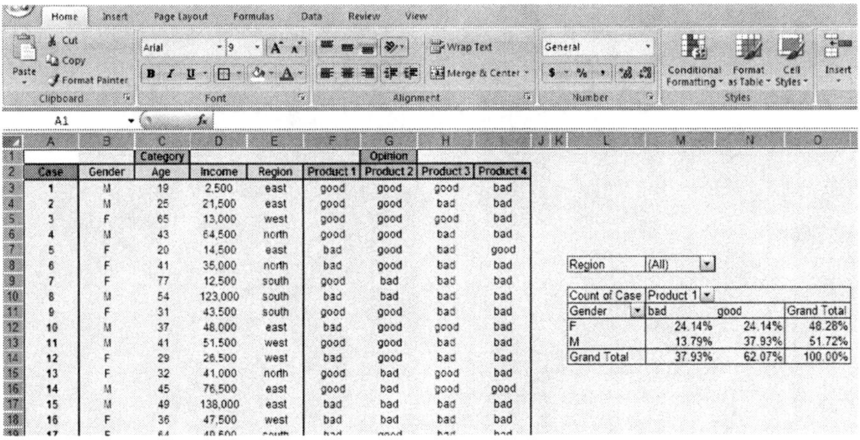

	Category			Opinion			
Case	Gender	Age	Income	Region	Product 1	Product 2	Product 3	Product 4
1	M	19	2,500	east	good	good	good	bad
2	M	25	21,500	east	good	good	bad	bad
3	F	65	13,000	west	good	good	good	bad
4	M	43	64,500	north	good	good	bad	bad
5	F	20	14,500	east	bad	good	bad	good
6	F	41	35,000	north	bad	good	bad	bad
7	F	77	12,500	south	good	bad	bad	bad
8	M	54	123,000	south	bad	bad	bad	bad
9	F	31	43,500	south	good	good	bad	bad
10	M	37	48,000	east	bad	good	good	bad
11	M	41	51,500	west	good	good	bad	bad
12	F	29	26,500	west	bad	good	bad	bad
13	F	32	41,000	north	good	bad	good	bad
14	M	45	76,500	east	good	bad	good	good
15	M	49	138,000	east	bad	bad	bad	bad
16	F	36	47,500	west	bad	bad	bad	bad

Region	(All)		
Count of Case	Product 1		
Gender	bad	good	Grand Total
F	24.14%	24.14%	48.28%
M	13.79%	37.93%	51.72%
Grand Total	37.93%	62.07%	100.00%

Exhibit 5.14 Region limited to east- resulting table

are seven respondents that are contained in the south region, and of the seven, three females responded that *Product 1* was *bad*. Additionally, the chart can be extended to include multiple fields, as shown by the addition of *Region* to the *Row* field, along with *Gender*. See Exhibit 5.20. This is equivalent to the *PivotTable* in Exhibit 5.15.

As with any Excel chart, the user can specify the type of chart and other options for presentation. In Exhibit 5.21 we show the data table associated with the chart; thus, the viewer has access to the chart and a table, simultaneously. Charts are powerful visual aids for presenting analysis, and are often more appealing and accessible than tables of numbers. The choice of a table versus a chart is a matter of preference. In Exhibit 5.21 we provide data presentation for all preferences—a chart and table of values.

5.4 TiendaMía.com Example—Question 1

Now back to the questions that the owners of TiendaMía.com asked earlier. But before we begin with the cross-tabulation analysis, a warning is in order. As with previous examples, this example has a relatively small number of respondents (29). It is dangerous to infer that the result of a small sample is indicative of the entire population of TiendaMía.com customers. We will say more about sample size in the next chapter, but needless to say, larger samples provide greater comfort in the generalization of results. For now, we can assume that this study is intended as a preliminary analysis, leading to more rigorous study later. Thus, we will also assume that our sample is large enough to be meaningful. And now, we consider the first question—Is there a webpage design that dominates others in terms of positive customer response?

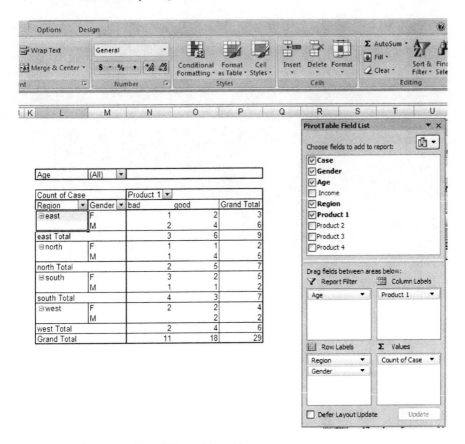

Exhibit 5.15 Extension of row field- resulting table

In order to answer this question, we need not use cross-tabulation analysis. Cross-tabulation provides insight into how the various characteristics of the respondents relate to preferences; our question is one that is concerned with summary data for respondents without regard to detailed characteristics. So, let us focus on how many respondents have a preference for each webpage design? Let's use the **COUNTIF(range, criteria)** cell function to count the number of *bad* and *good* responses that are found in our data table. For *Product 1* in Exhibit 5.22, the formula in cell F33 is *COUNTIF(F3:F31, "good")*. Thus, the counter will count a cell value *if* it corresponds to the criterion that is provided in the cell formula, in this case *good*. Note that a split screen is used in this exhibit (hiding most of the data), but it is obvious that *Product 1*, with 18 *good*, dominates all products. *Product 2* and *Product 3* are relatively close (15 and 13) to each other, but *Product 4* is significantly different, with only 5 *good* responses. Again, recall that this result is based on a relatively small sample size, so we must be careful to understand that if we require a high degree of assurance about the results, we may want a much larger sample size than 29 respondents.

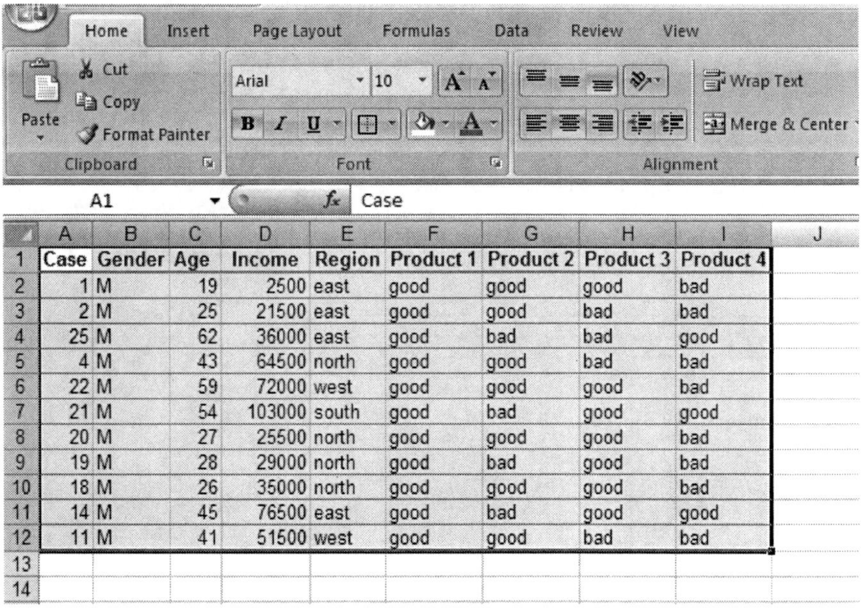

Case	Gender	Age	Income	Region	Product 1	Product 2	Product 3	Product 4
1	M	19	2500	east	good	good	good	bad
2	M	25	21500	east	good	good	bad	bad
25	M	62	36000	east	good	bad	bad	good
4	M	43	64500	north	good	good	bad	bad
22	M	59	72000	west	good	good	good	bad
21	M	54	103000	south	good	bad	good	good
20	M	27	25500	north	good	good	good	bad
19	M	28	29000	north	good	bad	good	bad
18	M	26	35000	north	good	good	good	bad
14	M	45	76500	east	good	bad	good	good
11	M	41	51500	west	good	good	bad	bad

Exhibit 5.16 Identify the respondents associated with a table cell

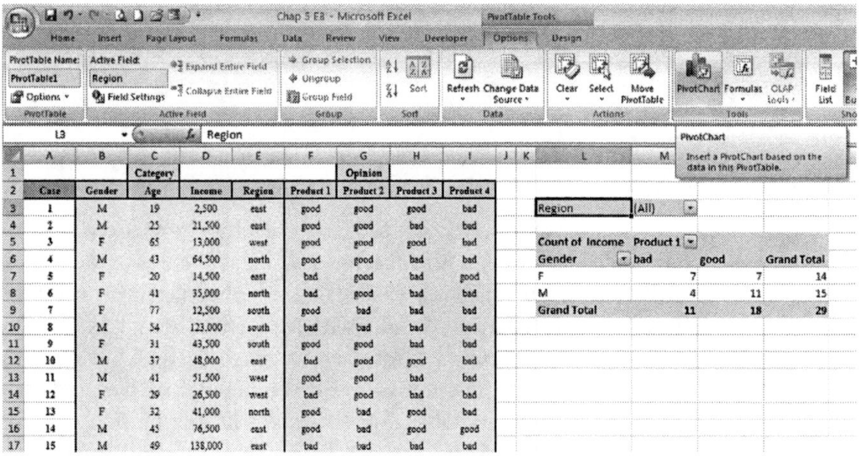

		Category				Opinion		
Case	Gender	Age	Income	Region	Product 1	Product 2	Product 3	Product 4
1	M	19	2,500	east	good	good	good	bad
2	M	25	21,500	east	good	good	bad	bad
3	F	65	13,000	west	good	good	good	bad
4	M	43	64,500	north	good	good	bad	bad
5	F	20	14,500	east	bad	good	bad	good
6	F	41	35,000	north	bad	good	bad	bad
7	F	77	12,500	south	good	bad	bad	bad
8	M	54	123,000	south	bad	bad	bad	bad
9	F	31	43,500	south	good	good	bad	bad
10	M	37	48,000	east	bad	good	good	bad
11	M	41	51,500	west	good	good	bad	bad
12	F	29	26,500	west	bad	good	bad	bad
13	F	32	41,000	north	good	bad	good	bad
14	M	45	76,500	east	good	bad	good	good
15	M	49	138,000	east	bad	bad	bad	bad

Region: (All)

Count of Income	Product 1		
Gender	bad	good	Grand Total
F	7	7	14
M	4	11	15
Grand Total	11	18	29

Exhibit 5.17 *PivotTable* ribbon

The strength of the preferences in our data is recorded as a simple dichotomous choice—*good* or *bad*. In designing the survey, there are other possible data collection options that could have been used to record preferences. For example, the respondents could be asked to rank the webpages from best to worse.

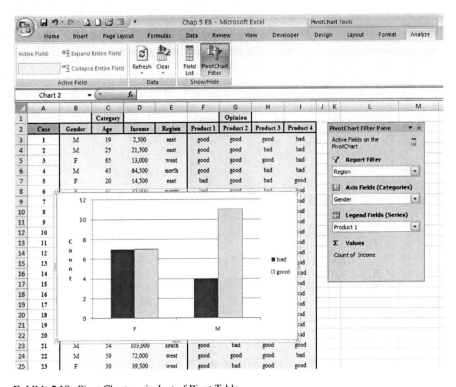

Exhibit 5.18 Pivot Chart equivalent of Pivot Table

This would provide information of the relative position (ordinal data) of the web-pages, but it would not determine if they were acceptable or unacceptable, as is the case with *good* and *bad* categories. An approach that brings both types of data together could create a scale with one extreme representing a *highly favorable* webpage, the center value a *neutral* position, and the other extreme as *highly unfavorable*. Thus, we can determine if a design is acceptable, and we can also determine the relative position of one design versus the others. For now, we can see that relative to Products 1, 2, and 3, Product 4 is by far the least acceptable option.

5.5 TiendaMía.com Example—Question 2

Question 2 asks how the demographic and financial data of the respondents relates to preferences. This is precisely the type of question that can easily be handled by using cross-tabulation analysis. Our demographic characteristics are represented by *Gender*, *Age*, *Income*, and *Region*. *Gender* and *Region* have two and four variable levels (categories), respectively, which are a manageable number of values. But

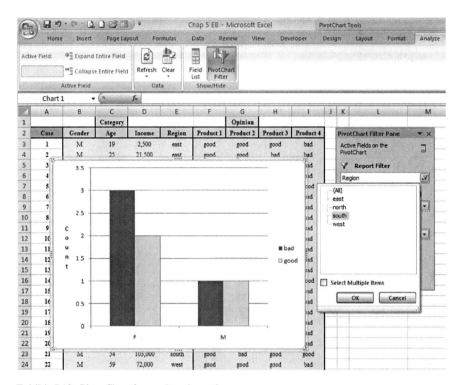

Exhibit 5.19 Pivot Chart for south region only

Income and *Age* are a different matter. The data has been collected in increments of $500 for income and units of years for age, resulting in many possible values for these variables. What is the value of having such detailed information? Is it absolutely necessary to have such detail for our goal of analyzing the connection between these demographic characteristics and preferences for webpages? Can we simplify the data by creating categories of contiguous values of the data and still answer our questions with some level of precision?

Survey studies often group individuals into age categories spanning multiple years (e.g. 17–21, 22–29, 30–37, etc.) that permit easier cross-tabulation analysis, with minor loss of important detail. The same is true of income. We often find with quantitative data that it is advantageous, from a data management point of view, to create a limited number of categories, and this can be done *after* the initial collection of detailed data. Thus, the data in Table 5.2 would be collected then *conditioned* or *scrubbed* to reflect categories for both *Age* and *Income*. In an earlier chapter we introduced the idea of collecting data that would serve multiple purposes, and even unanticipated purposes. Table 5.2 data is a perfect example of such data.

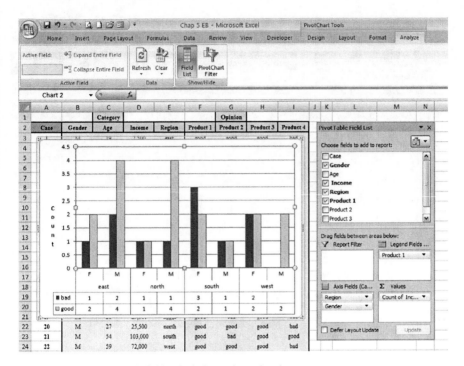

Exhibit 5.20 Extended axis field to include gender and region

So let us create categories for *Age* and *Income* that are easy[5] to work with
and simple to understand. For *Age* , we will use the following categories: 18–37;
38–older. Let us assume that these categories represent groups of consumers that
exhibit similar behavior: purchasing characteristics, visits to the site, level of expenditures per visit, etc. For *Income* , we will use \$0-38,000; \$38,001-above. Again,
assume that we have captured similar financial behavior for respondents in these
categories. Note that we have 2 categories for each dimension and we will apply
a numeric value to the categories—*1* for values in the lowest range and 2 in the
highest. The changes resulting for the initial Table 5.2 data are shown in Table 5.6.
The conversion to these categories can be accomplished with an *IF* statement. For
example, *IF(E3<=38000, 1,2)* returns *1* if the income of the first respondent is *less
than or equal to* \$38,000, otherwise 2 is returned.

Generally, the selection of the categories should be based on the expertise of
the data collector (TiendaMía.com) or their advisors. There are commonly accepted

---

[5] Since we are working with a very small sample, the categories have been chosen to reflect differences in the relationships between demographic/financial characteristics and preferences. In other
words, I have made sure the selection of categories results in interesting findings for the purpose
of this simple example.

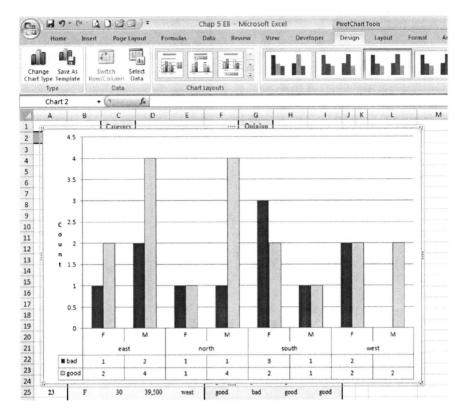

Exhibit 5.21   Data table options for PivotCharts

categories in many industries and they can be found by reading research studies
or the popular press associated with a business sector—e.g. the Census Bureau
often uses the following age categories for income studies—below 15, 15–24,
25–34, 35–44, etc. Other sources producing studies in specific areas of industry
or business, such as industry trade associations, can be invaluable as sources of
demographic/financial category standards.

Now back to the question related to the respondent's demographic characteristics
and how those characteristics relate to preferences. TiendaMía.com is interested in
targeting particular customers with particular products, and doing so with a particular web design. Great product offerings are not always enough to entice customers to
buy. TiendaMía.com understands that a great web design can influence customers
to buy more items and more expensive products. This is why they are concerned
with the attitudes of respondents toward the set of four webpage designs. So let us
examine *which* respondents prefer *which* webpages.

Assume that our management team at TiendaMía.com believes that *Income* and
Age are the characteristics of greatest importance; *Gender* plays a small part in preferences and *Region* plays an even lesser role. We construct a set of four *PivotTables*

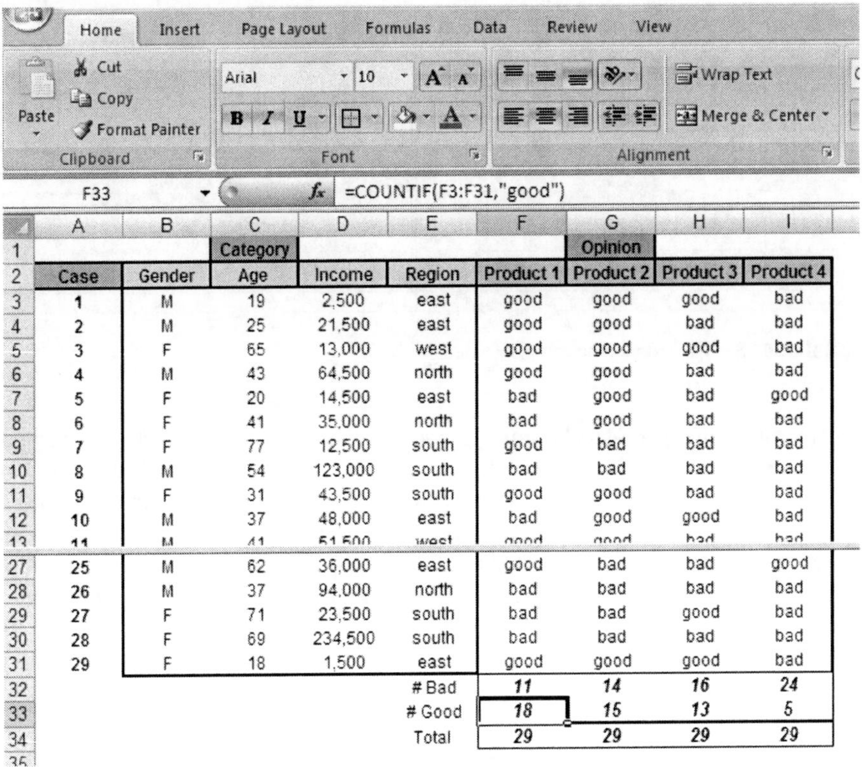

	Category			Opinion			
Case	Gender	Age	Income	Region	Product 1	Product 2	Product 3	Product 4
1	M	19	2,500	east	good	good	good	bad
2	M	25	21,500	east	good	good	bad	bad
3	F	65	13,000	west	good	good	good	bad
4	M	43	64,500	north	good	good	bad	bad
5	F	20	14,500	east	bad	good	bad	good
6	F	41	35,000	north	bad	good	bad	bad
7	F	77	12,500	south	good	bad	bad	bad
8	M	54	123,000	south	bad	bad	bad	bad
9	F	31	43,500	south	good	good	bad	bad
10	M	37	48,000	east	bad	good	good	bad
11	M	41	51,500	west	good	good	bad	bad
25	M	62	36,000	east	good	bad	bad	good
26	M	37	94,000	north	bad	bad	bad	bad
27	F	71	23,500	south	bad	bad	good	bad
28	F	69	234,500	south	bad	bad	bad	bad
29	F	18	1,500	east	good	good	good	bad
				# Bad	11	14	16	24
				# Good	18	15	13	5
				Total	29	29	29	29

Exhibit 5.22 Summary analysis of product preference

Table 5.6 Age and income category extension

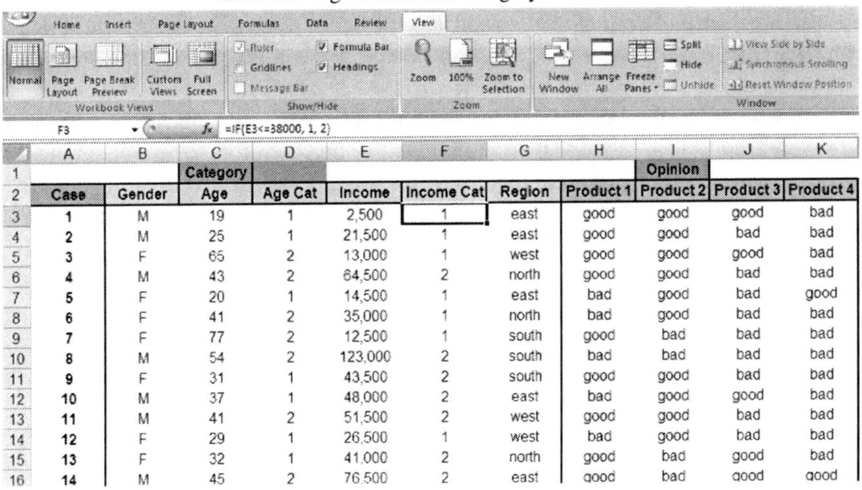

	Category					Opinion			
Case	Gender	Age	Age Cat	Income	Income Cat	Region	Product 1	Product 2	Product 3	Product 4
1	M	19	1	2,500	1	east	good	good	good	bad
2	M	25	1	21,500	1	east	good	good	bad	bad
3	F	65	2	13,000	1	west	good	good	good	bad
4	M	43	2	64,500	2	north	good	good	bad	bad
5	F	20	1	14,500	1	east	bad	good	bad	good
6	F	41	2	35,000	1	north	bad	good	bad	bad
7	F	77	2	12,500	1	south	good	bad	bad	bad
8	M	54	2	123,000	2	south	bad	bad	bad	bad
9	F	31	1	43,500	2	south	good	good	bad	bad
10	M	37	1	48,000	2	east	bad	good	good	bad
11	M	41	2	51,500	2	west	good	good	bad	bad
12	F	29	1	26,500	1	west	bad	good	bad	bad
13	F	32	1	41,000	2	north	good	bad	good	bad
14	M	45	2	76,500	2	east	good	bad	good	good

Count of Case		Product 1			Count of Case		Product 2		
AgeCat	IncomeCat	bad	good	Grand Total	AgeCat	IncomeCat	bad	good	Grand Total
1	1	2	6	8	1	1	1	7	8
	2	3	3	6		2	4	2	6
2	1	2	4	6	2	1	4	2	6
	2	4	5	9		2	5	4	9
Grand Total		11	18	29	Grand Total		14	15	29

Count of Case		Product 3			Count of Case		Product 4		
AgeCat	IncomeCat	bad	good	Grand Total	AgeCat	IncomeCat	bad	good	Grand Total
1	1	3	5	8	1	1	7	1	8
	2	3	3	6		2	5	1	6
2	1	4	2	6	2	1	5	1	6
	2	6	3	9		2	7	2	9
Grand Total		16	13	29	Grand Total		24	5	29

Exhibit 5.23   Age and income category extension

that contain the cross-tabulations for comparison of all the products and respondents in our study. All four products are combined in Exhibit 5.23, beginning with *Product 1* in the Northwest corner and *Product 4* in the Southeast—note the titles in the column field identifying the four products. One common characteristic of data in each cross-tabulation is the number of individuals that populate each combination of demographic/financial categories—e.g. there are 8 individuals in the combination of the 18–37 *Age* range and 0-$38,000 *Income* category; there are 6 that are in the 18–37 and $38,001–above categories, etc. These numbers are in the **Grand Totals** column in each *PivotTable*.

To facilitate the formal analysis, let us introduce a shorthand designation for identifying categories: *AgeCategory* \ *IncomeCategory*. We will use the category values introduced earlier to shorten and simplify the Age-Income combinations. Thus, 1\1 is the 18–37 *Age* and the 0-$38,000 *Income* combination. Now here are some observations that can be reached by examining Exhibit 5.23:

1. Category 1\1 has strong opinions about products. They are positive to very positive regarding *Products 1, 2,* and *3* and they strongly dislike *Product 4*. For example, for *Product 1*, category 1\1 rated it *bad* and *good*, 2 and 6, respectively.
2. Category 1\2 is neutral about *Products 1, 2,* and *3*, but strongly negative on *Product 4*. It may be argued that they are not neutral on *Product 2*. This is an important category due to their higher income and therefore their higher potential for spending. For example, for *Product 4*, category 1\2 rated it *bad* and *good*, 5 and 1, respectively.
3. Category 2\1 takes slightly stronger positions than 1\2, and they are only positive about *Product 1*. They also take opposite positions than 1\1 on *Products 2* and *3*, but agree on *Products 1* and *4*.
4. Category 2\2 is relatively neutral on *Product 1* and *2* and negative on *Product 3* and *4*. Thus, 2\2 is not particularly impressed with any of the products, but the category is certainly unimpressed with Products 3 and 4.
5. Clearly, there is universal disapproval for *Product 4*, and the disapproval is quite strong. Ratings by 1\1, 1\2, 2\1, 2\2 are far more negative than positive: 24 out 29 respondents rated it *bad*.

Table 5.7 Detailed view of respondent favorable ratings

Respondent category	1\1	1\2	2\1	2\2
Acceptable	P-2 (87.5%)			
	P-1 (75.0%)			
			P-1 (66.7%)	
	P-3 (62.5%)			
				P-1 (55.6%)
Neutral		P-1&3 (50%)		
				P-2 (44.4%)
		P-2 (33.3%)	P-2&3 (33.3%)	P-3 (33.3%)
				P-4 (22.2%)
		P-4 (16.7%)	P-4 (16.7%)	
Unacceptable		P-4 (0.125%)		

There is no clear consensus for a webpage design that is acceptable to all categories, but clearly Product 4 is a disaster. If TiendaMía.com decides to use a single webpage design, which one would be the most practical design to use? This is not a simple question to answer given the analysis above. TiendaMía.com may decide that the question requires further in-depth study. Why? Here are several important reasons why further study may be necessary:

1. if the conclusions from the data analysis are inconclusive
2. if the size of the sample is deemed to be too small, then the preferences reflected may not merit a generalization of results to the population of all possible website users
3. if the study reveals new questions of interest or guides us in new directions that might lead us to eventual answers for these new questions.

Let us consider number (3) above. We will perform a slightly different analysis by asking the following new question—is there a single measure that permits an overall ranking of products? The answer is yes. We can summarize the preferences shown in Exhibit 5.23 in terms of a new measure—*favorable rating*.

Table 5.7 organizes the respondent categories and their *favorable rating*—the ratio[6] of *good* responses relative to the total of *all responses*. From Exhibit 5.23 you see that respondents in category 1\1 have an 87.5% (7 of 8 rated the product *good*) favorable rating for *Product 2*. This is written as P-2 (87.5%). Similarly, category 2\1 has a favorable rating for P-2 and P-3 of 33.3% (2of 6). To facilitate comparison, the favorable ratings are arranged on the vertical axis of the table, with highest near the top of the table and lowest near the bottom, with a corresponding scale from acceptable to neutral to unacceptable. (Note this is simply a table and not a *PivotTable*.)

---

[6] [number *good*] ÷ [number *good* + number *bad*].

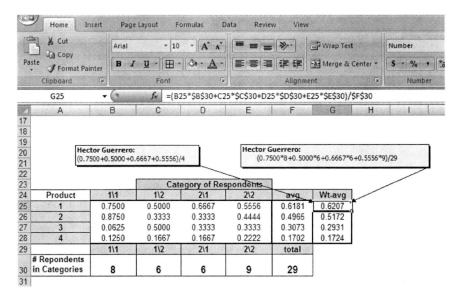

Exhibit 5.24 Calculation of average (avg) and weighted average (Wt-avg)

A casual analysis of the table suggests that *Product 1* shows a 50% or greater favorable rating for all categories. No other product can equal this favorable rating: *Product 1* is the top choice of all respondent categories except for 1\1 and it is tied with *Product 3* for category 1\2. Although this casual analysis suggests a clear choice, we can now do more formal analyses to arrive at the selection of a single website design.

First, we will calculate the average of all *favorable ratings* for each category (1\1, 1\2, 2\1, 2\2) as a single composite score. This is a simple calculation and it provides a straightforward method for TiendaMía.com to assess products. In Exhibit 5.24 the calculation of averages is found in F25:F28–0.6181 for *Product 1*, 0.4965 for *Product 2*, etc. *Product 1* has the highest average favorable rating. But there are some questions that we might ask about the fairness of the calculated averages. Should there be an approximately similar number of respondents in each category for this approach to be fair? Stated differently, is it fair to count an individual category average equally to others when the number of respondents in that category is substantially less than other categories?

In TiendaMía.com's study, there *are* different numbers of respondents in the various categories. This can be significant for the calculation of averages. The difference in the numbers can be due to the particular sample that we selected. A random sample of this small size can lead to wide variation in the respondents selected. One way to deal with this problem is to consciously sample customers to reflect the proportion of category members that shop at TiendaMía.com. There are many techniques and methods for formally **sampling**[7] data that we cannot study here.

---

[7] Sampling theory is a rich science that should be carefully considered prior to initiating a study.

For the moment, let's assume that the random sample has selected a proportionally fair representation of respondents, and this is precisely what TiendaMía.com desires. Thus, 28% (8÷29, 8 of 1\1 respondents out of 29) should be relatively close to the population of all 1\1's in TiendaMía.com's customer population. If we want to account for the difference in respondent category size in our analysis, then we will want to calculate a *weighted* average of favorable ratings, which reflects the relative size of the respondent categories. Note that the first average that we calculated is a special form of weighted average: one where all weights were assumed to be equal. In range G25:G28 of Exhibit 5.24 we see the calculation of the weighted average. Each average is multiplied by the fraction of respondents that it represents of the total sample.[8] This approach provides a proportional emphasis on averages. If a particular average is composed of many respondents, then it will receive a higher weight; if an average is composed of fewer respondents, then it will receive a lower weight.

So what do our respondent weighted averages (G25:G28) reveal about Products compared to the equally weighted averages (F25:F28)? The results are approximately the same for *Products 1* and *4*. The exceptions are *Product 2* with a somewhat stronger showing from 0.4965 to 0.5172 and Product 3 with a substantial drop in score from 0.3073 to 0.2931. Still, there is no change in the ranking of the products; it remains P-1, P-2, P-3, and P-4.

What has led to the increase in the *Product 2* score? Categories 1\1 and 2\2 are the highest favorable ratings for *Product 2*; they also happen to be the largest weighted categories (8/29=0.276 and 9/29=0.310). Larger weights applied to the highest scores will of course yield a higher weighted average. If TiendaMía.com wants to focus attention on these market segments, then a weighted average may be appropriate. **Market segmentation** is in fact a very important element in their marketing strategy.

There may be other ways to weight the favorable ratings. For example, there may be categories that are more important than others due to their higher spending per transaction or more frequent transactions at the site. So, as you can see, many weighting schemes are possible.

5.6 Summary

Cross-tabulation analysis through the use *PivotTables* and *PivotCharts* is a simple and effective way to analyze qualitative data, but to insure fair and accurate analysis the data must be carefully examined and prepared. Rarely is a data set of significant size exempt from errors. Although most errors are usually accidental, there may be some that are intentional. Excel provides many logical cell functions to determine if data have been accurately captured and fit the specifications that an analyst has imposed.

---

[8] $(0.7500 * 8 + 0.5000 * 6 + 0.6667 * 6 + 0.5556 * 9) / 29 = 0.6207$.

PivotTables and *PivotCharts* allow the analyst to view the interaction of several variables in a data set. To do so, it is often necessary to convert data elements that are collected in surveys into values that permit easier manipulation—e.g. we converted *Income* and *Age* into categorical data. This does not suggest that we have made an error in how we collected data; on the contrary, it is often advantageous to collect data in its *purest* form (e.g. 23 years of age) versus providing a category value (e.g. the 19-24 years of age category). This allows detailed uses of the data that may not be anticipated.

In the next chapter we will begin to apply more sophisticated statistical techniques to qualitative data. These techniques will permit us to not only study the interaction between variables, but they also allow us to quantify how confident we are that the conclusions we reach are indeed applicable to a population of interest. Among the techniques we will introduce are Analysis of Variance (ANOVA), tests of hypothesis with t-tests and z-tests, and chi-square tests. These are powerful statistical techniques that can be used to study the effect of *independent* variables on *dependent* variables, and determine similarity or dissimilarity in data samples. When used in conjunction with the techniques we have learned in this chapter, we are capable of uncovering the complex data interactions that are essential to successful decision making.

Key Terms

Data Errors	Collectively Exhaustive
Error Checking	Data Area
EXACT (text1, text2)	Count
TRUE/FALSE	Page, Column, Row, and Data
OR, AND, NOT, TRUE, FALSE	Sum, Count, Average, Min, Max
MOD (number, divisor)	COUNTIF (range, criteria)
Cross-tabulation	Grand Totals
PivotTable/PivotChart	Sampling
Data Scrubbing	Market Segmentation
Mutually Exclusive	

Problems and Exercises

1. Data errors are of little consequence in data analysis—T or F.
2. What does the term *data scrubbing* mean?
3. Write a *logical if* function for a cell (A1) that tests whether or not a cell contains a value larger than or equal to 15 or less than 15. Return phrases that say "15 or more" or "less than 15."
4. For Exhibit 5.1, write a *logical IF* function in the cells H2:I4 that calculates the difference between *Original Data Entry* and *Secondary Data Entry* for each

cell of the corresponding cells. If the difference is not 0, then return the phrase "Not Equal", otherwise return "Equal."

5. Use a *logical IF* function in cell A1 to test a value in B1. Examine the contents of B1and return "In" if the values are between and include the range 2 and 9. If the value is outside this range, return "Not In."

6. Use a *logical IF* function in cell A1 to test values in B1 and C1. If the contents of B1 *and* C1 are, 12 and 23, respectively, return "Values are 12 and 23", otherwise return "Values are not 12 and 23."

7. Use a *logical IF* function in cell A1 to test whether or not a value in B1 is an integer. Use the *Mod* function to make the determination. Return either "Integer" or "Not Integer."

8. What type of analysis does cross-tabulation allow a data analyst to perform?

9. What types of data (categorical, ordinal, etc.) will *PivotTables* and *PivotCharts* permit you to cross-tabulate?

10. Create a *PivotTable* from the data in Exhibit 5.2 (minus Case 13) that performs a cross-tabulation analysis for the following configuration: *Region* on the Row field; *Income* in the Values field; *Product 2* on the Column field; and *Age* on the Page field.

   a. What are the *counts* in the Values field?
   b. What are the *averages* in the Values field?
   c. What are the *maximums* in the value field?
   d. What Region has the maximum *count* ?
   e. For all regions taken as a whole, is there a clear preference, good or bad, for Product 2?
   f. What is the highest *average* income for a region/preference combination?
   g. What combination of region and preference has the highest *variation* in Income?

11. Create a cell for counting values in range A1:A25 if the cell contents are equal to the text "New."

12. Create a *PivotChart* from the *PivotTable* analysis in 10c above.

13. The *Income* data in Table 5.6 is organized into two categories. Re-categorize the *Income* data into 3 categories—0–24,000; 24,001–75,000; 75,001 and above? How will the Exhibit 5.23 change?

14. Perform the conversion of the data in Table 5.7 into a column chart that presents the same data graphically?

15. Create a weighted average based on the sum of incomes for the various categories. Hint—The weight should be related to the proportion of the category sum of income to the total of all income.

16. Your boss believes that the analysis in Exhibit 5.23 is interesting, but she would like to see the Age category replaced with Region. Perform the new analysis and display the results similarly to those of Exhibit 5.23.

17. *Advanced Problem*—A clinic that specializes in alcohol abuse has collected some data on their current clients. Their data for clients includes the number

of years of abuse have experienced, age, years of schooling, number of parents in the household as a child, and the perceived chances for recovery by a panel of experts at the clinic. Determine the following using *PivotTables* and *PivotCharts*:

a. Is there a general relationship between age and the number of years of abuse?
b. For the following age categories, what proportion of their lives have clients abused alcohol:

    i.   0–24
    ii. 25–35
    iii. 36–49
    iv. 49–over.

c. What factor is the most reliable predictor of perceived chances for recovery? Which is the least?
d. What is the co-relationship between number of parents in the household and years of schooling?
e. What is the average age of the clients with bad prospects?
f. What is the average number of years of schooling for clients with one parent in the household as a child?
g. What is the average number of parents for all clients that have poor prospects for recovery?

Case	Yrs abuse	Age	Years school	Number of parents	Prospects
1	6	26	12	1	G
2	9	41	12	1	B
3	11	49	11	2	B
4	5	20	8	2	G
5	6	29	9	1	B
6	8	34	13	2	B
7	12	54	16	2	G
8	7	33	16	1	G
9	9	37	14	1	G
10	7	31	10	2	B
11	6	26	7	2	B
12	7	30	12	1	G
13	8	37	12	2	B
14	12	48	7	2	B
15	9	40	12	1	B
16	6	28	12	1	G
17	8	36	12	2	G
18	9	37	11	2	B
19	4	19	10	1	B
20	6	29	14	2	G

Case	Yrs abuse	Age	Years school	Number of parents	Prospects
21	6	28	17	1	G
22	6	24	12	1	B
23	8	38	10	1	B
24	9	41	8	2	B
25	10	44	9	2	G
26	5	21	12	1	B
27	6	26	10	0	B
28	9	38	12	2	G
29	8	38	10	1	B
30	9	37	13	2	G

Chapter 6
Inferential Statistical Analysis of Data

Contents

H. Guerrero, *Excel Data Analysis*, DOI 10.1007/978-3-642-10835-8_6,
© Springer-Verlag Berlin Heidelberg 2010

6.1 Introduction

In Chap. 3, we introduced several statistical techniques for the analysis of data, most of which were descriptive or exploratory. But, we also got our first glimpse of another form of statistical analysis known as *Inferential Statistics*. Inferential statistics is how statisticians use inductive reasoning to move from the specific, the data contained in a sample, to the general, inferring characteristics of the population from which the sample was taken.

Many problems require an understanding of population characteristics; yet, it can be difficult to determine these characteristics because populations can be very large and difficult to access. So rather than throw our hands into the air and proclaim that this is an *impossible* task, we resort to a **sample**: a small slice or view of a population. It is not a perfect solution, but we live in an imperfect world and we must make the best of it. Mathematician and popular writer John Allen Paulos sums it up quite nicely—"Uncertainty is the only certainty there is, and knowing how to live with insecurity is the only security."

So what sort of imperfection do we face? Sample data can result in measurements that are not representative of the population from which they are taken, so there is always uncertainty as to how well the sample represents the population. We refer to these circumstances as **sampling error**: the difference between the measurement results of a sample and the true measurement values of a population. Fortunately, through carefully designed sampling methods and the subsequent application of statistical techniques, statisticians *are* able to infer population characteristics from results found in a sample. If performed correctly, the sampling design will provide a measure of reliability about the population inference we will make.

Let us carefully consider why we rely on inferential statistics:

1. The size of a population often makes it impossible to measure characteristics for every member of the population—often there are just too many members of populations. Inferential statistics provides an alternative solution to this problem.
2. Even if it is possible to measure characteristics for the population, the cost can be prohibitive. Accessing measures for every member of a population can be costly.
3. Statisticians have developed techniques that can quantify the uncertainty associated with sample data. Thus, although we know that samples are not perfect, inferential statistics provides a reliability evaluation of how well a sample measure represents a population measure.

This was precisely what we were attempting to do in the survey data on the four webpage designs in Chap. 5; that is, to make population inferences from the webpage preferences found in the sample. In the descriptive analysis we presented a numerical result. With inferential statistics we will make a statistical statement about our confidence that the sample data is representative of the population. For the numerical outcome, we *hoped* that the sample did in fact represent the population, but it was mere hope. With inferential statistics, we will develop techniques that allow us to *quantify* a sample's ability to reflect a population's characteristics, and this will all be done within Excel. We will introduce some often used and important inferential statistics techniques in this chapter.

6.2 Let the Statistical Technique Fit the Data

Consider the type of sample data we have seen thus far in Chaps. 1–5. In just about every case, the data has contained a combination of quantitative and qualitative data elements. For example, the data for teens visiting websites in Chap. 3 provided the number of page views for each teen, and also described the circumstances related to the page views—either *new* or *old* site. This was our first exposure to sophisticated statistics and to **cause and effect** analysis—one variable causing an effect on another. We can think of these categories, new and old, as experimental **treatments**, and the page views as a **response variable**. Thus, the treatment is the assumed cause and the effect is the number of views. In an attempt to determine if the sample means of the two treatments were different or equal, we performed an analysis called a **paired t-Test**. This test permitted us to consider complicated questions.

So when do we need this *more* sophisticated statistical analysis? Some of the answers to this question can be summarized as follows:

1. When we want to make a precise mathematical statement about the data's capability to infer characteristics of the population.
2. When we want to determine how closely these data fit some assumed model of behavior.
3. When we need a higher level of analysis to further investigate the preliminary findings of descriptive and exploratory analysis.

This chapter will focus on data that has both qualitative and quantitative components, but we will also consider data that is strictly qualitative (categorical), as you will soon see. By no means can we explore the exhaustive set of statistical techniques available for these data types; there are thousands of techniques available and more are being developed as we speak. But, we will introduce some of the most often used tools in statistical analysis. Finally, I repeat that it is important to remember that the type of data we are analyzing will dictate the technique that we can employ. The misapplication of a technique on a particular set of data is the most common reason for dismissing or ignoring the results of an analysis; the analysis just does not match the data.

6.3 χ²—Chi-Square Test of Independence for Categorical Data

Let us begin with a powerful analytical tool applied to a frequently occurring type of data—categorical variables. In this analysis, a test is conducted on sample data, and the test attempts to determine if there is an association, or relationship, between

Table 6.1 Results of mutual fund sample

Investor risk preference	Fund types frequency				
	Bond	Income	Income/Growth	Growth	Totals
Risk-taker	30	9	45	66	150
Conservative	270	51	75	54	450
Totals	300	60	120	120	600

two categorical (**nominal**) variables. Ultimately, we would like to know if the result can be extended to the entire population, or is due simply to chance. For example, consider the relationship between two variables: (1) an investor's self-perceived behavior toward investing, and 2) the selection of mutual funds made by the investor. This test is known as the **Chi-square**, or Chi-squared, **test of independence**. As the name implies, the test addresses the question of whether or not the two categorical variables are independent (not related).

Now let us consider a specific example. A mutual fund investment company samples a total of 600 potential investors who have indicated their intention to invest in mutual funds. The investors have been asked to classify themselves as either *risk-taking* or *conservative* investors. Then, they are asked to identify a single type of fund they would like to purchase. Four fund types are specified for possible purchase and only one can be selected—*bond, income, growth*, and *income and growth*. The results of the sample are shown in Table 6.1. This table structure is known as a **contingency table** and this particular contingency table happens to have 2 rows and 4 columns—a 2 by 4 contingency table. Contingency tables show the frequency of occurrence of the row and column categories. For example, 30 (first row /first column) of the 150 (*Totals* row for risk-takers) investors in the sample that identified themselves as risk-takers said they would invest in a bond fund, and 51 (second row/second column) investors considering themselves to be conservative said they would invest in an income fund. These values are **counts** or the frequency of observations associated with a particular cell.

6.3.1 Tests of Hypothesis—Null and Alternative

The mutual fund investment company is interested in determining if there is a relationship in an investor's perception of his own risk and the selection of a fund that the investor actually makes. This information could be very useful for marketing funds to clients and also for counseling clients on risk tailored investments. To make this determination, we perform an analysis of the data contained in the sample. The analysis is structured as a **test of the null hypothesis**. There is also an alternative to the *null hypothesis* called, quite appropriately, the **alternative hypothesis**. As the name implies, a test of hypothesis, either null or alternative, requires that a hypothesis is posited, and then a test is performed to see if the null hypothesis can be (1) rejected in favor of the alternative or (2) *not* rejected.

In this particular case, our null hypothesis assumes that self-perceived risk preference is **independent** of a particular mutual fund selection. That suggests that an investor's self-description as an investor is not related to the mutual funds he purchase, or more strongly stated, does not *cause* a purchase of a particular type of mutual fund. If our test suggests otherwise, that is, the test leads us to **reject the null hypothesis**, then we conclude that it is likely to be **dependent** (related).

This discussion may seem tedious, but if you do not have a firm understanding of tests of hypothesis, then the remainder of the chapter will be very difficult, if not impossible, to understand. Before we move on to the calculations necessary for performing the test, the following summarizes the general procedure we have just discussed:

1) an assumption (*null* hypothesis) that the variables under consideration are independent, or that they are *not* related, is made
2) an alternative assumption (*alternative* hypothesis) relative to the null is made that there *is* dependence between variables
3) the chi-square test is performed on the data contained in a contingency table to test the *null* hypothesis
4) the results, a statistical calculation, will be used to attempt to reject the null hypothesis
5) if the null *is* rejected, then this implies that the alternative is accepted; if the null is *not* rejected, then the alternative hypothesis is rejected

The chi-square test is based on a null hypothesis that assumes independence of relationships. If we believe the independence assumption, then the *overall* fraction of investors in a perceived risk category and fund type should be *indicative* of the entire investing population. Thus, an *expected* frequency of investors in each cell can be calculated. We will have more to say about this later in the chapter. The expected frequency, assuming independence, is compared to the actual (observed) and the variation of expected to actual is tested by calculating a statistic, the χ^2**statistic** (χ is the lower case Greek letter chi). The variation between what is actually observed and what is expected is based on the formula that follows. Note that the calculation squares the difference between the observed frequency and the expected frequency, divides by the expected value, and then sums across the two dimensions of the i by j contingency table:

$$\chi^2 = \sum_i \sum_j [(\text{obs}_{ij} - \text{expval}_{ij})^2 / \text{expval}_{ij}]$$

where:

obs_{ij} = frequency or count of observations in the ith row and jth column of the contingency table exp val$_{ij}$ = expected frequency of observations in the ith row and jth column of the contingency table, when independence of the variables is assumed.[1]

Once the χ^2 statistic is calculated, then it can be compared to a benchmark value of χ^2_α that sets a limit, or threshold, for rejecting the null hypothesis. The value of χ^2_α is the limit the χ^2 statistic can achieve before we reject the null hypothesis. These values can be found in most statistics books. To select a particular χ^2_α, the α (the **level of significance** of the test) must be set by the investigator. It is closely related to the *p-value*—the probability of obtaining a particular statistic value or more extreme by chance, when the null hypothesis is true. Investigators often set α to 0.05; that is, there is a 5% chance of obtaining this χ^2 statistic (or greater) when the null is true. So, in essence, our decision maker only wants a 5% chance of *erroneously* rejecting the null hypothesis. That is relatively conservative, but a more conservative (less chance of erroneously rejecting the null hypothesis) stance would be to set α to 1%, or even less.

Thus, if our χ^2 is greater than or equal to χ^2_α, then we *reject* the null. Alternatively, if the p-value is less than α we *reject* the null. These tests are equivalent. In summary, the rules for rejection are either:

$$\text{Reject the null hypothesis when } \chi^2 >= \chi^2_\alpha$$
or
Reject the null hypothesis when p-value <= α
(Note that these rules are equivalent)

Exhibit 6.1 shows a worksheet that performs the test of independence using the chi-square procedure. The exhibit also shows the typical calculation for contingency table expected values. Of course, in order to perform the analysis, both tables are needed to calculate the χ^2 statistic since both the observed frequency and the expected are used in the calculation. Using the Excel **CHITEST (actual range, expected range)** cell function permits Excel to calculate the data's χ^2 and then return a p-value (see cell F17 in Exhibit 6.1). You can also see from Exhibit 6.1 that the *actual range* is C4:F5 and does not include the marginal totals. The *expected range* is C12:F13 and the marginal totals are also omitted. The internally calculated χ^2 value takes into consideration the number of variables for the data, 2 in our case, and the possible levels within each variable—2 for risk preference and 4 for mutual fund types. These variables are derived from the range data information (rows and columns) provided in the *actual* and *expected* tables.

From the spreadsheet analysis in Exhibit 6.1 we can see that the calculated χ^2 value in F18 is 106.8 (a relatively large value), and if we assume α to be 0.05, then

---

[1]Calculated by multiplying the row total and the column total and dividing by total number of observations—e.g. in Exhibit 1 expected value for conservative/growth cell is 120 ∗ 450/600 = 90. Note that 120 is the marginal total Income/Growth and 450 is the marginal total for Conservative.

	F17		f_x	=CHITEST(C4:F5,C12:F13)			
A	B	C	D	E	F	G	H

1							
2	Investor Risk		Fund Types Frequency				
3	Preference	Bond	Income	Income/Growth	Growth	Totals	
4	Risk-taker	30	9	45	66	150	
5	Conservative	270	51	75	54	450	
6	Totals	300	60	120	120	600	
7							
8							
9							
10	Investor Risk		Fund Types Expected Values				
11	Preference	Bond	Income	Income/Growth	Growth	Totals	
12	Risk-taker	75	15	30	30	150	
13	Conservative	225	45	90	90	450	
14	Totals	300	60	120	120	600	
15							
16	Hector Guerrero: Cell Formula is:		Hector Guerrero : Cell Formula as:				
17					p-value =	5.35687E-23	
18	C6*G4 / G6		D6*G5 / G6		x^2 =	106.8	
19							
20							
21							

Exhibit 6.1 Chi-squared calculations via contingency table

χ^2_α is approximately 7.82. Thus, we can reject the null since $106.8 > 7.82$.[2] Also, the p-value from Exhibit 6.1 is extremely small (5.35687E-23) [3] indicating a very small probability of obtaining the χ^2 value of 106.8 when the null hypothesis is true. The p-value returned by the CHITEST function is shown in cell F17, and it is the only value that is needed to reject, or not reject, the null hypothesis. Note that the cell formula in F18 is the calculation of the χ^2 given in the formula above and is not returned by the CHITEST function. This result leads us to conclude that the null hypothesis is likely not true, so we reject the notion that the variables are independent. Instead, there appears to be a strong dependence given our test statistic.

Earlier, we summarized the general steps in performing a test of hypothesis. Now we describe in detail how to perform the test of hypothesis associated with the χ^2 test. The steps of the process are:

1. Organize the frequency data related to two categorical variables in a contingency table.

---

[2] Tables of χ^2_α can be found in most statistics texts. You will also need to calculate *the degrees of freedom* for the data: (number of rows–1) × (number of columns–1). In our example: (2–1) × (4–1) =3.

[3] Recall this is a form of what is known as "scientific notation". E-17 means 10 raised to the –17 power, or the decimal point moved 17 decimal places to the left of the current position for 3.8749. Positive (E+13 e.g.) powers of 10 moves the decimal to the right (13 decimal places).

2. From the contingency table values, calculate expected frequencies (see Exhibit 6.1 cell comments) under the assumption of independence. The calculation of χ^2 is relatively simple and performed by the *CHITEST(actual range, expected range)* function. The function returns the p-value of the calculated χ^2. Note that it does not return the χ^2 value, although it does calculate the value for internal use.

3. By considering an explicit level of α, the decision to reject the null can be made on the basis of determining if $\chi^2 > = \chi^2_\alpha$. Alternatively, α can be compared to the calculated p-value: p-value $< = \alpha$. Both rules are interchangeable and equivalent. It is often the case that an α of 0.05 is used by investigators.

6.4 z-Test and t-Test of Categorical and Interval Data

Now, let us consider a situation that is similar in many respects to the analysis just performed, but it is different in one important way. In the χ^2 test, the subjects in our sample were associated with two variables, both of which were categorical. The cells provided a count, or frequency, of the observations that were classified in each cell. Now we will turn our attention to sample data that contains categorical *and* interval or ratio data. Additionally, the categorical variable is dichotomous, and thereby can take on only two levels. The categorical variable will be referred to as the experimental *treatment* and the interval data as the *response* variable. In the next section, we consider an example problem related to the training of human resources that considers experimental treatments and response variables.

6.5 An Example

A large firm with 12,000 call center employees in two locations is experiencing explosive growth. One call center is located in South Carolina (SC) and the other is in Texas (TX). The firm has done its own *standard* internal training of employees for 10 years. The CEO is concerned that the quality of call center service is beginning to deteriorate at an alarming rate. They are receiving many more complaints from customers, and when the CEO disguised herself as a customer requesting call center information, she was appalled at the lack of courtesy and the variation of responses to a relatively simple set of questions. She finds this to be totally unacceptable and has begun to consider possible solutions. Among the solutions being considered is a training program to be administered by an outside organization with experience in the development and delivery of call center training. The hope is to create a systematic and predictable customer service response.

A meeting of high level managers is held to discuss the options and some skepticism is expressed about training programs in general: many ask the question—Is there really any value in these outside programs? Yet in spite of the skepticism, managers agree that something has to be done about the deteriorating quality of customer service. The CEO contacts a nationally recognized training firm, EB Associates. EB has considerable experience and understands the concerns of management. The CEO expresses her concern and doubts about training. She is not sure that training can be effective, especially for the type of unskilled workers they hire. EB listens carefully and has heard these concerns before. EB proposes a test to determine if the *special* training methods they provide can be of value for the call center workers. After careful discussion with the CEO, EB makes the following suggestion for testing the effectiveness of *special* (EB) versus *standard* (internal) training:

1. A test will be prepared and administered to all of the customer service representatives working in the call centers—4000 in SC and 8000 TX. The test is designed to assess the *current* competency of the customer service representatives. From this overall data, specific groups will be identified and a sample of 36 observations (test scores) for each group will be a taken–e.g. call center personnel with *standard* training in SC.
2. Each customer service representative taking the test will receive a score from 0 to 100. The results will form a database for the competency of the workers.
3. A *special* training course devised by EB will be offered to a selected group of customer service representatives in South Carolina: 36 incarcerated women. The competency test will be *re-administered* to this group after the special training program to detect changes in scores, if any.
4. Analysis of the difference in performance for the sample that is specially trained and those with the current standard training will be used to consider the application of the training to all employees. If the special training indicates significantly better performance on the exam after training, then EB will receive a large contract to perform training for all employees.

As mentioned above, the 36 customer service representatives selected to receive special training are a group of woman that are incarcerated in a low security prison facility in the state of South Carolina. The CEO has signed an agreement with the state of South Carolina to provide the SC women with an opportunity to work as customer service representatives and gain skills before being released to the general population. In turn, the firm receives significant tax benefits from South Carolina. Because of the relative ease with which these women can be trained, they are chosen for the special training. They are, after all, a captive audience. There is a similar group of customer service representatives that also are incarcerated woman. They are located in a similar low security Texas prison, but these women are not chosen for the special training.

The results of the tests for employees are shown in Table 6.2. Note that the data included in each of five columns is a sample of personnel scores of similar size (36): (1) non-prisoners in TX, (2) women prisoners in TX, (3) non-prisoners in SC, (4) women prisoners in SC before special training, and (5) women prisoners in SC after special training. All the columns of data, except the last, are scores for customer service representatives that have only had the internal standard training. The last column is the re-administered test scores of the SC prisoners that received

Table 6.2 Special training and no training scores

Observation	36 Non-prisoner scores TX	36 Women prisoners TX	36 Non-prisoner scores SC	36 Women SC (before special training)*	36 Women SC (with special training)*
1	81	93	89	83	85
2	67	68	58	75	76
3	79	72	65	84	87
4	83	84	67	90	92
5	64	77	92	66	67
6	68	85	80	68	71
7	64	63	73	72	73
8	90	87	80	96	98
9	80	91	79	84	85
10	85	71	85	91	94
11	69	101	73	75	77
12	61	82	57	62	64
13	86	93	81	89	90
14	81	81	83	86	89
15	70	76	67	72	73
16	79	90	78	82	84
17	73	78	74	78	80
18	81	73	76	84	85
19	68	81	68	73	76
20	87	77	82	89	91
21	70	80	71	77	79
22	61	62	61	64	65
23	78	85	83	85	87
24	76	84	78	80	81
25	80	83	76	82	84
26	70	77	75	76	79
27	87	83	88	90	93
28	72	87	71	74	75
29	71	76	69	71	74
30	80	68	77	80	83
31	82	90	86	88	89
32	72	93	73	76	78
33	68	75	69	70	72
34	90	73	90	91	93
35	72	84	76	78	81
36	60	70	63	66	68
Averages=	75.14	80.36	75.36	79.08	81.06
Variance=	72.12	80.47	78.47	75.11	77.31
Total TX Av=(8000 obs.)	74.29		Total TX VAR=	71.21	
Total SC Av= (4000 obs.)	75.72		Total SC VAR=	77.32	
Total Av= (12000 obs.)	74.77		TX&SC VAR=	73.17	

*Same 36 SC women prisoners that received training.

special training from EB. Additionally, the last two columns are the same individual subjects, matched as before and after special training, respectively. The sample sizes for the samples need not be the same, but it does simplify the analysis calculations. Also, there are important advantages to samples greater than 30 observations that we will discuss later.

Every customer service representative at the firm was tested at least once and the SC women prisoners twice. Excel can easily store these sample data and provide access to specific data elements using the filtering and sorting capabilities we learned in Chap. 5. The data collected by EB provides us with an opportunity for thorough analysis of the effectiveness of the special training.

So what are the questions of interest and how will we use inferential statistics to answer them? Recall that EB administered special training to 36 women prisoners in SC. We also have a standard trained non-prisoner group from SC. EB's first question might be—Is there any difference between the *average* score of a randomly selected SC non-prisoner sample with no special training and the SC prisoner's *average* score after special training? Note that our focus is on the aggregate statistic of *average* scores for the groups. Additionally, EB's question involves SC data exclusively. This is done to not confound results, should there be a difference between the competency of customer service representatives in TX and SC. We will study the issue of the possible difference between Texas and SC scores later in our analysis.

6.5.1 z-Test: 2 Sample Means

To answer the question of whether or not there is a difference between the average scores of SC non-prisoners *without* special training and prisoners *with* special training, we use the **z-Test: Two Sample for Means** option found in Excel's *Data Analysis* tool. This analysis tests the null hypothesis that there is *no* difference between the two sample means and is generally reserved for samples of 30 observations or more. Pause for a moment to consider this statement. We are focusing on the question of whether two means from sample data are different; different in statistics suggests that the samples come from different underlying populations with different means. For our problem, the question is whether the SC non-prisoner group and the SC prisoner group with special training have different population means for their scores. Of course, the process of calculating sample means will very likely lead to different values. If the means are relatively close to one another, then we will conclude that they came from the same population; if the means are relatively different, we are likely to conclude that they are from different populations. Once calculated, the sample means will be examined and a probability estimate will be made as to how likely it is that the two sample means came from the same population. But the question of importance in these tests of hypothesis is related to the populations—are the averages of the population of SC non-prisoners and of the population of SC prisoners with special training the same, or are they different?

If we reject the null hypothesis that there is no difference in the average scores, then we are deciding in favor of the training indeed leading to a difference in scores. As before, the decision will be made on the basis of a statistic that is calculated from the sample data, in this case the **z-Statistic**, which is then compared to a critical value. The critical value incorporates the decision maker's willingness to commit an error by possibly rejecting a true null hypothesis. Alternatively, we can use the p-value of the test and compare it to the level of significance which we have adopted—frequently assumed to be 0.05. The steps in this procedure are quite similar to the ones we performed in the chi-square analysis, with the exception of the statistic that is calculated, z rather than chi-square.

6.5.2 Is There a Difference in Scores for SC Non-Prisoners and EB Trained SC Prisoners?

The procedure for the analysis is shown in Exhibits 6.2 and 6.3. Exhibit 6.2 shows the *Data Analysis* dialogue box in the Analysis group of the Data ribbon used to perform the z-Test. We begin data entry for the z-Test in Exhibit 6.3 by identifying the range inputs, including labels, for the two samples: 36 SC non-prisoner standard trained scores (E1:E37) and 36 SC prisoners that receive special training (G1:G37). Next, the dialog box requires a hypothesized mean difference. Since we are assuming there is *no* difference in the null hypothesis, the input value is 0. This

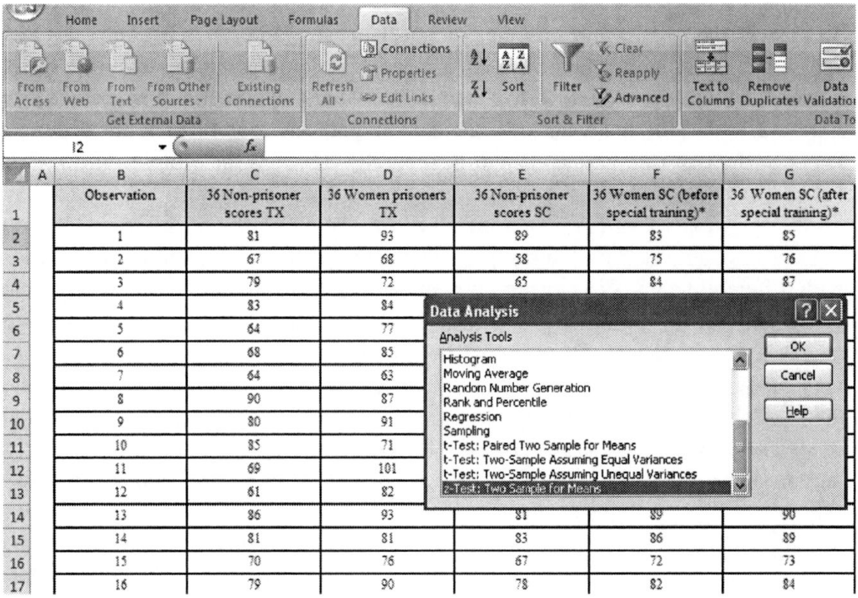

Exhibit 6.2 Data analysis tool for z-test

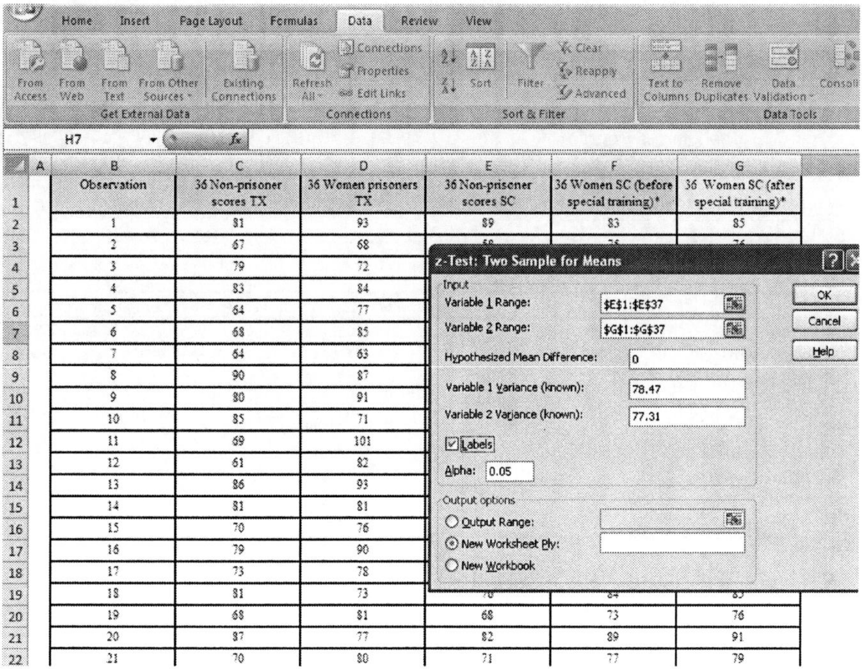

Exhibit 6.3 Selection of data for z-test

is usually the case, but you are permitted to designate other differences if you are hypothesizing a specific difference in the sample means. For example, consider the situation in which management is willing to purchase the training, but only if it results in some minimum increase in scores. The desired difference in scores could be tested as the *Hypothesized Mean Difference*.

The variances for the variables can be estimated to be the variances of the samples, as long as the samples are greater than approximately 30 observations. Recall earlier that I suggested that a sample size of at least 30 was advantageous—this is why! We can also use the variance calculated for the entire population at SC (Table 6.2—Total SC VAR $=77.31$) since it is available, but the difference in the calculated z-statistics is very minor: z-statistic using the sample variance is 2.7375 and 2.7475 for the known variance of SC. Next, we choose an α value of 0.05, but you may want to make this smaller if you want to be very cautious about rejecting true null hypotheses. Finally, this test of hypothesis is known as a *two-tail* test since we are not speculating on whether one specific sample mean will be greater than the other mean. We are simply positing a difference in the alternative. This is important in the application of a critical z-value for possible rejection of the null hypothesis. In cases where you have evidence that one mean is greater than another, then a *one-tail* test is appropriate. The critical z-values, **z-Critical one-tail** and **z-Critical two-tail**, and p-values, **P(Z<=z) one-tail** and **P(Z<=z) two-tail**, are provided when the analysis is complete. These values represent our thresholds for the test.

Table 6.3   Results of z-test for training of customer service representatives

	Home	Insert	Page Layout	Formulas	Data	Review	View	

	A	B	C
1	z-Test: Two Sample for Means		
2			
3		36 Non-prisoner scores SC	36 Women SC (after special training)*
4	Mean	75.36111111	81.05555556
5	Known Variance	78.47	77.31
6	Observations	36	36
7	Hypothesized Mean Difference	0	
8	z	-2.737453564	
9	P(Z<=z) one-tail	0.003095843	
10	z Critical one-tail	1.644853627	
11	P(Z<=z) two-tail	0.006191686	
12	z Critical two-tail	1.959963985	
13			
14			
15			
16			

The results of our analysis is shown in Table 6.3. Note that a z-statistic of approximately -2.74 has been calculated. We reject the null hypothesis if the test statistic (z) is either:

$$z >= \text{critical two-tail value } (1.959962787)\ldots\ldots\text{see cell B12}$$
$$or$$
$$z <= \text{- critical two-tail value } (-1.959962787).$$

Note that we have two rules for rejection since our test does not assume that one of the sample means is larger or smaller than the other. Alternatively, we can compare the p-value= 0.006191794 (in cell B11) to $\alpha = 0.05$ and reject if the p-value is $<= \alpha$. In this case the critical values (and the p-value) suggest that we *reject* the null hypothesis that the samples means are the same; that is, we have found evidence that the EB training program at SC has indeed had a significant effect on scores for the customer service representatives. EB is elated with this news since it suggests that the training does indeed make a difference, at least at the $\alpha = 0.5$ level of significance. This last comment is recognition that it is still possible, in spite of the current results, that our samples have led to the rejection of a true null hypothesis. If greater assurance is required, then run the test with a smaller α, for example 0.01. The results will be the same since a p-value 0.006191794 is less than 0.01.

6.5.3 t-Test: Two Samples Unequal Variances

A very similar test, but one that does not explicitly consider the variance of the population to be *known*, is the **t-Test**. It is reserved for small samples, less than 30 observations, although larger samples are permissible. The lack of knowledge of a population variance is a very common situation. Populations are often so large that it is practically impossible to measure the variance or standard deviation of the population, not to mention the possible change in the population's membership. We will see that the calculation of the *t-statistic* is very similar to the calculation of the *z-statistic*.

6.5.4 Do Texas Prisoners Score Higher Than Texas Non-Prisoners?

Now, let's consider a second, but equally important question that EB will want to answer—Is it possible that women prisoners, regardless of the state affiliation, normally score higher than others in the population, and that training is not the only factor in their higher scores? If we ignore the possible differences in state (SC or TX) affiliation of the prisoners for now, we can test this question by performing a test of hypothesis with the Texas data samples and form a general conclusion. Why might this be an important question? We have already concluded that there is a difference between the mean score of SC prisoners and that of the SC non-prisoners. Before we attribute this difference to the special training provided by EB, let us consider the possibility that the difference may be due to the affiliation with the prison group. One can build an argument that women in prison might be motivated to learn and achieve, especially if they are soon likely to rejoin the general population. As we noted above, we will not deal with state affiliation at this point, although it is possible that one state may have higher scores than another.

To answer this question we will use the **t-Test: Two Samples Unequal Variances** in the *Data Analysis* tool of Excel. In Exhibit 6.4 we see the dialog box associated with the tool. Note that it appears to be quite similar to the z-Test, except that rather than requesting values for known variances, the t-Test calculates the sample variances and uses the calculated values in the analysis. The results of the analysis are shown in Table 6.4, and the t-statistic indicates that we should reject the null hypothesis that the means are the same for prisoners and non-prisoners. This is so because the −2.53650023 (cell B9) is less than the negative of the critical two-tail t-value, −1.994435479 (negative of cell B13). Additionally, we can see that the p-value 0.013427432 (cell B12) is $<= \alpha$ (0.05). We therefore conclude that alternative hypothesis is true—there *is* a difference between the mean scores of the prisoners and non-prisoners. This could be due to many reasons and might require further investigation.

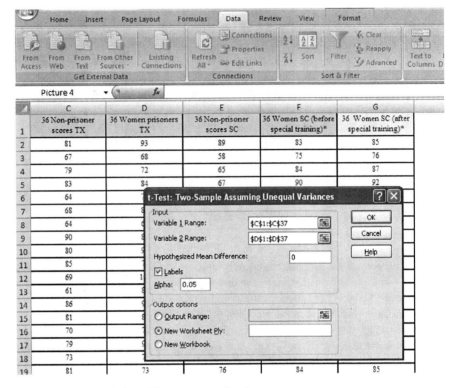

Exhibit 6.4 Data analysis tool for t-test unequal variances

6.5.5 Do Prisoners Score Higher Than Non-Prisoners Regardless of the State?

Earlier we suggested that the analysis did not consider state affiliation, but in fact our selection of data has explicitly done so—only Texas data was used. The data is *controlled* for the state affiliation variable; that is, the state variable is *held constant* since all observations are from Texas. What might be a more appropriate analysis if we do not want to hold the state variable constant and thereby make a statement that is not state dependent? The answer is relatively simple: combine the SC and Texas non-prisoner scores in Exhibit 6.2 columns C and E (72 observations; 36 + 36) and the SC and Texas Prisoner scores in column D and F (also 72). Note that we use Column F data rather than G since we are interested in the standard training only. Now we are ready to perform the analysis on these larger sample data sets, and fortuitously, more data is more reliable. The outcome is now independent of the state affiliation of the observations. In Table 6.5 we see that the results are similar to those in Table 6.4: we reject the null hypothesis in favor of the alternative that there is a difference. A t-statistic of approximately -3.085 (cell F11) and a p-value of 0.0025 (cell F14) is evidence of the need to reject the null hypothesis; -3.085 is less than the critical value -1.977 (cell D21) and 0.0025 is less than α (0.05).

Table 6.4 Results t-test scores of prisoner and non-prisoner customer service representatives in TX

	A	B	C
1	t-Test: Two-Sample Assuming Unequal Variances		
2			
3		*36 Non-prisoner scores TX*	*36 Women prisoners TX*
4	Mean	75.13888889	80.36111111
5	Variance	72.12301587	80.46587302
6	Observations	36	36
7	Hypothesized Mean Difference	0	
8	df	70	
9	t Stat	-2.536560023	
10	P(T<=t) one-tail	0.006713716	
11	t Critical one-tail	1.66691448	
12	P(T<=t) two-tail	0.013427432	
13	t Critical two-tail	1.994437086	
14			
15			
16			
17			

6.5.6   How do Scores Differ Among Prisoners of SC and Texas Before Special Training?

A third and related question of interest is whether the prisoners in SC and TX have mean scores (before training) that are significantly different. To test this question, we can compare the two samples of the prisoners, TX and SC, using the SC prisoners' scores prior to special training. To include EB trained prisoners would be an unfair comparison, given that the special training may have an effect on the scores. Table 6.6 shows the results of the analysis. Again, we perform the t-Test: two-samples unequal variances and find a t-statistic of 0.614666361 (cell B9). Given that the two-tail critical value is 1.994435479 (cell B13), the calculated t-statistic is not sufficiently extreme to reject the null hypothesis that there is no difference in mean scores for the prisoners of TX and SC. Additionally, the p-value, 0.540767979, is much larger than the α of 0.05. This is not an unexpected outcome given how similar the mean scores, 79.083 and 80.361, were for prisoners in both states.

Finally, we began the example with a question that focused on the viability of special training—Is there a significant difference in scores after special training?

Table 6.5 Results of t-test scores of prisoner (SC & TX) and non-prisoner (SC & TX)

	A	B	C	D	E	F	G
1	Observation	Non-prisoner SC & TX	Prisoners SC & TX				
2	1	81	93				
3	2	67	68		t-Test: Two-Sample Assuming Unequal Variances		
4	3	79	72				
5	4	83	84			Non-prisoner SC & TX	Prisoners SC & TX
6	5	64	77		Mean	75.25	79.72222222
7	6	68	85		Variance	74.24647887	77.10485133
8	7	64	63		Observations	72	72
9	8	90	87		Hypothesized Mean Dif	0	
10	9	80	91		df	142	
11	10	85	71		t Stat	-3.084583296	
12	11	69	101		P(T<=t) one-tail	0.001225206	
13	12	61	82		t Critical one-tail	1.655655173	
14	13	86	93		P(T<=t) two-tail	0.002450411	
15	14	81	81		t Critical two-tail	1.976810963	
16	15	70	76				
17	16	79	90				
18	17	73	78				

The analysis for this question can be done with a specific form of the t-statistic that makes a very important assumption: the samples are **paired** or **matched**. Matched samples simply imply that the sample data is collected from the same 36 observations, in our case the same SC prisoners. This form of sampling *controls* for individual differences in the observations by focusing directly on the special training as a level of treatment. It also can be thought of as a *before-and-after* analysis. For our analysis, there are two levels of training—standard training and special (EB) training. The tool in the Data Analysis menu to perform this type of analysis is **t-Test: Paired Two-Sample for Means**.

Exhibit 6.5 shows the dialog box for matched samples. The data entry is identical to that of the two-sample assuming unequal variances in Exhibit 6.4. Before we perform the analysis, it is worthwhile to consider the outcome. From the data samples collected in Table 6.2, we can see that the average score difference between the two treatments is about 2 points (79.08 before; 81.06 after). More importantly, if you examine the final two data columns in Table 6.2, it is clear that every observation for the prisoners with only standard training is improved when special training is applied. Thus, an informal analysis suggests that scores definitely have improved. We would not be as secure in our analysis if we achieved the same sample mean score improvement, but the individual matched scores were not consistently higher.

Table 6.6 Test of the difference in standard trained TX and SC prisoner scores

	A	B	C	D
1	t-Test: Two-Sample Assuming Unequal Variances			
2				
3		36 Women prisoners TX	36 Women SC (before special training)*	
4	Mean	80.36111111	79.08333333	
5	Variance	80.46587302	75.10714286	
6	Observations	36	36	
7	Hypothesized Mean	0		
8	df	70		
9	t Stat	0.614666361		
10	P(T<=t) one-tail	0.270383989		
11	t Critical one-tail	1.66691448		
12	P(T<=t) two-tail	0.540767979		
13	t Critical two-tail	1.994437086		

In other words, if we have an improvement in mean scores, but some individual scores improve and some decline, the perception of consistent improvement is far less compelling.

6.5.7 Does the EB Training Program Improve Prisoner Scores?

Let us now perform the analysis and review the results. Table 6.7 shows the results of the analysis. First, note the Pearson Correlation for the two samples is 99.62% (cell B7). This is a very strong positive correlation in the two data series, verifying the observation that the two scores move *together* in a very strong fashion—relative to the standard training score, the prisoner scores move in the same direction (positive) after special training. The t-statistic is −15.28688136 (cell B10), which is a very large negative value, and is much more negative[4] than the critical two-tail t-value

---

[4] −15.28688136 is a negative t-statistic because of the entry order of our data in the dialog box. If we reverse ranges for variable entry, the result is +15.28688136.

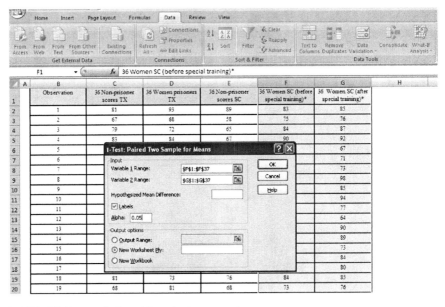

Exhibit 6.5   Data analysis tool for paired two-sample means

for rejection of the null hypothesis, 2.030110409 (cell B14). Thus, we reject the null and accept the alternative that the training does make a difference. The p-value is miniscule, 4.62055E-17 (cell B13), and far smaller than the 0.05 level set for α, which of course similarly suggests rejection of the null. The question remains

Table 6.7   Test of matched samples SC prisoner scores

	A	B	C	D
1	t-Test: Paired Two Sample for Means			
2				
3		36 Women SC (before special training)*	36 Women SC (after special training)*	
4	Mean	79.08333333	81.05555556	
5	Variance	75.10714286	77.31111111	
6	Observations	36	36	
7	Pearson Correlation	0.996172822		
8	Hypothesized Mean Difference	0		
9	df	35		
10	t Stat	-15.28688136		
11	P(T<=t) one-tail	2.31028E-17		
12	t Critical one-tail	1.68957244		
13	P(T<=t) two-tail	4.62055E-17		
14	t Critical two-tail	2.030107915		
15				
16				
17				
18				

whether or not an improvement of approximately 2 points is worth the investment in the training program.

6.5.8 What If the Observations Means Are Different, But We Do Not See Consistent Movement of Scores?

To see how the results will change if consistent improvement in matched pairs does not occur, while maintaining the averages, I will shuffle the data for training scores. In other words, the scores in the *36 Women prisoners SC (trained)* column will remain the same, but they will not be associated with the same values in the *36 Women prisoners SC (before training)* column. Thus, no change will be made in values; only the matched pairs will be changed. Table 6.8 shows the new (shuffled) pairs with the same mean scores as before. Table 6.9 shows the new results. Note that the means remain the same, but the Pearson Correlation value is quite different from before— −0.15617663 (cell E9). This negative value indicates that as one matched pair value increases there is generally a very mild decrease in the other value. Now the newly calculated t-statistic is −0.876116006 (cell E12). Given the critical t-value of 2.030107915 (cell E16), we *cannot* reject the null hypothesis that there is no difference in the means. The results are completely different than before, in spite of similar averages for the matched pairs. Thus, you can see that the consistent movement of matched pairs is extremely important to the analysis.

6.5.9 Summary Comments

In this section, we progressed through a series of hypothesis tests to determine the effectiveness of the EB special training program applied to SC prisoners. As you have seen, the question of the special training's effectiveness is not a simple one to answer. Determining statistically the true effect on the mean score improvement is a complicated task that may require several tests and some personal judgment. It is often the case that observed data can have numerous associated factors. In our example, the observations were identifiable by state (SC or TX), status of freedom (prisoner and non-prisoner), exposure to training (standard or EB special), and finally gender, although it was not fully specified for all observations. It is quite easy to imagine many more factors associated with our sample observations—e.g. age, level of education, etc.

In the next section, we will apply Analysis of Variance (ANOVA) to similar problems. ANOVA will allow us to compare the effects of multiple factors, with each factor containing several levels of treatment on a variable of interest, for example a test score. We will return to our call center example and identify 3 factors with 2 levels of treatment each. If gender could also be identified for each observation, the results would be 4 factors with 2 treatments for each. ANOVA will split our data into components, or groups, which can be associated with the various levels of factors.

Table 6.8 Scores for matched pairs that have been shuffled

36 women prisoners SC (before training)*	36 women prisoners SC (trained)
83	85
73	94
86	77
90	64
64	90
69	89
71	73
95	84
83	80
93	85
74	76
61	87
88	92
87	67
72	71
82	73
79	98
83	93
74	75
89	74
76	83
63	89
86	78
79	72
83	85
76	76
91	91
74	79
73	65
80	87
86	81
77	84
70	79
92	81
80	68
65	93
79.08	81.06

6.6 ANOVA

In this section, we will use **ANOVA** to find what are known as **main** and **interaction effects** of categorical (nominal) independent variables on an interval, dependent variable. The *main* effect of an independent variable is the *direct* effect it exerts on a dependent variable. The *interaction* effect is a bit more complex. It is the effect that results from the joint interactions of two or more independent variables on a dependent variable. Determining the effects of independent variables on dependent variables is quite similar to the analysis we performed in the section above. In that

Table 6.9 New matched pairs analysis

	A	B	C	D	E	F
1	36 Women prisoners SC (before training)*	36 Women Prisoners SC (trained)				
2	83	85				
3	73	94		t-Test: Paired Two Sample for Means		
4	86	77				
5	90	64			36 Women prisoners SC (before training)*	36 Women Prisoners SC (trained)
6	64	90		Mean	79.08333333	81.05555556
7	69	89		Variance	80.47857143	77.31111111
8	71	73		Observations	36	36
9	95	84		Pearson Correlation	-0.15617663	
10	83	80		Hypothesized Mean Difference	0	
11	93	85		df	35	
12	74	76		t Stat	-0.876116006	
13	61	87		P(T<=t) one-tail	0.193470266	
14	88	92		t Critical one-tail	1.68957244	
15	87	67		P(T<=t) two-tail	0.386940533	
16	72	71		t Critical two-tail	2.030107915	
17	82	73				
18	79	98				
19	83	93				

analysis, our independent variables were the state (SC or TX), status of freedom (prisoner and non-prisoner), and exposure to training (standard or special). These categorical independent variables are also known as **factors**, and depending on the **level** of the factor, they can affect the scores of the call center employees. Thus in summary, the levels of the various factors for the call center problem are: (1) *prisoner* and *non-prisoner* status for the freedom factor, (2) *standard* and *special* for the training factor, (3) *SC* and *TX* for state affiliation factor.

Excel permits a number of ANOVA analyses—*single factor, two-factor without replication*, and *two-factor with replication*. **Single factor ANOVA** is similar to the t-Tests we previously performed, and it provides an extension of the t-Tests analysis to more that two samples means; thus, the ANOVA tests of hypothesis permit the testing of equality of three or more sample means. It is also found in the Data Analysis tool in the Data Ribbon. This reduces the annoyance of constructing many pair-wise t-Tests to fully examine all sample relationships. The two-factor ANOVA, with and without replication, extends ANOVA beyond the capability of t-Tests. Now, let us begin with a very simple example of the use of single factor ANOVA.

6.6.1 ANOVA: Single Factor Example

A shipping firm is interested in the theft and loss of refrigerated shipping containers, commonly called *reefers*, that they experience at three similar sized terminal facilities at three international ports—Port of New York/New Jersey, Port of Amsterdam, and Port of Singapore. Containers, especially refrigerated, are serious investments of capital, not only due to their expense, but also due to the limited production capacity

Table 6.10 Reported missing reefers for terminals

Monthly obs.	NY/NJ	Amsterdam	Singapore	Security system
1	24	21	12	A
2	34	12	6	A
3	12	34	8	A
4	23	11	9	A
5	7	18	11	A
6	29	28	3	A
8	18	21	21	A
9	31	25	19	A
10	25	23	6	A
11	23	19	18	A
12	32	40	11	A
13	18	21	4	B
14	27	16	7	B
15	21	17	17	B
16	14	18	21	B
17	6	15	9	B
18	15	7	10	B
19	9	9	3	B
20	12	10	6	B
21	15	19	15	B
22	8	11	9	B
23	12	9	13	B
24	17	13	4	B
Average=	18.78	18.13	10.52	
Std =	8.37	8.15	5.66	

Note: Month 7 is missing

available for their manufacture. The terminals have similar security systems at all three locations and they were all updated one year ago. Therefore, the firm assumes the average number of missing containers at all the terminals should be relatively similar over time. The firm collects data over 23 months at the three locations to determine if the monthly means of lost and stolen reefers at the various sites are significantly different. The data for reefer theft and loss is shown in Table 6.10.

The data in Table 6.10 is in terms of reefers missing per month and represents a total of 23 months of collected data. A casual inspection of the data reveals that the average of missing reefers for Singapore is substantially lower than the averages for Amsterdam and NY/NJ. Also, note that the data includes an additional data element—the security system in place during the month. Security system A was replaced with system B in the beginning of the second year. In our first analysis of a single factor, we will only consider the Port factor with 3 levels—NY/NJ, Amsterdam, and Singapore. This factor is the independent variable and the number of missing reefers is the response, or dependent variable. It also is possible to consider the security system later as an additional factor with two levels, A and B. Here is our first question of interest.

6.6.2 Do the Mean Monthly Losses of Reefers Suggest That the Means are Different for the Three Ports?

Now we consider the application of ANOVA to our problem. Exhibit 6.6 shows the dialog box for initiating the single factor ANOVA. In Exhibit 6.7 we see the dialog box entries that permit us to perform the analysis. As before, we must identify the data range of interest; in this case, the three treatments of the Port factor (C2:E25), including labels. The α selected for comparison to the p-value is 0.05. Also, unlike the t-Test, where we calculate a t-statistic for rejection or acceptance of the null, in ANOVA we calculate an **F-Statistic** and compare it to a **critical F-value**. Thus, the statistic is different, but the general procedure is similar.

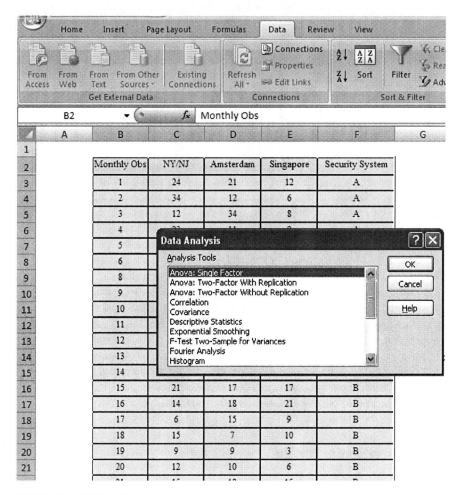

Exhibit 6.6 ANOVA: Single factor tool

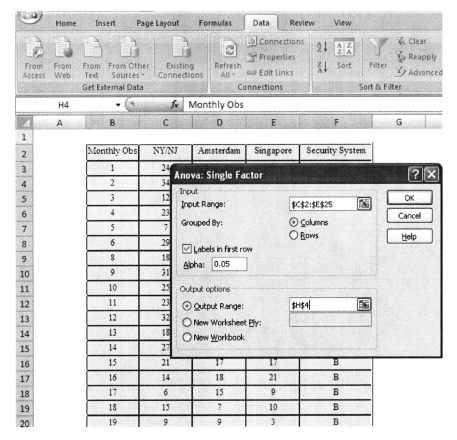

Exhibit 6.7 ANOVA: Single factor dialog box

Table 6.11 shows the result of the analysis. Note that the F-statistic, 8.63465822 (cell L15) is larger than the critical F-value, 3.13591793 (cell N15), so we can reject the null that all means come from the same population of expected reefer losses. Also, as before, if the p-value, 0.0004666 (cell M15) is less than our designated α (0.05), which is the case, we reject the null hypothesis. Thus, we have rejected the notion that the average monthly losses at the various ports are similar. At least one of the averages seems to come from a different distribution of monthly losses and is not similar to the averages of the other ports. We are not surprised to see this result given the much lower average at the Port of Singapore—about 10.5 versus 18.8 and 18.1 for the Port of NY/NJ and Amsterdam, respectively.

6.7 Experimental Design

There are many possible methods by which we conduct a data collection effort. Researchers are interested in carefully controlling and designing experimental studies, not only the analysis of data, but also its collection. The term used for

Table 6.11 ANOVA single factor analysis for missing reefers

Security System		Anova: Single Factor							
A									
A		**SUMMARY**							
A		*Groups*	*Count*	*Sum*	*Average*	*Variance*			
A		NY/NJ	23	432	18.7826087	70.0869565			
A		Amsterdam	23	417	18.1304348	66.4822134			
A		Singapore	23	242	10.5217391	31.9881423			
A									
A									
A		**ANOVA**							
B		*Source of Variation*	*SS*	*df*	*MS*	*F*	*P-value*	*F crit*	
B		Between Groups	970.2899	2	485.144928	8.63465822	0.0004666	3.13591793	
B		Within Groups	3708.261	66	56.1857708				
B									
B		Total	4678.551	68					
B									
B									

explicitly controlling the collection of observed data is **Experimental Design**. Experimental design permits researchers to refine their understanding of how factors affect the dependent variables in a study. Through the control of *factors* and their levels, the experimenter attempts to eliminate ambiguity and confusion related to the observed outcomes. This is equivalent to eliminating alternative explanations of observed results. Of course, completely eliminating alternative explanations is not possible, but attempting to control for alternative explanations is the hallmark of a well conceived study: a *good* experimental design.

There are some studies where we do not actively become involved in the manipulation of factors. These studies are referred to as **observational studies**. Our refrigerated container example above is best described as an observational study, since we have made no effort to manipulate the study's single factor of concern— Port. These ports simply happen to be where the shipping firm has terminals. If the shipping firm had many terminal locations and it had explicitly selected certain ports to study for some underlying reason, then our study would have been best described as an **experiment**. In experiments we have greater ability to influence *factors*. There are many types of experimental designs, some simple and some quite complex. Each design serves a different purpose in permitting the investigator to come to a scientifically focused and justifiable conclusion. We will discuss a small number of designs that are commonly used in analyses. It is impossible to exhaustively cover this topic in a small segment of a single chapter, but there are many good texts available on the subject if you should want to pursue the topic in greater detail.

Now, let us consider in greater detail the use of experimental designs in studies that are *experimental* and not *observational*. As I have stated, it is impossible to consider all the possible designs, but there are three important designs worth considering due to their frequent use. Below I provide a brief description that explains the major features of the three experimental designs:

- **Completely Randomized Design**: This experimental design is structured in a manner such that the treatments that are allocated to the **experimental units** (subjects or observations) are assigned completely at random. For example, consider 20 analysts (our experimental unit) from a population. The analysts will use 4 software products (treatments) for accomplishing a specific technical task. A response measure, the time necessary to complete the task, will be recorded. Each analyst is assigned a unique number from 1 to 20. The 20 numbers are written on 20 identical pieces of paper and placed into a container marked *subject*. These numbers will be used to allocate analysts to the various software products. Next, a number from 1 to 4, representing the 4 products, is written on 4 pieces of paper and repeated 5 times, resulting in 4 pieces of paper with the number 1, 4 pieces with the number 2, etc. These 20 pieces of paper are placed in a container marked *treatment*. Finally, we devise a process where we pick a single number out of each container and record the number of the analyst (subject or experimental unit) and the number of the software product (treatment) they will use. Thus, a couplet of an analyst and software treatment is recorded; for example, we might find that analyst 14 and software product 3 form a couplet. After the selection of each couplet, discard the selected pieces of paper (do not return to the containers) and repeat the process until all pieces of paper are discarded. The result is a completely randomized experimental design. The analysts are randomly assigned to a randomly selected software product, thus the description—completely randomized design.
- **Randomized Complete Block Design**: This design is one in which the experimental subjects are grouped (blocked) according to some variable which the experimenter wants to control. The variable could be intelligence, ethnicity, gender, or any other characteristic deemed important. The subjects are put into groups (blocks), with the same number of subjects in a group as the number of treatments. Thus, if there are 4 treatments, then there will be 4 subjects in a block. Next, the constituents of each block are then randomly assigned to different treatment groups, one subject per treatment. For example, consider 20 randomly selected analysts that have a recorded historical average time for completing a software task. We decide to organize the analysts into blocks according to their historical average times. The 4 lowest task averages are selected and placed into a block, the next 4 lowest task averages are selected to form the next block, and the process continues until 5 blocks are formed. Four pieces of paper with a unique number (1, 2, 3, or 4) written on them are placed in a container. Each member of a block randomly selects a single number from the container and discards the number. This number represents the treatment (software product) that the analyst will receive. Note that the procedure accounts for the possible individual differences

in analyst capability through the blocking of average times; thus, we are controlling for individual differences in capability. As an extreme case, a block can be
comprised of a single analyst. In this case, the analysts will have all four treatments (software products) administered in randomly selected order. The random
application of the treatments helps eliminate the possible interference (learning,
fatigue, loss of interest, etc) of a fixed order of application. Note this randomized
block experiment with a single subject in a block (20 blocks) leads to 80 data
points (20 blocks × 4 products), while the first block experiment (5 blocks) leads
to 20 data points (5 blocks × 4 products).

- **Factorial Design**: A factorial design is one where we consider more than one
  factor in the experiment. For example, suppose we are interested in assessing the
  capability of our customer service representatives by considering both training
  (standard and special) and their freedom status (prisoners or non-prisoners) for
  SC. Factorial designs will allow us to perform this analysis with two or more factors, simultaneously. Consider the customer representative training problem. It
  has 2 treatments in each of 2 factors, resulting in a total of 4 unique treatment
  combinations, sometimes referred to as a cell: prisoner/special training, pris-
  oner/standard training, non-prisoner/special training, and non-prisoner/standard
  training. To conduct this experimental design, we randomly select an equal number of prisoners and non-prisoners and subject equal numbers to special training
  and standard training. So, if we randomly choose 12 prisoners and 12 non-
  prisoners from SC (a total of 24 subjects), we then allocate equal numbers of
  prisoners and non-prisoners to the 4 treatment combinations—6 observations in
  each treatment. This type of design results in **replications** for each cell, 6 to be
  exact. Replication is an important factor for testing the adequacy of models to
  explain behavior. It permits testing for *lack-of-fit*. Although it is an important
  topic in statistical analysis, it is beyond the scope of this introductory material.

There are many, many types of experimental designs that are used to study specific experimental effects. We have covered only a small number, but these are some
of the most important and commonly used designs. The selection of a design will
depend on the goals of the study that is being designed. Now for some examples of
experimental design.

6.7.1 Randomized Complete Block Design Example

Now, let us perform one of the experiments discussed above in the *Randomized
Complete Block Design*. Our study will collect data in the form of task completion
times from 20 randomly selected analysts. The analysts will be assigned to one of
five blocks (A–E) by considering their *average task* performance times in the past
6 months. The consideration (blocking) of their *average task* times for the previous
6 months is accomplished by sorting the analysts on the *6 Month Task Average* key
in Table 6.12. Groups of 4 analysts (A–E) will be selected and blocked until the list

Table 6.12 Data for four software products experiment

Obs. (Analysts)	6 month task average	Block assignment	Software treatment	Task time
1	12	A	d	23
2	13	A	a	14
3	13	A	c	12
4	13	A	b	21
5	16	B	a	16
6	17	B	d	25
7	17	B	b	20
8	18	B	c	15
9	21	C	c	18
10	22	C	d	29
11	23	C	a	17
12	23	C	b	28
13	28	D	c	19
14	28	D	a	23
15	29	D	b	36
16	31	D	d	38
17	35	E	d	45
18	37	E	b	41
19	39	E	c	24
20	40	E	a	26

is exhausted, beginning with the top 4, and so on. Then analysts will be randomly assigned to one of 4 software products, within each block. Finally, a score will be recorded on their task time and the Excel analysis **ANOVA: Two-Factor without Replication** will be performed. This experimental design and results is shown in Table 6.13.

Although we are using the Two-Factor procedure, we are interested only in a single factor—the four software product treatments. Our blocking procedure is more an attempt to focus our experiment by eliminating unintended influences (the skill of the analyst prior to the experiment), than it is to explicitly study the effect of more capable analysts on task times. Table 6.12 shows the 20 analysts, their previous 6 month average task scores, the 5 blocks the analysts are assigned to, the software product they are tested on, and the task time scores they record in the experiment. Exhibit 6.8 shows the data that Excel will use to perform the ANOVA. Note that *analyst no. 1 in Block A (see Table 6.12) was randomly assigned *product d*. In Exhibit 6.8 the cell (C8) associated with the cell comment represents the score of *analyst no. 20 on product a*.

We are now prepared to perform the ANOVA on the data and we will use the Excel tool *ANOVA: Two-Factor without Replication* to test the null hypothesis that the task completion times for the various software products are no different.

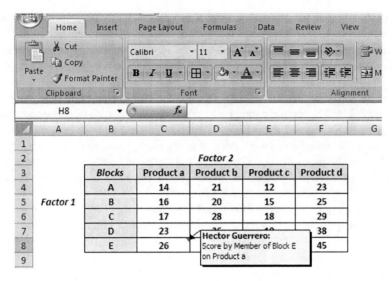

Exhibit 6.8 Randomized complete block design analyst example

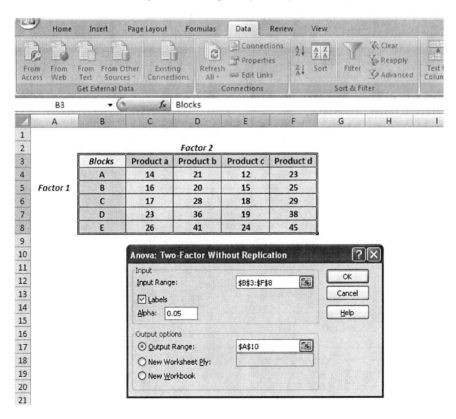

Exhibit 6.9 Dialog box or ANOVA: Two-factor without replication

Exhibit 6.9 shows the dialog box to perform the analysis. The *Input Range* is the entire table, including labels, and the level of significance, α, is 0.05

The results of the ANOVA are shown in Table 6.13. The upper section of the output, entitled *SUMMARY*, shows descriptive statistics for the two factors in the analysis—Groups (A–E) and Products (a–d). Recall that we will be interested only in the single factor, Products, and have used the blocks to mitigate the extraneous effects of skill. The section entitled ANOVA provides the statistics we need to either accept or reject the null hypothesis: there is no difference in the task completions times of the 4 software products. All that is necessary for us to reject the hypothesis is for one of the four software products task completion times to be significantly different from the others. Why do we need ANOVA for this determination? Recall we used the t-Test procedures for comparison of pair-wise differences—two software products with one compared to another. Of course, there are 6 exhaustive pair-wise comparisons possible in this problem—a/b, a/c, a/d, b/c, b/d, and c/d. Thus, 6 tests would be necessary to exhaustively cover all possibilities. It is much easier to use ANOVA to accomplish the same analysis as the t-Tests, especially as the number of pairwise comparisons begins to grow large.

Table 6.13 ANOVA for analyst example

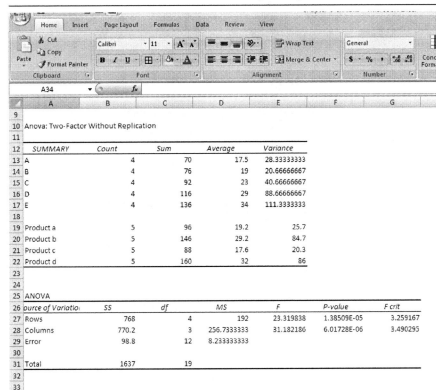

A	**B**	**C**	**D**	**E**	**F**	**G**	
9							
10 Anova: Two-Factor Without Replication							
11							
12 *SUMMARY*	*Count*	*Sum*	*Average*	*Variance*			
13 A	4	70	17.5	28.33333333			
14 B	4	76	19	20.66666667			
15 C	4	92	23	40.66666667			
16 D	4	116	29	88.66666667			
17 E	4	136	34	111.3333333			
18							
19 Product a	5	96	19.2	25.7			
20 Product b	5	146	29.2	84.7			
21 Product c	5	88	17.6	20.3			
22 Product d	5	160	32	86			
23							
24							
25 ANOVA							
26 *ource of Variatio*	*SS*	*df*	*MS*	*F*	*P-value*	*F crit*	
27 Rows	768	4	192	23.319838	1.38509E-05	3.259167	
28 Columns	770.2	3	256.7333333	31.182186	6.01728E-06	3.490295	
29 Error	98.8	12	8.233333333				
30							
31 Total	1637	19					
32							
33							

What is our verdict for the data? Do we reject the null? We are interested in the statistics associated with the sources of variation entitled *columns*. Why? Because in the original data used by Excel, the software product factor was located in the *columns* of the table. Each treatment, Product a–d, contained a column of data for the 5 block groups that were submitted to the experiment. According to the analysis in Table 6.13, the F-Statistic, 31.182186 (cell E28) is much larger than the critical F, 3.490295 (cell G28) and our p-value is 0.00000601728 (cell F28) which is much smaller than the assumed α of 0.05. Given the results, we clearly must reject the null hypothesis in favor of the alternative—*at least one* of the mean task completion times is significantly different from the others. If we reexamine the summary statistics in D19:D22 of Table 6.13, we see that at least two of our averages, 29.2 (b) and 32 (d), are much larger than the others, 19.2 (a) and 17.6 (c). So, this casual examination substantiates the results of the ANOVA.

6.7.2 Factorial Experimental Design Example

Now, let us return to our *prisoner/non-prisoner* and *special training/ no special training* two factors example. Suppose we collect a *new* set of data for an experimental study—24 observations of equal numbers of prisoners/non-prisoners and standard trained/special trained. This implies a selection of two factors of interest: prisoner status and training. The treatments for prisoner status are *prisoner* and *non-prisoner*, while the treatments for training are *trained* and *not-trained*. The four cells formed by the treatments each contain 6 replications (unique individual scores) and lead to another type of ANOVA—**ANOVA: Two-Factor with Replication**.

Table 6.14 shows the 24 observations in the two-factor format and Table 6.15 shows the result of the ANOVA. The last section in Table 6.15, entitled *ANOVA*, provides the F-Statistics (E34:E36) and p-values (cells F34:F36) to reject the null hypotheses related to the effect of both factors. In general, the null hypotheses states that the various treatments of the factors do not lead to significantly different averages for the scores.

Factor A (Training) and Factor B (Prisoner Status) are represented by the sources of variation entitled *Sample* and *Columns*, respectively. Factor A has an F-Statistic of 1.402199 (cell E34) and a critical value of 4.35125 (cell G34), thus we cannot reject the null. The p-value, 0.250238 (cell F34), is much larger than the assumed α of 0.05.

Factor B has an F-Statistic of 4.582037 (cell E35) that is slightly larger than the critical value of 4.35124 (cell G35). Also, the p-value, 0.044814 (cell F35), is slightly smaller than 0.05. Therefore, for Factor B we can reject the null hypothesis, but not with overwhelming conviction. Although the rule for rejection is quite clear, a result similar to the one we have experienced with Factor B might suggest that further experimentation is in order. Finally, the interaction of the factors does not lead us to reject the null. The F-Statistic is rather small, 0.101639 (cell E36), compared to the critical value, 4.35124 (cell G36).

Table 6.14 Training data revisited

		Factor B	
	Observations	Non-Prisoners	Prisoners SC
	Trained	74	85
		68	76
		72	87
		84	92
		77	96
Factor A		85	78
	Not-Trained	63	73
		77	88
		91	85
		71	94
		67	77
		72	64

Table 6.15 ANOVA: Two-Factor with Replication

	A	B	C	D	E	F	G
16	Anova: Two-Factor With Replication						
17							
18	SUMMARY	Non-Prisoners	Prisoners SC	Total			
19	Count	6	6	12			
20	Sum	460	514	974			
21	Average	76.66666667	85.66666667	81.16666667			
22	Variance	45.46666667	60.26666667	70.15151515			
23	Count	6	6	12			
24	Sum	441	481	922			
25	Average	73.5	80.16666667	76.83333333			
26	Variance	95.9	119.7666667	110.1515152			
27		Total					
28	Count	12	12				
29	Sum	901	995				
30	Average	75.08333333	82.91666667				
31	Variance	66.99242424	90.08333333				
32	ANOVA						
33	Source of Variation	SS	df	MS	F	P-value	F crit
34	Sample	112.6666667	1	112.6666667	1.402199	0.250238	4.35124
35	Columns	368.1666667	1	368.1666667	4.582037	0.044814	4.35124
36	Interaction	8.166666667	1	8.166666667	0.101639	0.753178	4.35124
37	Within	1607	20	80.35			
38							
39	Total	2096	23				
40							

6.8 Summary

The use of inferential statistics is invaluable in analysis and research. Inferential statistics allows us to infer characteristics for a population from the data obtained in a sample. We are often forced to collect sample data because the cost and time required in measuring the characteristics of a population can be prohibitive. In addition, inferential statistics provides techniques for quantifying the inherent uncertainty associated with using samples to specify population characteristics. It does not eliminate the uncertainty due to sampling, but it can provide a quantitative measure for the uncertainty we face about our conclusions for the data analysis.

Throughout Chap. 6 we have focused on analyses that involve a variety of data types—categorical, ordinal, interval, and rational. Statistical studies usually involve a rich variety of data types that must be considered simultaneously to answer our questions or to investigate our beliefs. To this end, statisticians have developed a highly structured process of analysis known as tests of hypothesis to formally test the veracity of a researcher's beliefs about behavior. A hypothesis and its alternative are posited and then tested by examining data collected in observational or experimental studies. We then construct a test to determine if we can reject the null hypothesis based on the results of the analysis.

Much of this chapter focused on the selection of appropriate analyses to perform the tests of hypothesis. We began with the chi-squared test of independence of variables. This is a relatively simple, but useful, test performed on categorical variables. The z-Test and t-Test expanded our view of data from strictly categorical, to combinations of categorical and interval data types. Depending on our knowledge of the populations we are investigating, we execute the appropriate test of hypothesis, just as we did in the chi-squared. The t-Test was then extended to consider more complex situations through ANOVA. Analysis of variance is a powerful family of techniques for focusing on the effect of independent variables on some response variable. Finally, we discussed how design of experiments helps reduce ambiguity and confusion in ANOVA by focusing our analyses. A thoughtful design of experiments can provide an investigator with the tools for sharply focusing a study, so that the potential of confounding effects can be reduced.

Although application of these statistics appears to be difficult, it is actually very straight forward. Table 6.16 below provides a summary of the various tests presented in this chapter and the rules for rejection of the null hypothesis.

In the next chapter we will begin our discussion of *Model Building* and *Simulation*—these models represent analogs of realistic situations and problems that we face daily. Our focus will be on *what-if* models. These models will allow us to incorporate the complex uncertainty related to important business factors, events, and outcomes. They will form the basis for rigorous experimentation. Rather than strictly gather empirical data, as we did in this chapter, we will collect data from our models that we can submit to statistical analysis. Yet, the analyses will be similar to the analyses we have performed in this chapter.

Table 6.16 Summary of test statistics used in inferential data analysis

Test statistic	Application	Rule for *Rejecting* null hypothesis
x^2 – Test of Independence	Categorical Data	χ^2 (calculated) $> = \chi^2{}_\alpha$ (critical) *or* p-value $< = \alpha$
z Test	Two Sample Means of Categorical and Interval Data Combined	z stat $> = $ z Critical Value z stat $< = -$ z Critical Value *or* p-value $< = \alpha$
t Test	Two Samples of Unequal Variance; Small Samples ($<$ 30 observations)	t stat $> = $ t Critical Value t stat $< = -$ t Critical Value *or* p-value $< = \alpha$
ANOVA: Single Factor	Three or More Sample Means	F Stat $> = $ F Critical Value *or*
ANOVA: Two Factor without Replication	Randomized Complete Block Design	p-value $< = \alpha$
ANOVA: Two Factor with Replication	Factorial Experimental Design	

Key Terms

Sample
Sampling Error
Cause and Effect
Treatments
Response Variable
Paired t-Test
Nominal
Chi-square
Test of Independence
Contingency Table
Counts
Test of the Null Hypothesis
Alternative Hypothesis
Independent
Reject the Null Hypothesis
Dependent
χ^2 Statistic
$\chi^2{}_\alpha$, α–level of significance
CHITEST(actual range, expected range)
z-Test: Two Sample for Means
z-Statistic
z-Critical one-tail
z-Critical two-tail
P(Z<=z) one-tail and P(Z<=z) two tail

t-Test
t-Test: Two-Samples Unequal Variances
Paired or Matched
t-Test: Paired Two-Sample For Means
ANOVA
Main and Interaction Effects
Factors
Levels
Single Factor ANOVA
F-Statistic
Critical F-Value
Experimental Design
Observational Studies
Experiment
Completely Randomized Design
Experimental Units
Randomized Complete Block Design
Factorial Design
Replications
ANOVA: Two-Factor without Replication
ANOVA: Two-Factor with Replication

Problems and Exercises

1. Can you ever be totally sure of the *cause and effect* of one variable on another by employing *sampling*?—Y or N
2. Sampling errors can occur naturally, due to the uncertainty inherent in examining less than all constituents of a population—T or F?
3. A sample mean is an estimation of a population mean—T or F?
4. In our webpage example, what represents the treatments, and what represents the response variable?
5. A coffee shop opens in a week and is considering a choice among several brands of coffee, Medalla de Plata and Startles, as their single offering. They hope their choice will promote visits to the shop. What are the treatments and what is the response variable?
6. What does the Chi-square test of independence for categorical data attempt to suggest?
7. What does a contingency table show?
8. Perform a Chi-squared test on the following data. What do you conclude about the null hypothesis?

Customer Type	Coffee Drinks				
	Coffee	Latte	Cappuccino	Soy-based	Totals
Male	230	50	56	4	
Female	70	90	64	36	
Totals					600

9. What does a particular level of significance, $\alpha = 0.05$, in a test of hypothesis suggest?
10. In a chi-squared test, if you calculate a p-value that is smaller than your desired α, what is concluded?
11. Describe the basic calculation to determine the expected value for a contingency table cell.
12. Perform tests on Table E1 data. What do you conclude about the test of hypothesis?

    a. z-Test: Two Sample for Means
    b. t-Test: Two Sample Unequal Variances
    c. t-Test: Paired Two Sample for Means

13. Perform an ANOVA: Two-Factor Without Replication test of the blocked data in Table E2. What is your conclusion about the data?

Table E1

Sample 1	Sample 2
83	85
73	94
86	77
90	64
84	90
69	89
71	73
95	84
83	80
93	91
74	76
72	87
88	92
87	67
72	71
82	73
79	98
83	90
74	75
81	74
76	83
63	89
86	78
71	72
83	85
76	76
96	91
77	79
73	65
80	87
86	81
77	84
70	79
92	81
80	68
65	93

14. *Advanced Problem*—A company that provides network services to small business has three locations. In the past they have experienced errors in their accounts receivable systems at all locations. They decide to test two systems for detecting accounting errors and make a selection based on the test results. The data in Table E3 represents samples of errors (columns 2–4) detected in accounts receivable information at three store locations. Column 5 shows the system used to detect errors. Perform an ANOVA analysis on the results. What is your conclusion about the data?

Table E2

	Blocks	Factor 2			
		W	X	Y	Z
	A	14	21	12	23
Factor 1	B	12	20	15	25
	C	17	18	23	19
	D	23	36	19	38
	E	26	21	24	32

Table E3

Observations	Location 1	Location 2	Location 3	Type of System
1	24	21	17	A
2	14	12	6	A
3	12	24	8	A
4	23	11	9	A
5	17	18	11	A
6	29	28	3	A
8	18	21	21	A
9	31	25	19	A
10	25	23	9	A
11	13	19	18	A
12	32	40	11	A
13	18	21	4	B
14	21	16	7	B
15	21	17	17	B
16	14	18	11	B
17	6	15	9	B
18	15	13	10	B
19	9	9	3	B
20	12	10	6	B
21	15	19	15	B
22	12	11	9	B
23	12	9	13	B
24	17	13	9	B

15. *Advanced Problem*—A transportation and logistics firm, Mar y Tierra (MyT), hires seamen and engineers, foreign and domestic, to serve on board its container ships. The company has in the past accepted the worker's credentials without an official investigation of veracity. This has led to problems with workers lying about or exaggerating their service history, a very important concern for MyT. MyT has decided to hire a consultant to design an experiment to determine the extent of the problem. Some managers at MyT believe that the foreign workers may be exaggerating their service, since it is not easily verified. A test for first-class engineers is devised and administered to 24 selected workers. Some of the workers are foreign and some are domestic. Also, some have previous experience with MyT and some do not. The consultant randomly

selects 6 employees to test in each of the four categories—Domestic/Experience with MyT, Domestic/No Experience with MyT, etc. A Proficiency exam is administered to all the engineers and it is assumed that if there is little difference between the workers scores then their concern is unfounded. If the scores are significantly different (0.05 level), then their concern is well founded. What is your conclusion about the exam data and differences among workers?

Table E4

Observations	Foreign	Domestic
	72	82
Previous Employment	67	76
with MyT	72	85
	84	92
	77	96
	85	78
	63	73
No Previous Experience	77	88
with MyT	91	85
	71	94
	67	77
	72	64

Chapter 7
Modeling and Simulation: Part 1

Contents

7.1 Introduction

The previous six chapters have provided us with a general idea of what a model is and how it can be used. Yet, if we are to develop good model building and analysis skills, we still need a more formal description, one that permits us a precise method for discussing models. Particularly, we need to understand a model's structure, its capabilities, and its underlying assumptions. So let us carefully re-consider

H. Guerrero, *Excel Data Analysis*, DOI 10.1007/978-3-642-10835-8_7,
© Springer-Verlag Berlin Heidelberg 2010

the question—what is a model? This might appear to be a simple question, but as is often the case, simple questions can often lead to complex answers. Additionally, we need to walk a fine line between an answer that is simple, and one that does not over-simplify our understanding. Albert Einstein was known to say—"Things should be made as simple as possible, but not any simpler." We will heed his advice.

Throughout the initial chapters, we have discussed models in various forms. Early on, we broadly viewed models as an attempt to capture the behavior of a system. The presentation of quantitative and qualitative data in Chaps. 2 and 4 provided visual models of the behavior of a system for a number of examples: sales data of products over time, payment data in various categories, and auto sales for sales associates. Each graph, data sort, or filter modeled the outcome of a focused question. For example, we determined which sales associates sold automobiles in a specified time period and we determined the types of expenditures a college student made on particular days of the week. In Chaps. 3 and 5 we performed data analysis on both quantitative and qualitative data leading to models of general and specific behavior, like summary statistics and *PivotTables*. Each of these analyses relied on the creation of a model to determine behavior. For example, our paired t-Test for determining the changes in average page views of teens modeled the number of views before and after website changes. In all these cases, the model was the way we *arranged*, *viewed*, and *examined* data.

Before we proceed with a formal answer to our question, let's see where this chapter will lead. The world of modeling can be described and categorized in many ways. One important way to categorize models is related to the circumstances of their *data availability*. Some modeling situations are **data rich**; that is, data for modeling purposes exists and is readily available for model development. The data on teens viewing a website was such a situation, and in general, the models we examined in Chaps. 2, 3, 4, 5, and 6 were all data rich. But what if there is little data available for a particular question or problem—a **data poor** circumstance? For example, what if we are introducing a new product that has no reasonable equivalent in a particular sales market? How can we model the potential success of the product if the product has no sales history and no related product exists that is similar in potential sales? In these situations modelers rely on models that *generate* data based on a set of underlying assumptions. Chaps. 7 and 8 will focus on these models that can be analyzed by the techniques we have discussed in our early chapters.

Since the academic area of Modeling and Simulation is very broad, it will be necessary to divide the topics into two chapters. Chapter 7 will concentrate on the basics of modeling. We will learn how models can be used and how to construct them. Also, since this is our first formal view of models, we will concentrate on models that are less complex in their content and structure. Although uncertainty will be modeled in both Chaps. 7 and 8, we will deal explicitly with uncertainty in Chap. 8. Yet, for both chapters, considering the uncertainty associated with a process will help us analyze the risk associated with overall model results.

Chapter 8 will also introduce methods for constructing *Monte Carlo* simulations, a powerful method for modeling uncertainty. Monte Carlo simulation uses random numbers to model the probability distributions of outcomes for uncertain variables in our problems. This may sound complicated, and it can be, but we will take great care in understanding the fundamentals—simple, but not too simple.

7.1.1 What is a Model?

So, now back to our original question—what is a model? To answer this question, let us begin by identifying a broad variety of model types.

1. **Physical model**: a physical replica that can be operated, tested, and assessed— e.g. a model of an aircraft that is placed in a wind-tunnel to test its aerodynamic characteristics and behavior.
2. **Analog model**: a model that is analogous (shares similarities)—e.g. a map is analogous to the actual terrestrial location it models.
3. **Symbolic model**: a model that is more abstract than the two discussed above and that is characterized by a symbolic representation—e.g. a financial model of the US economy used to predict economic activity in a particular economic sector.

Our focus will be on symbolic models: models constructed of mathematical relationships that attempt to mimic and describe a process or phenomenon. Of course, this should be of no surprise since this is exactly what Excel does, besides all its clerical uses like storing, sorting, manipulating, and querying data. Excel, with its vast array of internal functions, is used to represent phenomenon that can be translated into mathematical and logical relationships.

Symbolic models also permit us to observe how our decisions will perform under a particular set of model conditions. We can build models where the conditions within which the model operates are assumed to be known with certainty. Then the specific assumptions we have made can be changed and the changed conditions applied to the model. Becoming acquainted with these models is the goal of this chapter.

We can also build models where the behavior of model elements is uncertain, and the range of uncertainty is built directly into the model. This is the goal of Chap. 8. The difference between the two approaches is subtle, but under the first approach, the question that is addressed is—if we impose these specific conditions, what is the resulting behavior of our model? It is a very focused approach. In the latter approach we incorporate the full array of possible conditions into the model and ask—if we assume these possible conditions, what is the full array of outcomes for the model? Of course, this latter approach is much broader in its scope.

The models we will build in this chapter will permit us to examine complex decisions. Imagine you are considering a serious financial decision. Your constantly scheming neighbor has a business idea, which for the first time you can recall, appears to have some merit. But the possible outcomes of the idea can result either in a huge financial success or a colossal financial loss, and thus the venture is very risky. You have a conservatively invested retirement portfolio that you are considering liquidating and reinvesting in the neighbor's idea, but you are cautious and you wonder how to analyze your decision options carefully before committing your hard-earned money. In the past you have used intuition to make choices, but now the stakes are extremely high because your neighbor is asking you to invest the *entire* value of your retirement portfolio. The idea could make you a multi-millionaire or a penniless pauper at retirement.

Certainly in this situation it is wise to rely on *more* that intuition! Chapters 7 and 8 will describe procedures and tools to analyze the risk in decision outcomes, both good and bad. As we have stated, this chapter deals with risk by answering questions related to *what* outcome occurs *if* certain conditions are imposed. In the next chapter we will discuss a related, but more powerful, method for analyzing risk— **risk profiles**. Risk profiles are graphical representations of the risk associated with decision strategies or choices. They make explicit the many possible outcomes of a complex decision problem, and their estimated probability of occurrence, explicit. For example, consider the risk associated with the purchase of a one dollar lottery ticket. There is a very high probability, 99%, that you will *lose* the dollar invested; there is also a very small probability, 1%, that you will *win* one million dollars. This risk profile is shown in Exhibit 7.1. Note that the *win* outcome, $999,999, is the $1 million net of your $1 investment for the lottery ticket. Now let's turn our attention to classifying models.

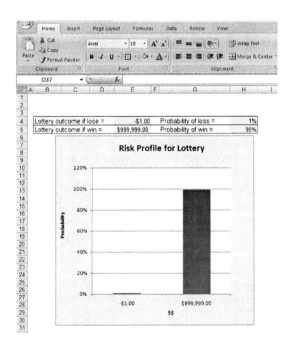

Exhibit 7.1 Lottery risk profile

7.2 How Do We Classify Models?

There are ways to classify models other than by the circumstances within which they exist. For example, earlier we discussed the circumstances of data rich and data poor models. Another fundamental classification for models is as either **deterministic** or **probabilistic**. A deterministic model will generally ignore, or assume away, any uncertainty in its relationships and variables. Even in problems where uncertainty exists, if we reduce uncertain events to some *determined* value, for example an *average* of various outcomes, then we refer to these models as deterministic. Suppose you are concerned with a particular task in a project that you believe to have a 20% probability of requiring 2 days, a 60% probability of 4 days, and a 20% probability of 6 days. If we reduce the uncertainty of the task to a single value of 4 days, the average and the most likely outcome, then we have converted an uncertain outcome into a deterministic outcome. Thus, in deterministic models, all variables are assumed to have a specific value, which for the purpose of analysis remains constant.

Even in deterministic models, if conditions change we can adjust the current values of the model and assume that a new value is known with certainty, at least for the purpose of analysis. For example, suppose that you are trying to calculate equal monthly payments due on a mortgage with a particular term (30 years or 360 monthly payments), an annual interest rate (6.5%), a loan amount ($200 K), and a down-payment ($50 K). The model used to calculate a constant payment over the life of the mortgage is the **PMT()** financial function in Excel. The model returns a precise value that corresponds to the deterministic conditions assumed by the modeler. In the case of the data provided above, the resulting payment is $948.10, calculated by the function PMT(0.065/12,360,150,000). See Exhibit 7.2 for this calculation.

Now, what if we would like to impose a new set of conditions, where all PMT() values remain the same, except that the annual interest rate is now 7%, rather than 6.5%. This type of *what-if* analysis of deterministic models helps us understand the potential variation in a deterministic model, variation that we have assumed away. The value of the function with a new interest rate of 7% is $997.95 and is shown in Exhibit 7.3. Thus, deterministic models can be used to study uncertainty, but only through the manual change of values.

Unlike deterministic models, probabilistic models explicitly consider uncertainty; they incorporate a technical description of how variables can change and the uncertainty is embedded in the model structure. It is generally the case that probabilistic models are more complex and difficult to construct because the explicit consideration of the uncertainty must be accommodated. But in spite of the complexity, these models provide great value to the modeler; after all, almost all important problems contain some elements of uncertainty.

Uncertainty is an ever present condition of life and it forces the decision maker to face a number of realities:

1. First and foremost, we usually make decisions based on what we currently know, or think we know. We also base decisions and actions on the outcomes we expect

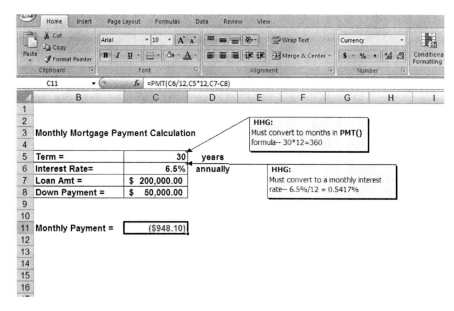

Exhibit 7.2 Model of mortgage payments with rate 6.5%

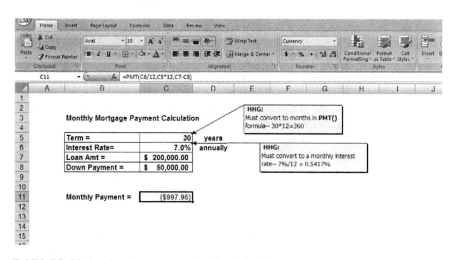

Exhibit 7.3 Model of mortgage payments with rate 7.0%

to occur. Introducing uncertainty for both existing conditions and the outcomes resulting from actions can severely complicate decision making.

2. It is not unusual for decision makers to delay or abandon decision making because they feel they are unable to deal with uncertainty. Decision makers often believe that taking *no* action is a superior alternative to making decisions with highly uncertain problem elements. Of course, there is no guarantee of this.

Not acting can be just as damaging as acting under difficult to model uncertain circumstances.
3. Decision makers who incorporate a better understanding of uncertainty in their modeling and how uncertainty is related to the elements of their decision problems are far more likely to achieve better results than those who do not.

So how do we deal with these issues? We do so by *systematically* dealing with uncertainty. This suggests that we need to understand a number of important characteristics about the uncertainty that surrounds our problem. In Chap. 8 we will see precisely how to deal with these problems.

7.3 An Example of Deterministic Modeling

Now let us consider a relatively simple problem from which we will create a deterministic model. We will do so by taking the uncertain problem elements and converting them into deterministic elements. Thus, in spite of the uncertainty the problem contains, our approach will be to develop a deterministic model.

A devoted parish priest, Fr. Moses Efia, has an inner city parish in a large American city, Baltimore, MD. Fr. Efia is faced with a difficult situation. His poor parish church, Our Lady of Perpetual Succor (OLPS), is scheduled for removal from the official role of Catholic parishes in the Baltimore dioceses. This means that the church and school that served so many immigrant populations of Baltimore for decades will no longer exist. A high-rise condominium will soon replace the crumbling old structure. Fr. Efia is from Ghana and understands the importance of a community that tends to the needs of immigrants. Over the decades the church has ministered to German, Irish, Italian, Filipino, Vietnamese, Cambodian, and most recently Central American immigrants. These immigrants have arrived in waves, each establishing their neighborhoods and communities near the church, and then moving to other parts of the city as economic prosperity has taken hold.

Fr. Efia knows that these *alumni* of OLPS have a great fondness and sense of commitment to the church. He has decided to save the church by holding a fund raising event that he calls *Vegas Night at OLPS*. His boss, the Archbishop has strictly forbidden Fr. Efia to solicit money directly from past and present parishioners. Thus the event, appropriately named to evoke Las Vegas style gambling, is the only way to raise funds without a direct request of the parish alumni. The event will occur on a Saturday afternoon after evening mass, and it will feature a number of games of fortune. The Archbishop, a practical and empathetic man, has allowed Fr. Efia to invite alumni, but he has warned that if he should notice anything that suggests a direct request for money, he will cancel the event. Many of the alumni are now very prosperous and Fr. Efia hopes that they will attend and open their pockets to the event's games of chance.

7.3.1 A Preliminary Analysis of the Event

Among one of his strongest supporters in this effort is a former parishioner who has achieved considerable renown as a risk analyst, Voitech Schwartzman. Voitech has volunteered to provide Fr. Efia with advice regarding the design of the event. This is essential since an event based on games of chance offers no absolute guarantee that OLPS will make money; if things go badly and lady-luck frowns on OLPS, the losses could be disastrous. Voitech and Fr. Efia decide that the goal of their *design* and *modeling* effort should be to construct a tool that will provide a forecast of the revenues associated with the event. In doing so, the tool should answer several important questions. Will *Vegas Night at OLPS* make money? Will it make too little revenue to cover costs and cause the parish a serious financial problem? Will it make too much revenue and anger the Archbishop?

Voitech performs a simple, preliminary analysis to help Fr. Efia determine the design issues associated with *Vegas Night at OLPS* in Table 7.1. It is a set of questions that he addresses to Fr. Efia regarding the event and the resolution of the issues raised by the questions. You can see from the nature of these questions, Voitech is attempting to prompt Fr. Efia to think carefully about how he will design the event. The questions deal specifically with the types of games, the sources of event revenues, and the turn-out of alumni he might expect.

This type of interview process is typical of what a consultant might undertake to develop a model of the events. It is a preliminary effort to *define* the problem that a client wants to solve. Fr. Efia will find the process useful for understanding the choices he must make to satisfy the Archbishop's concerns—the games to be played, the method for generating revenues, the attendees that will participate, etc. In response to Fr. Efia's answers, Voitech notes the resolution of the question or the steps needed to achieve resolution. These appear in the third column of Table 7.1. For example, in question 4 of Table 7.1, it is resolved to consider a number of possible attendance fees and their contribution to overall revenues. This initial step is critical to the design and modeling process, and is often referred to as the **model or problem definition phase**.

In the second step of the model definition phase, a *flow diagram* for planning the OLPS event process is generated. This diagram, which provides a view of the related steps of the process, is shown in Exhibit 7.4. Question 1 was resolved by creating this preliminary diagram of the process, including all its options. Since the event in our example is yet to be fully designed, the diagram must include the options that Fr. Efia believes are available. This is not always the case. It is possible that in some situations you will be provided a pre-determined process that is to be modeled, and as such, this step will not include *possible* design options. The answers to Voitech's questions and the discussion about the unsettled elements of the game permit Voitech to construct a relatively detailed process flow map of the event.

The process flow model, at this point, does not presume to have all questions related to the design of the event answered, but by creating this diagram Fr. Efia can begin to comprehend the decisions he must make to execute *Vegas Night at OLPS*. This type of diagram is usually referred to as a **process flow map** because of

Table 7.1 Simple analysis of Fr. Efia's risk related to Vegas night at OLPS

Voitech's question to Fr. Efia	Fr. Efia's answer	Resolution (if any)
1. How do you envision the event?	I'm not sure. What do you think? There will be food and gambling. The Archbishop is not happy with the gambling, but he is willing to go along for the sake of the event.	Let's create a diagram of the potential process—see Exhibit 7.4.
2. What games will be played?	• The Bowl of Treachery • Omnipotent Two-Sided Die • Wheel of Outrageous Destiny	We will have to think about the odds of wining and losing in the games. We can control the odds.
3. Will all attendees play all games?	I don't know. What do you think? But I do know that I want to make things simple. I will only allow attendees to play a game once. I don't really approve of gambling myself, but under these special circumstances a little won't hurt.	Let's consider a number of possibilities–attendees playing all games only once at one end of the spectrum, and at the other end, attendees having a choice of the games they play and how often they play. I am just not sure about the effect on revenue here.
4. Will the games be the only source of income?	No. I am also going to charge a fee for attending. It will be a cover charge of sorts. But, I don't know what to charge. Any ideas?	Let's consider a number of choices and see how it affects overall revenues.
5. How many alumni of OLPS will attend?	It depends. Usually the weather has a big effect on attendance.	Let's think carefully about how the weather will affect attendance.
6. Will there be any other attraction to induce the OLPS alumni to attend?	I think that we will have many wonderful ethnic foods that will be prepared by current and past parishioners. This will not cost OLPS anything. It will be a contribution by those who want to help. We will make the food concession *all-you-can-eat*. In the past this has been a very powerful inducement to attend our events. The local newspaper has even called it the *Best Ethnic Food Festival* in the city!	I urge you to do so. This will make an entry fee a very justifiable expense for attendees. They will certainly see great value in the excellent food and the *all-you-can-eat* format. Additionally, this will allow us to explore the possibility of making the entry fee a larger part of the overall revenue. The Archbishop should not react negatively to this.

the directed flow (or steps) indicated by the arrows. The rectangles represent steps in the process. For example, the *Revenue Collected* process indicates the option to collect an attendance or entry fee to supplement overall revenues. The diamonds in the diagram represent decision points for Fr. Efia's design. For example, the *Charge*

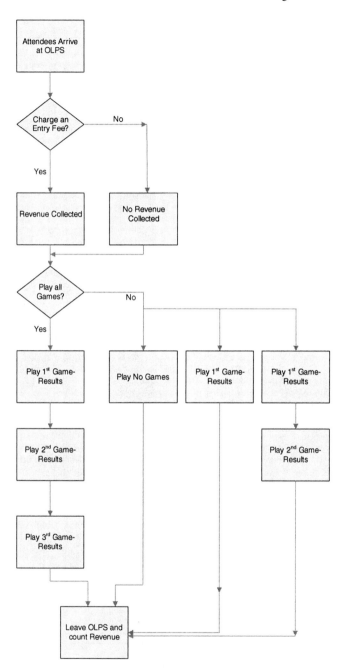

Exhibit 7.4 Simple process flow planning model of OPLS

an Entry Fee? diamond suggests that to finalize the event, Fr. Efia must either decide whether he will collect an entry fee or allow free admission.

From this preliminary analysis, we can also learn where the risk related to uncertainty occurs. Fr. Efia can see that uncertainty is associated with a number of event processes: (1) the number of parishioners attending *Vegas Night at OLPS* which is likely to be associated with weather conditions and the entry fee charged, and (2) the outcomes of the games (players winning or losing) which are associated with the odds that Fr. Efia and Voitech will set for the games. The question of setting the odds of the games is not included at this point, but could be a part of the diagram. In this example, it is assumed that after these preliminary design issues are resolved we can return to the question of the game odds. The design process is usually iterative due to the complexity of the design task, so you may return to a number of the resolved issues to investigate possible changes. Changes in one design issue can, and will, affect the design of other event elements. We will return to this problem later and see how we can incorporate uncertainty deterministically in an Excel based decision model.

7.4 Understanding the Important Elements of a Model

As we can see from the brief discussion of the OLPS event, understanding the processes and design of a model is not an easy task. In this section we will create a framework for building complex models. Let us begin by considering *why* we need models. First, we use models to help us analyze problems and eventually make decisions. If our modeling is accurate and thorough, we can greatly improve the quality of our decision making. As we determined earlier in our investment example, intuition is certainly a valuable personal trait, but one that may not be sufficient in *complex* and *risky* decision situations. So what makes a problem complex? **Complexity** comes from:

1. the need to consider the interaction of many factors
2. the difficulty in understanding the nature and structure of the interactions
3. the uncertainty associated with problem variables and structure
4. the potentially evolving and changing nature of a problem.

To deal with complexity, we need to develop a formal approach to the modeling process; that is, how we will organize our efforts for the most effective and efficient modeling. This does not guarantee success in understanding complex models, but it contributes mightily to the possibility of a *better* understanding. It is also important to realize that the modeling process occurs in stages, and that one iteration through the modeling process may not be sufficient for completely specifying a complex problem. It may take a number of iterations with progressively more complex modeling approaches to finally arrive at an understanding of our problem. This will become evident as we proceed through the OLPS example.

So let us take what we have learned thus far and organize the steps that we need to follow in order to perform effective and efficient modeling:

1. A **pre-modeling or design phase** that contributes to our preliminary understanding of the problem. This could, and often is, called the *problem definition* phase. This step can take a considerable proportion of the entire modeling effort. After all, if you define the problem poorly, no amount of clever analysis will be helpful. At this stage, the goal of the modeling effort should be made clear. What are we expecting from the model? What questions will it help answer? How will it be used and by whom?

2. A **modeling phase** where we build and implement a model that emerges from the pre-modeling phase. Here we refine our specification of the problem sufficiently to explore the model's behavior. At this point the model will have to be populated with very specific detail.

3. An **analysis phase** where we test the behavior of the model developed in steps (1) and (2), and we analyze the results. In this phase, we collect data that the model produces under controlled experimental conditions and analyze the results.

4. A **final acceptance phase** where we reconsider the model specification, if the result of the analysis phase suggests the need to do so. At this point, we can return to the earlier phases until the decision maker achieves desired results. It is, of course, also possible to conclude that the desired results are *not* achievable.

7.4.1 Pre-Modeling or Design Phase

In the pre-modeling or design phase it is likely that we have not settled on a precise definition of our problem, just as Fr. Efia has not decided on the detailed design of his event. I refer to this step as the pre-modeling phase, since the modeling is generally done on paper and does not involve the use of a computer based model. Fr. Efia will use this phase to make decisions about the activities that he will incorporate into *Vegas Night at OLPS*; thus, as we stated earlier, he is still defining the event's design. Voitech used the interview exercise in Table 7.1 to begin this phase. The resulting actions of Table 7.1 then led to the preliminary process flow design in Exhibit 7.4. If the problem is already well defined, this phase may not be necessary. But more often than not, the problem definition is not easy to determine without considerable work. It is not unusual for this step to be the longest phase of the process. And why not! There is nothing worse than realizing that you have developed an elegant model that solves the *wrong* problem.

7.4.2 Modeling Phase

Now it is time to begin the second phase—*modeling*. At this point Fr. Efia has decided on the basic structure of the events—the games to be played and their odds, the restrictions, if any, on the number of games played by attendees, whether an entry fee will be required, etc. A good place to begin the modeling phase is to create of an **Influence Diagram** (IFD). IFDs are diagrams that are connected by directed arrows, much like those in the preliminary process flow planning diagram of Exhibit 7.4. An IFD is a powerful tool that is used by decision analysts to specify *influences* in decision models. Though the concept of an IFD is relatively simple, the theoretical underpinnings can be complicated. For our example, we will develop two types of IFDs: one very simple and one a bit more complex.

We begin by identifying factors—processes, decisions, outcomes of decisions, etc. —that constitute the problem. In our first IFD, we will consider the links between these factors and determine the type of influence between them, either positive (+) or negative (−). A **positive influence** (+) suggests that if there is an increase in a factor, the factor that it influences also has an increase; it is also true that as a factor decreases so does the factor it influences. Thus, they move in the same direction. For example, if I increase marketing efforts for a product, we can expect that sales will also increase. This suggests that we have a positive influence between marketing efforts and sales. The opposite is true for **negative influence** (−): factors move in opposite directions. A negative influence can easily exist between the quality of employee training and employee errors—the higher the quality of training for employees the lower the number of errors committed by employees. Note that the IFD does not suggest the intensity of the influence, only the direction.

Not all models lend themselves to this simple form of IFD, but there will be many cases where this approach is quite useful. Now, let's apply the IFD to Fr. Efia's problem. Voitech has helped Fr. Efia to create a simple IFD of revenue generation for *Vegas Night at OLPS*. It is shown in Exhibit 7.5. Voitech does so by conducting another interview and having Fr. Efia consider more carefully the structure of the event, and how elements of the event are related. In order to understand the movement of one factor due to another, we first must establish a scale for each factor, from negative to positive. The negative to positive scale used for the levels of *weather quality* and *attendee good fortune* is *bad to good*. For *attendance* and *revenue*, the scale is quite direct: higher levels of attendance or revenue are positive and lower levels are negative. The IFD in Exhibit 7.5 provides an important view of how revenues are generated, which of course is the goal of the event. Fr. Efia has specified six important factors: Weather Quality, Attendance, Attendee Luck or Fortune in Gambling, Entry Admission Revenue, Gambling Proceeds Revenue, and Total Revenue. Some factors are uncertain and others are not. For example, weather and attendee fortune are uncertain, and obviously he hopes that the *weather quality* will be good (+) and that *attendee good fortune* will be bad (−). The effect of these two conditions will eventually lead to greater revenues (+). Entry Admission Revenues are known with certainty once we know the attendance, as is the Total Revenue once we determine Entry Admission Revenue and Gambling Proceeds Revenue.

Note that the model is still quite general, but it does provide a clear understanding of the factors that will lead to either success or failure for the OLPS event. There is no final commitment, yet, to a number of the important questions in Table 7.1, for example, questions 2, the *odds of the games*, and 3, *will all attendees play all games*. But, it has been decided that the three games mentioned in question 2 will

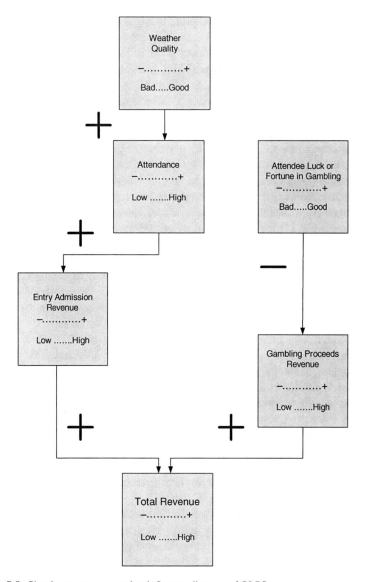

Exhibit 7.5 Simple revenue generation influence diagram of OLPS

all be a part of the event, and that an entry admission fee *will* be charged. The admission fee will supplement the revenues generated by the games. This could be important given that if the games generate a loss for OLPS, then entry admission revenues could offset them. Since Fr. Efia can also control the odds of these games, he eventually needs to consider how the odds will be set.

So in summary, what does the IFD tell us about our problem? If Fr. Efia wants the event to result in larger revenues, he now knows that he will want the following conditions:

1. Good weather to attract a higher attendance

    a. we have little control of the weather
    b. we can schedule events in time periods where the likelihood of good weather is more likely
    c. the exact effect of weather on attendance is uncertain

2. Poor attendee luck leading to high gambling revenues

    a. we do have control of an attendee's *luck* by setting the odds of games
    b. a fair game has 50–50 odds and an example of a game favoring OLPS is 60–40 odds (60% of the time OLPS wins and 40% the attendee wins), or the odds in favor of OLPS could possibly be higher depending on what attendees will tolerate

3. Charge an entry admission to supplement gambling revenue

    a. entry fee is a guaranteed form of revenue based on attendance, assuming there are attendees, unlike the gambling revenue which is uncertain
    b. entry fee is also a delicate matter because charging too much might diminish attendance or have it appear to be a direct request for funds that the Archbishop is firmly against—the entry fee can be justified on the basis of a fee for food and refreshments

As you can see, this analysis is leading to a formal design of the event (a formal problem definition). Just what will the event look like? At this point in the design, Voitech has skillfully directed Fr. Efia to consider all of the important issues related to revenue generation. Fr. Efia must make some difficult choices at this point if he is going to eventually create a model of the event. Note that he will still be able to change his choices in the course of testing the model, but he does need to settle on a particular event configuration to proceed with the modeling effort. Thus, we are moving toward the final stages of phase 2, the *modeling phase*.

Voitech is concerned that he must get Fr. Efia to make some definitive choices for specifying the model. Both men meet at a local coffee shop, and after considerable discussion, Voitech determines the following final details for *Vegas Night at OLPS*:

1. There will be an entry fee of $10 for all those attending. He feels this is a reasonable charge that will not deter attendance. Given the array of wonderful ethnic foods that will be provided by parishioners, this is really quite a modest charge for entry. Additionally he feels that weather conditions are the most important determinant for attendance.
2. The date for the event is scheduled for October 6th. He has weather information forecasting the weather conditions for that October date: 15% chance of rain,

40% chance of cloudy, and a 45% chance of sunshine. Note that these weather outcomes are **mutually exclusive** and **collectively exhaustive**. They are mutually exclusive in that there is *no overlap* in events; that is, it is either rainy or cloudy or sunny. They are collectively exhaustive in that the sum of their probabilities of occurrence is equal to 1; that is, these are *all* the outcomes that can occur.

3. Since weather determines attendance, Voitech interviews Fr. Efia with the intent to determine his estimates for attendance given the various weather conditions. Based on his previous experience with parish events, Fr. Efia believes that if weather is *rainy*, attendance will be 1500 people; if it is *cloudy*, attendance is 2500; if the weather is *sunshine*, attendance is 4000. Of course these are subjective estimates, but he feels confident that they closely represent likely attendance.

4. The selection of the games remains the same—Omnipotent Two-Sided Die (O2SD), Wheel of Outrageous Destiny (WOD), and the Bowl of Treachery (BT). To simplify the process and to correspond with Fr. Efia's wishes to limit gambling (recall he does not approve of gambling), he will insist that every attendee must play all three games and play them *only* once. Later he may consider relaxing this condition to permit individuals to do as they please—play all, some, none, of the games, and to possibly repeat games. This relaxation of play will cause Voitech to model a much more complex event by adding another factor of uncertainty: the unknown number and type of games each attendee will play.

5. He also has set the odds of attendees winning at the games as follows: probabilities of winning in O2SD, WOD, and BT, are 20, 35, and 55%, respectively. The structure of the games is quite simple. If an attendee, wins Fr. Efia gives the attendee the value of the bet (more on this in 6.); if the attendee, loses then the attendee gives Fr. Efia the value of the bet. The logic behind having a single game (BT with 55%) that favors the attendees is to avoid having attendees feel as if they are being exploited. He may want to later adjust these odds a bit to determine the sensitivity of gambling revenues to the changes.

6. All bets at all games are $50 bets, but he would also like to consider the possible outcomes of other quantities, for example $100 bets. This may sound like a substantial amount of money, but he believes that the very affluent attendees will not be sensitive to these levels of bets.

7.4.3 Resolution of Weather and Related Attendance

Now that the *Vegas Night at OLPS* is precisely specified, we can begin to model the behavior of the event. To do so, let us first use another form of influence diagram, one that considers the *uncertain events* associated with a process. This diagramming approach is unlike our initial efforts in Exhibit 7.5 and it is quite useful for identifying the complexities of uncertain outcomes. One of the advantages of this approach is its simplicity. Only two symbols are necessary to diagram a process: a rectangle and a circle. The rectangle represents a step or decision in the process, e.g. the arrival of attendees or the accumulation of revenue. The circle represents an *uncertain* event and the outputs of the circle are the anticipated results of the event. These are the symbols that are also used in **decision trees**, but our use of the symbols is slightly different from those of decision trees. Rectangles in decision trees represent decisions, actions, or strategies. In our use of these symbols, we will allow rectangles to also represent some state or condition, for example the collection of entry fee revenue or the occurrence of some weather condition like rain. Exhibit 7.6 shows the model for this new IFD modeling approach.

In Exhibit 7.6, the flow of the IFD proceeds from top to bottom. The first event that occurs in our problem is the resolution of the uncertainty related to weather. How does this happen? Imagine that Fr. Efia awakens early on October 6th and looks out his kitchen window. He notices the weather for the day. Then he assumes that the weather he has observed will persist for the entire day. All of this is embodied in the circle marked *Weather Condition* and the resulting arrows. The three arrows represent the possible resolution of *weather condition* uncertainty, each of which leads to an assumed, deterministic number of participants. In turn, this leads to a corresponding entry fee revenue varying from a low of $15,000 to a high of $40,000. For example, suppose Fr. Efia observes *sunshine* out of his kitchen window. Thus, *weather condition* uncertainty is resolved and 4000 attendees are expected to attend *Vegas Night at OLPS*, resulting in $40,000 in entry fees.

7.4.4 Attendees Play Games of Chance

Next, the number of attendees determined earlier will participate in each of the three games. The attendees either win or lose in each game; an attendee win is bad news for OLPS and a loss is good news for OLPS. Rather than concerning ourselves with the outcome of each individual attendee's gaming results, an *expected* outcome of revenues can be determined for each game and for each weather/attendee situation. An **expected value** in decision analysis has a special meaning. Consider an individual playing the WOD. On each play the player has a 35% chance of winning. Thus, the average or expected winnings on any single play are $17.50 ($50 * 0.35) and the losses are $32.50 ($50 * [1–0.35]). Of course, we know that an attendee either wins or loses and that the outcomes are either $50 or $0. The expected values represent a weighted average; outcomes weighted by the probability of winning or losing. Thus, if a player plays WOD 100 times, the player can *expect* to win $1750 (100 * $17.50) and Fr. Efia can *expect* to collect $3250 (100 * 32.50). The expected values should be relatively accurate measures of long term results, especially given the large quantity of attendees, and this should permit the averages for winning (or losing) to be relatively close to odds set by Fr. Efia.

At this point we have converted some portions of our probabilistic model into a deterministic model; the probabilistic nature of the problem has not been abandoned, but it has been modified to permit the use of a deterministic model. The weather

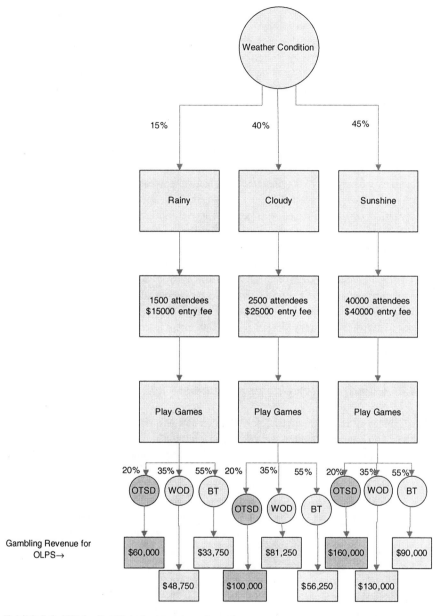

Exhibit 7.6 IFD for Fr. Efia's final event configuration

remains probabilistic because we have a distribution of probabilities and outcomes that specify weather behavior. The outcomes of attendee gambling also have become deterministic. To have a truly probabilistic model we would simulate the outcome of every play for every player. We have chosen not to simulate each uncertain event, but rather, to rely on what we *expect* to happen as determined by a weighted average of outcomes. Imagine the difficulty of simulating the specific fortune, or misfortune, of each game for each of the thousands of attendees.

These assumptions simplify our problem greatly. We can see in Exhibit 7.6 that the gambling revenue results vary from a low of $33,750[1] for the BT in rainy weather to a high of $160,000 for OTSD in sunshine. The range of *total revenue* (entry fee and gambling revenue for a given weather condition) varies from a low of $157,500[2] for rainy weather and a high of $420,000[3] for sunshine.

7.4.5 Fr. Efia's What-if Questions

In spite of having specified the model quite clearly to Voitech, Fr. Efia is still interested in asking numerous what-if questions. He feels secure in the basic structure of the games, but there are some questions that remain and they may lead to adjustments that enhance the event's revenue generation. For example, what if the entry fee is raised to $15, $20, or even $50? What if the value of each bet is changed from $50 to $100? What if the odds of the games are changed to be slightly different from the current values? These are all important questions because if the event generates too little revenue it may cause serious problems with the Archbishop. On the other hand, the Archbishop has also made it clear that the event should not take advantage of the parishioners. Thus, Fr. Efia is walking a fine line between too little revenue and too much revenue. Fr. Efia's what-if questions should provide insight on *how* fine that revenue line might be.

Finally, Voitech and Fr. Efia return to the goals they originally set for the model. The model should help Fr. Efia determine the revenues he can expect. Given the results of the model analysis, it will be up to him to determine if revenues are too low to halt the event or too high and attract the anger of the Archbishop. The model also should allow him to experiment with different revenue generating conditions. This important use of the model must be considered as we proceed to model building. Fortunately, there is a technique that allows us to examine the questions Fr. Efia faces. The technique is known as **sensitivity analysis**. The name might conjure an image of a psychological analysis that measures an individual's emotional response to some stimuli. This image is in fact quite similar to what we would like to accomplish with our model. Sensitivity analysis examines how sensitive the model *output* (revenues) is to changes in the model *inputs* (odds, bets, attendees, etc.). For example, if I change the entry fee, how will the revenue generated by the model change; how will gambling revenues change if the attendee winning odds of the WOD are changed to 30% from the original 35%? One of these changes could contribute to revenue to a greater degree than the other—hence the term *sensitivity* analysis.

---

[1] $1500 * \$50 * (1-0.55) = \$33,750.\ldots\ldots$and$.\ldots\ldots.4000 * \$50 * (1-0.20) = \$160,000.$

[2] $\$60,000 + \$48,750 + \$33,750 + \$15,000 = \$157,500$ (game revenue plus attendance fee).

[3] $\$160,000 + \$130,000 + \$90,000 + \$40,000 = \$420,000$ (game revenue plus attendance fee).

Through sensitivity analysis Fr. Efia can direct his efforts toward those changes that make the greatest difference in revenue.

7.4.6 Summary of OLPS Modeling Effort

Before we proceed, let us step back for a moment and consider what we have done thus far and what is yet to be done in our study of modeling:

- *Model categorization*—we began by defining and characterizing models as deterministic or probabilistic. By understanding the type of model circumstances we are facing we can determine the best approach for modeling and analysis.
- *Problem/Model definition*—we introduced a number of *paper modeling* techniques that allow us to refine our understanding of the problem or problem design. Among these were process flow diagrams that describe the logical steps that are contained in a process, Influence Diagrams (IFD) that depict the influence of and linkage between model elements, and even simple interview methods to probe the understanding of issues and problems related to problem definition and design.
- *Model building*—the model building phase has not been described yet, but it includes the activities that transform the paper models, diagrams, and results of the interview process into Excel based functions and relationships.
- *Sensitivity analysis*—this provides the opportunity to ask what-if questions of the model. These questions translate into input parameter changes in the model and the resulting changes in outputs. They also allow us to focus on parameter changes that have a significant effect on model output.
- *Implementation of model*—once we have studied the model carefully we can make decisions related to execution and implementation. We may decide to make changes to the problem or the model that fit with our changing expectations and goals. As the modeling process advances we may gain insights into new questions and concerns, heretofore not considered.

7.5 Model Building with Excel

In this section we will finally convert the efforts of Voitech and Fr. Efia into an Excel based model. Excel will serve as the programming platform for model implementation. All of their work, thus far, has been aimed at conceptualizing the problem design and understanding the relationships between problem elements. Now it is time to begin translating the model into an Excel workbook. Exhibit 7.6 is the model we will use to guide our modeling efforts. The structure of the IFD in Exhibit 7.6 lends itself quite nicely to an Excel based model. We will build the model with several requirements in mind. Clearly, it should permit Fr. Efia flexibility in revenue analysis; to be more precise, one that permits sensitivity analysis. Additionally, we want to use what we have learned in earlier chapters to help Fr. Efia understand the congruency of his decisions and the goals he has for *Vegas Night at OLPS*. In other words, the model should be user friendly and useful for those decisions relating to his eventual implementation of *Vegas Night at OLPS*.

Let's examine Exhibit 7.6 and determine what functions will be used in the model. Aside from the standard algebraic mathematical functions, there appears to be little need for highly complex functions. But, there are numerous opportunities in the analysis to use functions that we have not used or discussed before, for example control buttons that can be added to the **Quick Access Toolbar** menu via the Excel Options Customize tool menu—**Scroll Bars, Spinners, Combo Boxes, Option Buttons**, etc. We will see later that these buttons are a very convenient way to provide users with control of spreadsheet parameter values such as attendee entry fee and the value of a bet. Thus, they will be useful in sensitivity analysis.

So how will we begin to construct our workbook? The process steps shown in Exhibit 7.6 represent a convenient layout for our spreadsheet model. It also makes good sense that spreadsheets should flow either left-to-right or top-to-bottom in a manner consistent with process steps. I propose that left-to-right is a useful orientation and that we should follow all of our Feng Shui inspired best practices for workbook construction. The uncertain weather conditions will be dealt with deterministically, so the model will provide Fr. Efia outcomes for the event *given* one of the three weather conditions: rainy, cloudy, or sunshine. In other words, the model will not generate a weather event; a weather event will be *assumed* and the results of that event can then be analyzed. The uncertainty associated with the games also will be handled deterministically through the use of *expected values*. We will assume that precisely 20% of the attendees playing OTSD will win, 35% of the attendees playing WOD will win, and 55% of those playing BT will win. Note that in reality these winning percentages will rarely be exactly, 20, 35, and 55%, but if there are many attendees, the percentages should be close to these values.

Exhibit 7.7 shows the layout of the model. For the sake of simplicity, I have placed all analytical elements—brain, calculations, and sensitivity analysis on a single worksheet. If the problem were larger and more complex, it probably would be necessary to place each major part of the model on a separate worksheet. We will discuss aspects of the spreadsheet model in the following order: (1) the basic model and its calculations, (2) the sensitivity analysis that can be performed on the model, and (3) the controls that have been used in the spreadsheet model (scroll bars and options buttons) for user ease of control.

7.5.1 Basic Model

Let us begin by examining the general layout of Exhibit 7.7. The *Brain* is contained in the range B1 to C13. The *Brain* for our spreadsheet model contains the values that will be used in the analysis: Entry Fee, Attendance, Player (odds), and Bets.

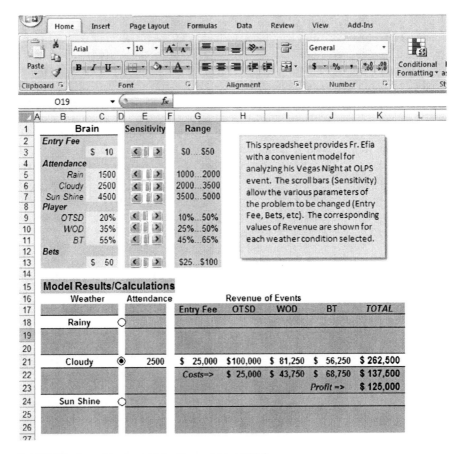

Exhibit 7.7 Spreadsheet model for Vegas night at OPLS

Note that besides the nominal values that Fr. Efia has agreed upon, on the right there is a *Range* of values that provide an opportunity to examine how the model revenue varies with changes in nominal values. These ranges come from Voitech's discussion with Fr. Efia regarding his interest in the model's sensitivity to change. The values currently available for calculations are in column C and they are referenced in the Model Results/Calculations section. The values in cells marked *Range* are text entries that are not meant as direct input data. They appear in G1:G13. As changes are made to the nominal values, they will appear in column C. Later I will describe how the scroll bars, in column E, are used to control the level of the parameter input without having to key-in new values, much like you would use the volume control scroll bars to set the volume of your radio or stereo.

The bottom section of the spreadsheet is used for calculations. The Model Results/Calculations area is straight forward. A weather condition (one of three) is selected by depressing the corresponding Option Button—the circular button next to the weather condition which contains a black dot when activated. Later we will discuss the operation of option buttons, but for know, it is sufficient to say that these buttons, when grouped together, result in a unique number to be placed in a cell. If there are 3 buttons grouped, the numbers will range from 1 to 3, each number representing a button. This provides a method for a specific condition to be used in calculation: Rainy $= 1$, Cloudy $= 2$ and Sunshine $= 3$. Only one button can be depressed at a time and the Cloudy condition, row 21, is the current condition selected. All this occurs in the area entitled Weather.

Once the Weather is selected, the Attendance is known, given the direct relationship Fr. Efia has assumed for Weather and Attendance. Note that the number in the Attendance cell, E21 of Exhibit 7.7, is 2500. This is the number of attendees for a Cloudy day. As you might expect, this is accomplished with a logical *IF* function and is generally determined by the following logical *IF* conditions: *IF* value of button $= 1$ then 1500, else *IF* value of button $= 2$ then 2500, else 4500. Had we selected the Rainy option button then the value for attendees, cell E18, will be 1500. As stated earlier, we will see later how the buttons are created and controlled.

Next, the number of attendees is translated into an *EntryFee* revenue (E21 * C3 $= \$2500 * 10 = \$25,000$) in cell G21. The various game revenues also are determined from the number of attendees. For example, OTSD revenue is the product of the number of attendees in cell E21 (2500), the value of each bet in cell C13 (\$50), and the probability of an OLPS win (1–C9 $= 0.80$), which results in \$100,000 (2500 * \$50 * 0.80) in cell H21. The calculations for WOB and BT are \$81,250[4] and 56,250,[5] respectively.

Of course, there are also payouts to the players that win, and these are shown as *Costs* on the line below revenues. Each game will have payouts, either to OLPS or the players, which when summed equal the total amount of money that is bet. In the case of *Cloudy*, each game has total bets of \$125,000 (\$50 * 2500). You can see that if you combine the revenue and cost for each game, the sum is indeed \$125,000, the total amount bet for each game. As you would expect, the only game where the costs (attendee's winnings) are greater than the revenues (OLPS's winnings) is the BT game. This game has odds that favor the attendee. The cumulative profit for the event is the difference between the revenue earned by OLPS in cell K21 (\$262,500) and the costs incurred in cell K22 (\$137,500). In this case, the event yields a profit of \$125,000 in cell K23. This represents the combination of *Entry Fee*, \$25,000, and *Profit* from the games, \$100,000.[6]

The model in Exhibit 7.7 represents the basic layout for problem analysis. It utilizes the values for entry fees, attendance, player odds, and bets that were agreed to by Voitech and Fr. Efia. In the next section we address the issues of sensitivity analysis that Fr. Efia has raised.

---

[4]2500 * \$50 * (1–0.35) = \$81,250.
[5]2500 * \$50 * (1–0.55) = \$56,250.
[6](\$100,000–\$25,000) + (\$81,250–\$43,750) + (\$56,250–\$68,750) = \$100,000.

7.5.2 Sensitivity Analysis

Once the basic layout of the model is complete, we can begin to explore some of the what-if questions that were asked by Fr. Efia. For example, what change in revenue occurs if we increase the value of a bet to $100? Obviously, if all other factors remain the same, revenue will increase. But will all factors remain the same? It is conceivable that a bet of $100 will dissuade some of the attendees from attending the event; after all, this doubles an attendee's total exposure to losses from $150 (three games at $50 per game) to $300 (three games at $100 per game). What percentage of attendees might not attend? Are there some attendees that would be more likely to attend if the bet increases? Will the Archbishop be angry when he finds out that the value of a bet is so high? The answers to these questions are difficult to know. The current model will not provide information on how attendees will respond since there is no economic model included to gauge the attendee's sensitivity to the value of the bet, but Fr. Efia can posit a guess as to what will happen with attendance and personally gauge the Archbishop's response. Regardless, with this model Fr. Efia can begin to explore the effects of these changes.

We begin sensitivity analysis by considering the question that Fr. Efia has been struggling with—how to balance the event revenues to avoid the attention of the Archbishop. If he places the odds of the games greatly in favor of OLPS, the Archbishop may not sanction the event. As an alternative strategy to setting *poor* player odds, he is considering increasing the entry fee to offset losses from the current player odds. He believes he can defend this approach to the Archbishop, especially in light of the *all-you-can-eat* format for ethnic foods that will be available to attendees. But of course, there are limits to the entry fee that OLPS alumni will accept as reasonable. Certainly a fee of $10 can be considered very modest for the opportunity to feast on at least 15 varieties of ethnic food.

So what questions might Fr. Efia pose? One obvious question is—How will an increase in the entry fee offset an improvement in player odds? Can an entry fee increase make up for lower game revenues? Finally, what Entry Fee increase will offset a change to fair odds: 50-50 for bettors and OLPS? Let us consider the Cloudy scenario in Exhibit 7.7 for our analysis. In this scenario total game revenue is $262,500 and cost is $137,500, resulting in overall profit of $125,000. Clearly, the entry fee will have to be raised to offset the lost gaming revenue if we improve the attendee's winning odds.

If we set the gaming odds to fair odds (50–50), we expect that the distribution of game funds to OLPS and attendees will be exactly the same, since the odds are now fair. Note that the odds have been set to 50% in cells C9, C10, and C11 in Exhibit 7.8. Thus, the *only* benefit to Fr. Efia is Entry Fee, which is $25,000 as shown in cell K23. The fair odds scenario has resulted in a $100,000 profit reduction. Now, let us increase the Entry Fee to raise the level of profit to the desired $125,000. To achieve such a profit we will have to set our Entry Fee to a significantly higher value. Exhibit 7.9 shows this result in cell K23. An increase to $50 per person in cell C3 eventually achieves the desired result. Although this analysis may seem trivial since

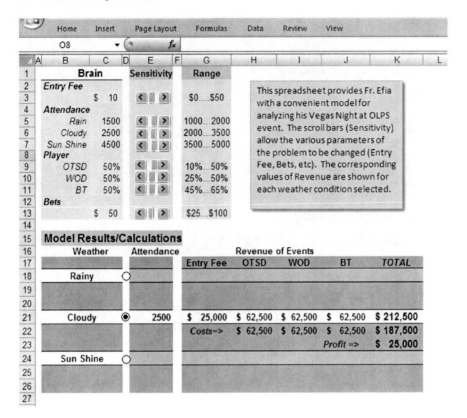

Exhibit 7.8   Fair (50–50) odds for OPLS and attendees

the fair odds simply mean that profit will only be achieved through Entry Fee, in more complex problems the results of the analysis need not be as simple.

What if $50 is just too large a fee for Fr. Efia or the Archbishop? Is there some practical combination of a larger (greater than $10), but reasonable, Entry Fee, and some *nearly* fair odds that will result in $125,000? Exhibit 7.10 shows that if we set odds cell C9 to 40%, C10 to 40%, and C11 to 50% for OTSD, WOD, and BT, respectively, and we also set the Entry Fee cell B3 to $30, we can achieve the desired results of $125,000 in cell K23. You can see that in this case the analysis is not as simple as before. There are many, many combinations of odds for three games that will result in the same profit. This set of conditions may be quite reasonable for all parties, but if they are not, then we can return to the spreadsheet model to explore other combinations.

As you can see, the spreadsheet model provides a very convenient system for performing sensitivity analysis. There are many other questions that Fr. Efia could pose and study carefully by using the capabilities of his spreadsheet model. For example, he has not explored the possibility of also changing the cost of a bet from the current

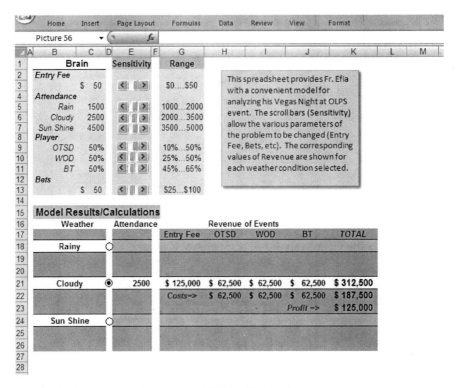

Exhibit 7.9 Entry fee needed to achieve $125,000 with fair odds

$50 to some higher value. In complex problems there will be many changes in variables that will be possible. Sensitivity analysis provides a starting point for dealing with these difficult *what-if* questions. Once we have determined areas of interest, we can take a more focused look at specific changes we would like to consider. To study these specific changes, we can use the **DataTable** feature in the Data ribbon. It is located in the What-If Analysis sub-group in the Data Tools group. The Data Table function allows us to select a variable (or two) and find the corresponding change in formula results for a given set of input values of the variable(s). For example, suppose we would like to observe the changes in the formula for *Profit* associated with our *Cloudy* scenario in Exhibit 7.10, by changing the value of *Bets* in a range from $10 to $100. Exhibit 7.11 shows a single-variable Data Table based on $10 increment changes in *Bets*. Note that for $10 increments in Bets the corresponding change in *Profit* is $10,000. For example, the difference in the *Profit* when *Bet Value* is changed from $30 in cell N5 to $40 in cell N6 is $10,000 ($115,000–$105,000).

What about simultaneous changes in *Bets* and *Entry Fee*? Which of the changes will lead to greater increases in *Profit*? A two-variable data Table is also shown in Exhibit 7.11, where values of *Bets* and *Entry Fee* are changed simultaneously. The cell with the rectangle border in the center of the table reflects the profit generated by the nominal values of *Bets* and *Entry Fee*, $50 and $30 respectively. From this

	B	C	D	E	F	G	H	I	J	K	L	M
1	**Brain**			**Sensitivity**		**Range**						
2	Entry Fee							This spreadsheet provides Fr. Efia				
3		$ 30		< ‖ >		$0...$50		with a convenient model for				
4	Attendance							analyzing his Vegas Night at OLPS				
5	Rain	1500		< ‖ >		1000...2000		event. The scroll bars (Sensitivity)				
6	Cloudy	2500		< ‖ >		2000...3500		allow the various parameters of				
7	Sun Shine	4500		< ‖ >		3500...5000		the problem to be changed (Entry				
8	Player							Fee, Bets, etc). The corresponding				
9	OTSD	40%		< ‖ >		10%...50%		values of Revenue are shown for				
10	WOD	40%		< ‖ >		25%...50%		each weather condition selected.				
11	BT	50%		< ‖ >		45%...65%						
12	Bets											
13		$ 50		< ‖ >		$25...$100						
14												
15	**Model Results/Calculations**											
16		Weather		Attendance				**Revenue of Events**				
17							Entry Fee	OTSD	WOD	BT	TOTAL	
18		Rainy	○									
19												
20												
21		Cloudy	◉	2500		$ 75,000	$ 75,000	$ 75,000	$ 62,500	$ 287,500		
22						Costs=>	$ 50,000	$ 50,000	$ 62,500	$ 162,500		
23									Profit =>	$ 125,000		
24		Sun Shine	○									
25												
26												
27												
28												

Exhibit 7.10 Entry fee and less than fair odds to achieve $125,000

analysis, it is clear that an increase in *Bets*, regardless of *Entry Fee*, will result in an increase of $10,000 in *Profits* for each $10 increment. We can see this by subtracting any two adjacent values in the column. For example, the difference between 125,000 in cell Q20 and 135,000 in cell Q21, for an *Entry Fee* of $30 and *Bets* of $50 and $60, is $10,000. Similarly, a $10 increment for *Entry Fee* results in a $25,000 increase in *Profits*, regardless of the level of the *Bet*. For example, 125,000 in cell Q20 and 150,000 in cell R20, for a *Bet* of $50 and *Entry Fee* of $30 and $40, is $25,000. This simple sensitivity analysis suggests that increasing *Entry Fee* is a more effective source of *Profit* than increasing *Bet*, for similar increments of change ($10).

Now, let us see how we create a single-variable and two-variable Data Table. The Data Table tool is a convenient method to construct a table for a particular cell formula calculation and record the formula's variation as one, or two, variables in the formula change. The process of creating a table begins by first selecting the formula that is to be used as table values. In our case the formula is *Profit* calculation. In Exhibit 7.12 we see the *Profit* formula reference, cell K23, placed in the cell N2. The *Bet Values* that will be used in the formula are entered in cells M3 through M12. This is a one variable table that will record the changes in *Profit* as *Bet Value* is varied from 10 to 100 in increments of 10. A one variable *Table* can take

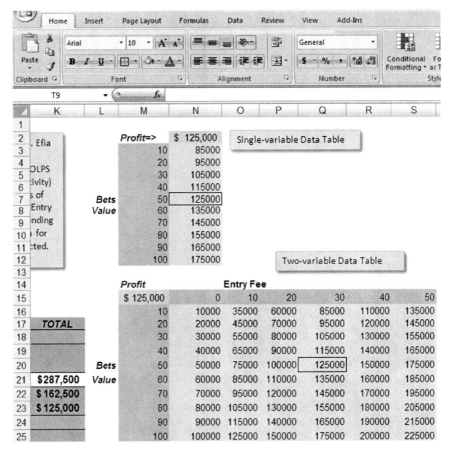

Exhibit 7.11 One-variable and two-variable data table

either a vertical or horizontal orientation. I have selected a vertical orientation that requires the *Bet Value* be placed in the column (M) immediately to the left of the formula column (N); for a horizontal orientation, the *Bet Value* would be in the row above the formula values. These are conventions that must be followed. The empty cells, N2:N12, will contain repetitive calculations of *Profit*. If a two variable table is required, the variables are placed in the column to the left and the row above the calculated values. Also, the variables used in the formula must have a cell location in the worksheet that the formula references. In other words, the variable cannot be entered directly into the formula as a number, but must reference a cell location; for example, the cell references that are in the Brain worksheet.

Once the external structure of a table is constructed (the variable values and the formula cell) we select the table range: M2:N12. See Exhibit 7.13. This step simply identifies the range that will contain the formula calculations and the value of the variable to use. Next, you will find in the Data ribbon the Data Table tool in the

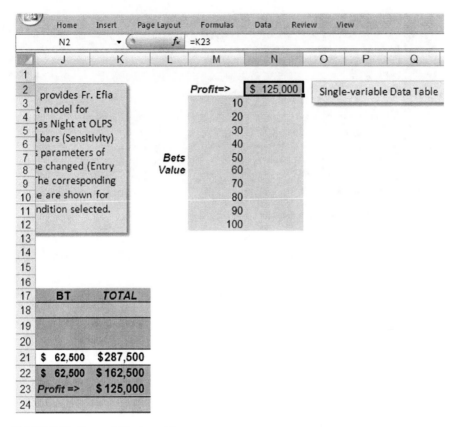

Exhibit 7.12 One-variable data table

What-If Analysis. This step utilizes a wizard that requests the location of the Row input cell and Column input cell. See Exhibit 7.14. In the case of the one variable table, in vertical orientation, the relevant choice is the Column input cell because our variable values appear in *column* M. This wizard input identifies where the variable is located that will be used by the formula. For the one-variable table, the wizard input is cell C13; it has a current value of $50. In essence, the Data Table is being told where to make the changes in the formula. In Exhibit 7.15 we see the two-variable Data Table with Row input cell as C3 and Column input cell as C13, the cell location of the formula input for *Entry Fee* and *Bets*, respectively. The results for both the one-variable and the two-variable Data Table are shown in the previously introduced Exhibit 7.11.

Of course in more complex problems, the possible combination of variables for sensitivity analysis could be numerous. Even in our simple problem, there are 8 possible variables that we can examine individually or in combination (2, 3, 4, . . ., and 8 at a time); thus, there are literally thousands of possible sensitivity analysis scenarios we can study.

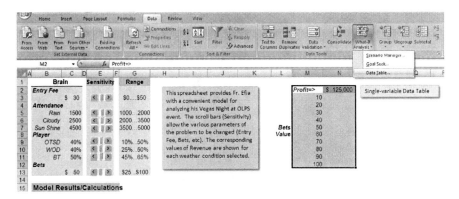

Exhibit 7.13   Data table tool in data ribbon

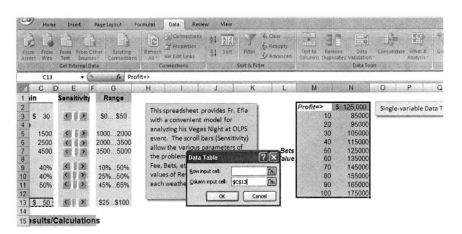

Exhibit 7.14   Table wizard for one-variable data table

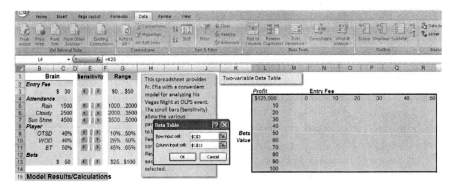

Exhibit 7.15   Table wizard for two-variable data table

The spreadsheet model has permitted an in-depth analysis of Fr. Efia's event. It has met his initial goal of providing a model that allows him to analyze the revenues generated by the event. Additionally, he is able to ask a number of important what-if questions by varying individual values of model inputs. Finally, the formal use of sensitivity analysis through the Data Table tool provides a systematic approach to variable changes. All that is left is to examine some of the convenient devices that he has employed to control the model inputs—Scroll Bars and Option Buttons from the Forms Control.

7.5.3 Controls from the Forms Control Tools

Now let us consider the devices that we have used for input control in Fr. Efia's spreadsheet model. These devices make analysis and collaboration with spreadsheets convenient and simple. We learned above that sensitivity analysis is one of the primary reasons we build spreadsheet models. In this section we consider two simple tools that aid sensitivity analysis: (1) one to change a variable through incremental change control, and (2) the other, a switching device to select a particular model condition or input. Why do we need such devices? Variable changes can be handled directly by selecting a cell and keying in new information, but this can be very tedious, especially if there are many changes to be made. So how will we implement these activities to efficiently perform sensitivity analysis and what tools will we use? The answer is the Forms Control. This is an often neglected tool that can enhance spreadsheet control and turn a pedestrian spreadsheet model into a user friendly and powerful analytical tool.

The Forms Control provides a number of functions that are easily inserted in a spreadsheet. They are based on a set of instructions that are bundled into a **Macro**. Macros as you will recall are a set of programming instructions that can be used to perform tasks by executing the macro. To execute a macro it must be assigned a name, keystroke, or symbol. For example, macros can be represented by *buttons* (a symbol) that launch their instructions. Consider the need to copy a column of numbers on a worksheet, perform some manipulation of the column, and move the results to another worksheet in the workbook. You can perform this task manually, but if this task has to be repeated many times it could easily be automated as a macro and attached to a button. By depressing the macro button we can execute multiple instructions with a single key stroke and a minimum of effort. Additionally, macros can serve as a method of control for the types of interactions a user may perform. It is often very desirable to control user interaction, and thereby the potential errors and the misuse of a model that can result.

To fully understand Excel Macros, we need to understand the programming language used to create them, Microsoft Visual Basic for Applications (**VBA**). Although this language can be learned through disciplined effort, Excel has anticipated that the majority of users will *not* be interested, or need, to make this effort. Incidentally, the VBA language is also available in MS Word and MS Project,

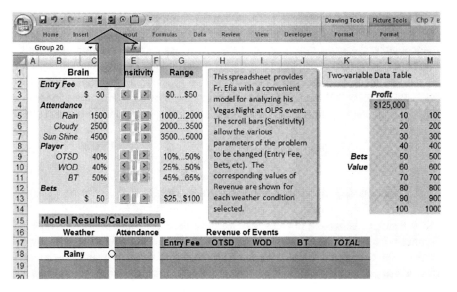

Exhibit 7.16 Use of forms tools for control

making it very attractive to use across programming platforms. Excel has provided a shortcut that permits some of the important uses of macros, without the need to learn VBA. Some of these shortcuts are found in the Forms Control. In Excel 2003 the Forms Control menu was found by engaging the pull-down menu View and selecting Toolbars. In Excel 2007 Forms Control is not available in the standard ribbons but can be placed into the Quick Access Toolbar. At the bottom of the Excel button there is an options Excel Options button. Upon entering the options you can select Customize to add tools to the Quick Access Toolbar. One of the options is *Commands Not in the Ribbon*. This is where Spin Button (Form Control) and Scroll Bar (Form Control) can be found and added to the Quick Access Toolbar. The arrow in Exhibit 7.16 shows four icons: (1) List Box where an item can be selected from a list, (2) the Scroll Bar which looks like a divided rectangle containing two opposing arrow heads, (3) Spin Button which looks quite similar to the Scroll Bar, (4) the Option Button which looks like a circle containing a large black dot, and (5) Group Box for grouping buttons.

7.5.4 Option Buttons

Let us begin with the Option Button. Our first task is to consider how many buttons we want to use. The number of buttons will depend on the number of options you will make available in a spreadsheet model. For weather in the OLPS model, we have 3 options to consider and each option triggers several calculations. Additionally, for the sake of clarity, a single option will be made visible at a time.

For example, in Exhibit 7.7 the cloudy option is shown and all others are hidden. The following are the detailed steps necessary to create a *group* of 3 options for which only one option will be displayed:

1. Creating a **Group Box**—We select the Group Box, the last icon in the Quick Access Toolbar in Exhibit 7.16. Drag-and-drop the Group Box onto the worksheet. See Exhibit 7.17 for an example. Once located, a right click will allow you to move a Group Box. The grouping of Option Buttons in a Group Box alerts Excel that any Option Buttons placed in the box will be connected or associated with each other. Thus, by placing three buttons in the box, each button will be assigned a specific output value (1, 2, or 3), and those values can be assigned to a cell of your choice on the worksheet. You can then use this value in a logical function to indicate the option selected. (If four buttons are used then the values will be 1, 2, 3, and 4.)

2. Creating Option Buttons—Drag-and-drop the Option Button found in the Quick Access Toolbar into the Group Box. When you click on the Option Button and move your cursor to the worksheet, a cross will appear. Left click your cursor and drag the box that appears into the size you desire. This box becomes an Option Button. Repeat the process in the Group Box for the number of buttons needed. A right click will allow you to reposition the button and text can be added to identify the button.

3. Connecting button output to functions—Now we must assign a location to the buttons that will indicate which of the buttons is selected. Remember, that only one button can be selected at a time. Place the cursor on any button and right click. A menu will appear and the submenu of interest is Format Control. See Exhibit 7.17. Select the submenu and then select the Control tab. At the bottom you will see a dialogue box requesting a **Cell Link**. In this box place the cell location where the buttons will record their unique identifier to indicate the single button that is selected. In this example D22 is the cell chosen, and by choosing this location for one button, *all* grouped buttons are assigned the same Cell Link. Now the cell can be used to perform worksheet tasks.

4. Using the Cell Link value—In Fr. Efia's spreadsheet model, the Cell Link values are used to display or hide calculations. For example, in Exhibit 7.18 cell E21, the Attendance for the Cloudy scenario, contains: $= IF(B29=2, C6,0)$. Cell C6 is currently set to 2500. This logical function examines the value of B29, the Cell Link that has been identified in step 3. If B29 is equal to 2, it returns 2500 as the value cell E21. If it is not equal to 2, then a zero is inserted. A similar calculation is performed for the Entry Fee in cell G21 by using the following cell function: $=IF(E21=0,0,C3 * E21)$. In this case, cell E21, the Attendance for Cloudy, is examined and if it is found to be zero, then a zero is inserted in E21. If it is *not* zero, then a calculation is performed to determine the Entry Fee revenue (2,500 * 30=75,000). The *Revenues of Events* are calculated in a similar manner. Note that it also is possible to show *all* values for *all* scenarios (Sunshine, Cloudy, and Rainy) and eliminate the logical aspect of the cell functions, and then the Option Buttons would not be needed. The buttons allow us to focus strictly on a single

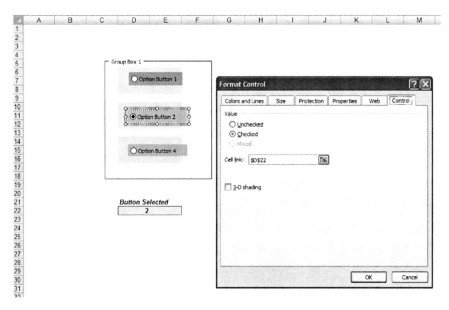

Exhibit 7.17 Assigning a cell link to grouped buttons

scenario. This makes sense, since only one weather condition will prevail for a particular day of *Vegas Night atOLPS*. Of course, these choices are often a matter of taste and of the specific application.

7.5.5 Scroll Bars

In the Brain, we also have installed 8 Scroll Bars. See Exhibit 7.18. They control the level of the variable in the column C. For example, the Scroll Bar in cell E5 controls C5 (1500), the attendance on a rainy day. These bars can be operated by clicking the left and right arrows, or by grabbing and moving the center bar. Scroll Bars can be designed to make incremental changes of a specific amount. For example, we can design an arrow click to result in an increase (or decrease) of five units, and the range of the bars can also be set to a specific maximum and minimum. Like the Option Button, a Cell Link needs to be provided to identify where cell values will be changed. Although Scroll Bars provide great flexibility and function, changes are restricted to be integer valued, for example, 1, 2, 3, etc. This will require additional minor effort if we are interested in producing a cell change that employs fractional values, for example percentages.

Consider the Scroll Bar located on E5 in Exhibit 7.19. This bar, as indicated in the formula bar above, controls C5. By right clicking the bar and selecting the Format Control tab, one can see the various important controls available for the bar: Minimum value (1000), Maximum value (2000), Incremental change (50-the

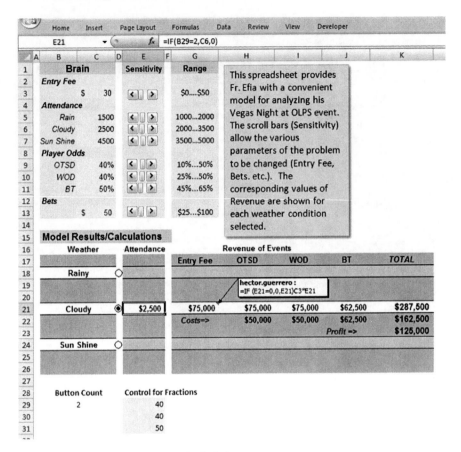

Exhibit 7.18   Button counter logic and calculations

change due to clicking the arrows), and Page change (10-the change due to clicking between the bar and arrow). Additionally, the Cell Link must also be provided, and in this case it is C5. Once the link is entered, a right click of the button will show the Cell Link in the formula bar, $C$5 in this case.

Now, consider how we might use a Scroll Bar to control fractional values in Exhibit 7.20. As mentioned above, since only integer values are permitted in Scroll Bar control, we will designate the Cell Link as cell E29. You can see the cell formula for C9 is E29/100. Dividing E29 by 100 will produce a fractional value, which we can use as a percentage. Thus, we can suggest the range of the Scroll Bar in G9 to range between 10 and 50, and this will result in a percentage in cell C9 from 10% to 50% for the Player Odds for OTSD.[7] You can also see that the other fractional odds

---

[7] E29/100 = 40/100 = 0.40 or 40% ...currently the value of the OTSD Player Odds.

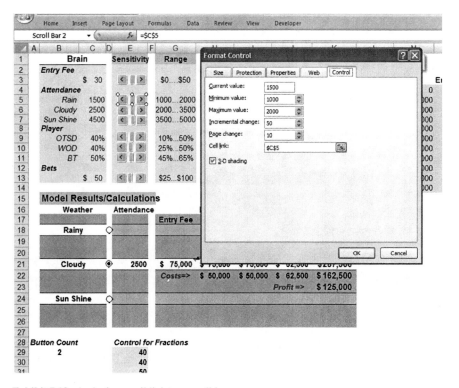

Exhibit 7.19 Assigning a cell link to a scroll bar

are linked to Scroll Bars in cells E30 and E31. This inability to directly assign fractional values to a Scroll Bar is a minor inconvenience that can be easily managed.

7.6 Summary

This chapter has provided a foundation for modeling complex business problems. Although modeling contains an element of art, a substantial part of it is science. We have concentrated on the important steps for constructing sound and informative models to meet a modeler's goals for analysis. It is important that however simple a problem might appear, a rigorous set of development steps must be followed to insure a successful model outcome. Just as problem definition is the most important step in problem solving, model conceptualization is the most important step in modeling. In fact, I suggest that problem definition and model conceptualization are essentially the same.

In early sections of this chapter we discussed the concept of models and their uses. We learned the importance of classifying models in order to develop appropriate strategies for their construction, and we explored tools (Flow and Influence Diagrams) that aid in arriving at a preliminary concept for design.

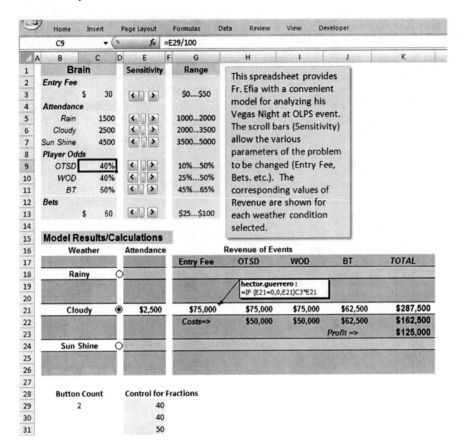

Exhibit 7.20 Using a scroll bar for fractional values

All our work in this chapter has been related to deterministic models. Although Fr. Efia's model contained probabilistic elements (game odds, uncertainty in weather conditions, etc.), we did not explicitly model these uncertain events. We accepted their deterministic equivalents: *expected value* of gambling outcomes. Thus, we have yet to explore the world of probabilistic simulation, in particular Monte Carlo simulation. Monte Carlo simulation will be the focus of Chap. 8. It is a powerful tool that deals explicitly with uncertain events. The process of building Monte Carlo simulations will require us to exercise our current knowledge and understanding of probabilistic events. This may be daunting, but as we learned in previous chapters on potentially difficult topics, special care will be taken to provide thorough explanations and I will present numerous examples to guide your learning.

This chapter has provided us with the essentials of creating effective models. In Chap. 8 we will rely on what we have learned in Chaps. 1–7 to create complex probabilistic models and analyze the results of our model experimentation.

Key Terms

Data rich

Data poor

Physical Model

Analog Model

Symbolic Model

Risk Profiles

Deterministic

Probabilistic

PMT() Function

Model or Problem Definition Phase

Process Flow Map

Complexity

Pre-Modeling or Design Phase

Modeling Phase

Analysis Phase

Final Acceptance Phase

Influence Diagram (IFD)

Positive Influence

Negative Influence

Mutually Exclusive

Collectively Exhaustive

Decision Trees

Expected Value

Sensitivity Analysis

Quick Access Toolbar

Scroll Bars

Spinners

Combo Boxes

Option Buttons

Data Table

Macro

VBA

Group Box

Cell Link

Problems and Exercises

1. Data rich data is expensive to collect—T or F?
2. What type of model is a site map that is associated with a website?
3. The x-axis of a risk profile is associated with potential outcomes—T of F?
4. Deterministic is to probabilistic as point estimate is to range—T or F?
5. What is a single annual payment for the PMT() function for the following data: 6.75% annual interest rate; 360 months term; and $100,000 principal?
6. Draw a Process Flow Map of your preparation to leave your house, dormitory, or apartment in the morning. Use a rectangle to represent process steps like, *brush teeth*, and diamonds to represent decisions, like *wear warm weather clothes (?).*
7. Create a diagram of a complex decision or process of your choice by using the structure of an influence diagram.
8. An investor has 3 product choices in a year long investment with forecasted outcomes—bank deposit (2.1% guaranteed); a *bond* mutual fund (0.35 probability of a 4.5% return; 0.65 probability of 7%), and a *growth* mutual fund (0.25 probability of –3.5% return, 50% probability of 4.5%, and remaining probability of 10.0%).

   a. Draw the decision tree and calculate the expected value of the three investment choices. You decide that the maximum expected value is how you will choose an investment. What is your investment choice?

b. What will the guaranteed return for the bank deposit have to be to change your decision in favor of the bank deposit?

c. Create a spreadsheet that permits you to perform the following sensitivity analysis: What must the value of the largest return (currently 7%) for the bond fund be for the expected value of the bond fund to be equal to the expected value of the growth fund?

9. For Fr. Efia's OLPS problem perform the following changes:

a. Introduce a 4th weather condition, *Absolutely Miserable*, where the number of alumni attending is a point estimate of only 750.

b. Perform all the financial calculations in a separate area below the others.

c. Add the scroll bar (range of 500–900) and the option button associated with the new weather condition, such that the look of the spreadsheet is consistent.

d. What will the entry fee for the new weather condition have to be in order for the profit to equal that in Exhibit 7.7?

e. Find a different combination of Player odds that leads to the same Profit ($125,000) in Exhibit 7.10.

f. Create a two-variable Data Table for cloudy weather, where the variables are Bet Value ($10 to $100 in $10 increments) and OTSD player odds (10–80% in 10% increments).

10. Create a set of 4 buttons that when a specific button is depressed (X) it provides the following message in a cell: *Button X is Selected* (X can take on values 1–4). Also, add conditional formatting for the cell that changes the color of the cell for each button that is depressed.

11. Create a calculator that asks a person their current weight and permits them to chose, by way of a scroll bar, only one of 5 percentage reductions 5, 10, 15, 20, and 25%. The calculator should take the value of the percentage reduction and calculate their *desired* weight.

12. For the same calculator in 11, create a one-variable Data Table that permits the calculation of the desired weight for weight reduction from 1 to 25% in 1% increments.

13. *Advanced Problem*—Income statements are excellent mechanisms for modeling the financial feasibility of projects. Modelers often choose a level of revenue, a percent of the revenue as COGS (Cost of Goods Sold), and a percent of revenue as variable costs.

a. Create a deterministic model of a simple income statement for the data elements shown below (d-i)–(d-iv). The model should permit selection of various data elements through the use of option buttons and scroll bars, as needed.

b. Produce a risk profile of the numerous combinations of data elements assuming that all data element combinations are of equal probability. (Recall the vertical axis of a risk profile is the probability of occurrence of the outcomes on the horizontal axis, and in this case, all probabilities are equal).

    c. Also, provide summary data for all the profit combinations for the problem–average, max, min, and standard deviation.

    d. Data elements for the problem:

        i. Revenue $100 k and $190 k (Option Button)

        ii. COGS % of Revenue with outcomes of 24 and 47% (Option Button)

        iii. Variable costs % of Revenue with outcome 35 and 45% (Option Button)

        iv. Fixed costs $20 k to $30 k in increments of $5 k (Scroll bar)

    e. Create a Data Table that will permit you to change (with a scroll bar) the fixed cost in increments of $1 k that will result in instantaneous changes in the graph of the risk profile. Hint: combine (d-ii) and (d-iii) as a single variable and as a single dimension of a two variable Data Table, while using revenue as the second dimension. Fixed cost will act as a third dimension in the sensitivity analysis, but will not appear on the borders of the two variable Data Table.

Chapter 8
Modeling and Simulation: Part 2

Contents

8.1 Introduction

Chapter 8 continues with our discussion of modeling. In particular, we will discuss modeling in the context of *simulation*, a term that we will soon discuss in detail. The terms **model** and **simulation** can be a bit confusing because they are often used

H. Guerrero, *Excel Data Analysis*, DOI 10.1007/978-3-642-10835-8_8,
© Springer-Verlag Berlin Heidelberg 2010

interchangeably; that is, simulation *as* modeling and vice versa. We will make a distinction between the two terms and we will see that in order to simulate a process we must first create a model of the process. Thus, modeling precedes simulation, and simulation is an activity that depends on *exercising* a model. This may sound like a great deal of concern about the minor distinctions between the two terms, but as we discussed in Chap. 7, being systematic and rigorous in our approach to modeling helps insure that we don't overlook critical aspects of a problem. Over many years of teaching and consulting, I have observed very capable people make serious modeling errors, simply because they felt that they could approach modeling in a casual manner, thereby abandoning a systematic approach.

So why do we make the distinction between modeling and simulation? In Chap. 7 we developed deterministic models and then exercised the model to generate outcomes based on simple *what-if* changes. We did so with the understanding that not all models require sophisticated simulation. For example, Fr. Efia's problem was a very simple form of simulation. We exercised the model by *imposing* a number of conditions: weather, an expected return on bets, and an expected number of attendees. Similarly, we imposed requirements (rate, term, principal) in the modeling of mortgage payments. But models are often not this simple, and can require considerable care in conducting simulations; for example, modeling the process of patients visiting a hospital emergency department. The arrival of many types of injury and illness, the staffing required to deal with the cases, and the updating of current bed and room capacity based on the state of conditions in the emergency department make this a complex model to simulate.

The difference between the mortgage payment and a hospital emergency department simulation, aside from the model complexity, is how we deal with uncertainty. For the mortgage payment model, we used a manual approach to managing uncertainty by changing values and asking *what-if* questions individually: what-if the interest rate is 7% rather than 6%, what if I change the principal amount I borrow, etc. In the OLPS model we reduced uncertainty to **point estimates** (specific values) and then used a manual approach to exercise a specific model configuration; for example, we set the number of attendees for *Cloudy* weather to exactly 2500 people and we considered a what-if change to *Entry Fee* from $10 to $50. This approach was sufficient for our simple what-if analysis, but with models containing more elements of uncertainty and even greater complexity due to the interaction of uncertainty elements, we will have to devise complex simulation approaches for managing uncertainty.

The focus of this chapter will be on a form of simulation that is often used in modeling of complex problems—a methodology called **Monte Carlo Simulation**. Monte Carlo simulation has the capability of handling the more complex models that we will encounter in this chapter. This does not suggest that all problems are destined to be modeled as Monte Carlo simulations, but many can and should. In the next section, I will briefly discuss several types of simulation. Emphasis will be placed on the differences between approaches and on the appropriate use of techniques. Though there are many commercially available simulation software packages for a variety of applications, remarkably, Excel is a very capable tool that can be useful with many simulation techniques. In cases where a commercially available package is necessary, Excel can still have a critical role to play in the early or **rapid prototyping** of problems. Rapid prototyping is a technique for quickly creating a model that need not contain the level of detail and complexity that an end-use model requires. It can save many, many hours of later programming and modeling effort by determining the feasibility and direction an end-use model should take.

Before we proceed, I must caution you about an important concern. Building a Monte Carlo simulation model must be done with great care. It is very easy to build faulty models due to careless consideration of processes. As such, the chapter will move methodically toward the goal of constructing a useful and thoroughly conceived simulation model. At critical points in the modeling process, we will discuss the options that are available and why some may be better than others. There will be numerous tables and figures that build upon one another, so I urge you to read all material with great care. At times the reading may seem a bit tedious and pedantic, but such is the nature of producing a high quality model—these things cannot be rushed. Try to avoid the need to get to the *punch-line* too soon.

8.2 Types of Simulation and Uncertainty

The world of simulation is generally divided into two categories—**continuous event simulation** and **discrete event simulation**. The difference in these terms is related to how the process of simulation evolves—how results change and develop over some dimension, usually time. For example, consider the simulation of patient arrivals to the local hospital emergency room. The patient arrivals, which we can consider to be **events**, occur sporadically and trigger other events in a *discrete* fashion. For example, if a cardiac emergency occurs at 1:23 pm on a Saturday morning, this might lead to the need of a defibrillator to restore a patient's heartbeat, specialized personnel to operate the device, as well as a call to a physician to attend to the patient. These circumstances require a simulation that triggers random events at *discrete* points in time and we need not be concerned with tracking model behavior when events are not *occurring*. The arrival of patients at the hospital is not continuous over time, as might be the case for the flow of daytime traffic over a busy freeway in Southern California. It is not unusual to have modeling phenomenon that involves both discrete and continuous events. The importance of making a distinction relates to the techniques that must be used to create suitable simulation models. Also, commercial simulation packages are usually categorized as having either continuous, discrete, or both modeling capabilities.

8.2.1 Incorporating Uncertain Processes in Models

Now, let us reconsider some of the issues we discussed in the Chap. 7, particularly those in Fr. Efia's problem of planning the events of *Vegas Night at OLPS* and let us focus on the issue of uncertainty. The problem contained several elements of uncertainty—the weather, number of attendees, and the outcome of games of chance. We simplified the problem analysis by assuming **deterministic values** (specific and unchanging) for these uncertainties. In particular, we considered only a single result for each of the uncertain values, for example *rainy* weather as the weather condition. We also reduced uncertainty to a value determined as an average, for example the winning odds for the game of chance, WOD. In doing so, we fashioned the analysis to focus on various scenarios we *expected* to occur. On the face of it, this is not a bad approach for analysis. We have scenarios in which we can be relatively secure that the deterministic values represent what Fr. Efia will experience, conditional on the specific weather condition being investigated. This provides a simplified picture of the event and it can be quite useful in decision making, but in doing so, we may miss the richness of all the possible outcomes, due to the *condensation* of uncertainty that we have imposed.

What if we have a problem in which we desire a greater degree of accuracy, and a more complete view of possible outcomes? How can we create a model to allow simulation of such a problem and how do we conceptualize such a form of analysis? To answer these questions, let me remind you of something we discussed earlier in Chap. 6—sampling. As you recall, we use sampling when it is difficult, or impossible, to investigate every possible outcome in a population. If Fr. Efia had 12 uncertain elements in his problem and if each element had 10 possible outcomes, how many distinct outcomes are possible; that is, if we want to consider *all* combinations of the uncertain outcomes, how many will Fr. Efia face? The answer is 10^{12} ($10 \times 10 \times 10 \dots$ etc.) possible outcomes, which is a whopping 1 trillion (1,000,000,000,000). For complex problems, 12 elements that are uncertain with 10 or more possible outcome values each are not at all unusual. In fact, this is a relatively small problem. Determining 1 trillion distinct combinations of possible outcome values is a daunting task, and I further suggest that it may be impossible.

This is where sampling comes to our rescue. If we can perform carefully planned sampling, we can arrive at a reasonably good estimate of the variety of outcomes we face: not a complete view, but one that is useful and manageable. By this, I mean that we can determine enough outcomes to produce a reasonably complete profile of the entire set of outcomes. This profile will become one of our most important tools for analysis and decision making. We call it a **risk profile** of the problem outcomes, but more on this later. Now, how can we organize our efforts to accomplish efficient and accurate sampling?

8.3 The Monte Carlo Sampling Methodology

In the 1940s Stanislaw Ulam, working with famed mathematician John von Neumann and other scientists, formalized a methodology for arriving at approximate solutions to difficult quantitative problems which came to be called Monte Carlo methods. Monte Carlo methods are based on *stochastic processes*, or the study of mathematical probability and the resolution of uncertainty. The reference to Monte

Carlo is due to the games of chance that are common in the gambling establishments of Monte Carlo, in the Principality of Monaco. Ulam and his colleagues determined that by using statistical sampling and performing the sampling repeatedly, they were able to arrive at solutions to problems that would be impossible, or at a minimum, very difficult by standard analytical methods. For the types of complex simulations we are interested in performing in Chap. 8, this approach will be extremely useful. It will require knowledge of a number of well known probability distributions and an understanding of the use of random numbers. The probability distributions will be used to describe the behavior of the uncertain events, and the random numbers will become input for the functions generating sampling outcomes for the distributions.

8.3.1 Implementing Monte Carlo Simulation Methods

Now, let us consider the basics of Monte Carlo simulation (MCS) and how we will implement them in Excel. MCS relies on sampling through the generation of random events. The cell function which is absolutely essential to our discussion of MCS, **RAND()**, is contained in the *Math* and *Trig* functions of Excel. In the following, I present six steps that utilize the RAND() function to implement MCS models:

1. *Uncertain events are modeled by sampling from the distribution of the possible outcomes for each uncertain event.* A sample is the random selection of a value(s) from a distribution of outcomes, where the distribution specifies the outcomes that are possible and their related probabilities of occurrence. For example, the **random sampling** of a coin toss is an experiment where a coin is tossed a number of times (the sample size) and the distribution of the individual coin outcomes is heads with a 50% probability and tails with a 50% probability. Exhibit 8.1 shows the probability distribution for the fair (50% chance of head or tail) coin toss. If I toss the coin and record the *outcome*, this value is referred to as the **resolution of an uncertain event**—the coin toss. In a model where there are many uncertain events and many uncertain outcomes are possible for each event, the process is repeated for all relevant uncertain events and the resulting values are then used as a fair (random) representation of the model's behavior. Thus, these resolved uncertain events tell us what the condition of the model is at a point in time. Here is a simple example of how we can use sampling to provide information about a distribution of unknown outcomes. Imagine a very large bowl containing a distribution of one million colorful stones: 300,000 white, 300,000 blue, and 400,000 red. You, as an observer, do not know the number of each colorful stone contained in the bowl. Your task is to try to determine what the true distribution of colorful stones is in the very large bowl. Of course, we can use the colors to represent other outcomes. For example, each color could represent one of the three weather conditions of the *OLPS* problem in the previous chapter. We can physically perform the selection of a colorful stone by randomly reaching into the bowl and selecting a stone, or we can use a convenient analogy.

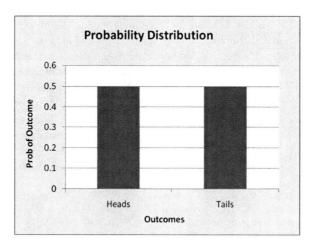

Exhibit 8.1 Probability distribution of a fair coin toss

The analogy will produce random outcomes of the uncertain events and can be extended to all the uncertain elements in the model.

2. *The RAND( ) function in Excel is the tool we will use to perform sampling in MCS.* By using RAND() we can create a *virtual* bowl from which to sample. The output of the RAND() function is a Continuous *Uniform* distribution, with output greater than or equal to 0 and less than 1; thus, numbers from 0.000000 to 0.999999 are possible values. The RAND() function results in up to sixteen digits to the right of the decimal point. A **Uniform distribution** is one where every outcome in the distribution has the same probability of being randomly selected. Therefore the sample outcome 0.831342 has exactly the same probability of being selected as the sample outcome of 0.212754. Another example of a Uniform distribution is our fair coin toss example, but in this case the outcomes are discrete (only heads or tails) and not continuous. Is the distribution of colorful stones a Uniform distribution? The answer is *no* since the blue stones have a higher probability of being selected in a random sample than white or red.

We now turn to a spreadsheet model of our sampling of colorful stones. This model will allow us to discuss some of the basic tenants of sampling. Exhibit 8.2 shows a table of 100 RAND() functions in cell range B2:K11. We will discuss these functions in greater detail in the next section, but for now, note that cell K3 contains a RAND() function that results in a randomly selected value of 0.2570. Likewise, every cell in the range B2:K11 is the RAND() function, and importantly, every cell has a different outcome. This a key characteristic of the RAND(): each time it is used in a cell, it is independent of other cells containing the RAND() function.

3. *How do we use the RAND() function to sample from the distribution of 30% Red, 30% White, and 40% Blue?* We've already stated that the RAND() function returns Uniformly distributed values from 0 up to, but not including, 1. So how do we use RAND() to model our bowl of colorful stones? In Exhibit

| K3 | | ▼ | | f_x | =RAND() | | | | | | | | | | |

	A	B	C	D	E	F	G	H	I	J	K	L	M	N	O	P
1		1	2	3	4	5	6	7	8	9	10		**Table for Various Sample Sizes**			
2	1	0.6896	0.9636	0.2993	0.4248	0.0139	0.6225	0.5296	0.7244	0.0008	0.6080		Sample Size	Red	White	Blue
3	2	0.3313	0.6392	0.7580	0.0673	0.6435	0.0929	0.5785	0.0029	0.7023	0.2570		10	3	2	5
4	3	0.5269	0.0427	0.9626	0.8250	0.4500	0.3197	0.1058	0.0580	0.1752	0.9426		%	30%	20%	50%
5	4	0.9130	0.4994	0.2788	0.6619	0.5619	0.4190	0.8554	0.6816	0.0550	0.8856		20	8	5	7
6	5	0.1023	0.4591	0.2630	0.4476	0.0278	0.6985	0.2361	0.2993	0.4342	0.2378		%	40%	25%	35%
7	6	0.6916	0.7518	0.9981	0.1630	0.9020	0.5683	0.0183	0.7129	0.5079	0.5843		30	10	9	11
8	7	0.0452	0.9726	0.0252	0.4482	0.5528	0.0342	0.5316	0.7395	0.8310	0.0988		%	33%	30%	37%
9	8	0.6035	0.3076	0.6143	0.2003	0.7761	0.5845	0.7719	0.7045	0.0603	0.4297		50	11	17	22
10	9	0.8971	0.9144	0.3351	0.3568	0.7021	0.8783	0.4424	0.9463	0.5020	0.3229		%	22%	34%	44%
11	10	0.4751	0.5628	0.7779	0.2022	0.0866	0.1036	0.7376	0.5732	0.6289	0.9555		100	30	30	40
12					**Random Numbers Table**								%	30%	30%	40%
13		1	2	3	4	5	6	7	8	9	10					
14	1	Blue	Blue	Red	White	Red	Blue	White	Blue	Red	Blue					
15	2	White	Blue	Blue	Red	Blue	Red	White	Red	Blue	Red					
16	3	White	Red	Blue	Blue	White	White	Red	Red	Red	Blue					
17	4	Blue	White	Red	Blue	White	White	Blue	Blue	Red	Blue					
18	5	Red	White	Red	White	Blue	Red	Red	White	Red						
19	6	Blue	Blue	Blue	Red	Blue	White	Red	Blue	White	White					
20	7	Red	Blue	Red	White	White	Red	White	Blue	Blue	Red					
21	8	Blue	White	Blue	Red	Blue	White	Blue	Blue	Red	White					
22	9	Blue	Blue	White	White	Blue	Blue	White	Blue	White	White					
23	10	White	White	Blue	Red	Red	Red	Blue	White	Blue	Blue					
24				**Translation of Random Numbers to Outcomes**												

Hector.Guerrero:
=COUNTIF(Results100,"Red")

Hector.Guerrero:
=IF(K3<0.3,"Red",IF(K3<0.6,"White","Blue")

Exhibit 8.2 RAND( ) function example

8.2 you can see two tables entitled *Random Numbers Table* and *Translation of Random Numbers to Outcomes*. Each cell location in *Random Numbers Table* has an equivalent location in the *Translation of Random Numbers to Outcomes*; for example, K15 is the equivalent of K3. Every cell in the translation table references the random numbers produced by RAND() in the random number table. An IF statement is used to compare the RAND( ) value in K3 with a set of values, and based on the comparison, the IF() assigns a color to a sample. The formula in cell K15 is—(=IF (K3<0.3, "Red", IF (K3<0.6,"White","Blue")). Thus, if the K3 value is less than 0.3 *Red* is returned. If the value in K3 is greater than 0.3, but less than 0.6, then *White* is returned. If neither of these conditions is satisfied, then *Blue* is returned. This logical function insures that 30% (0.3–0) of the randomly selected values are red; 30% (0.6–0.3=0.3) are white, and the remainder (1.0–[0.3+0.3]=0.4) are blue. Since K3 is 0.2570, the first condition is met and the value returned is the color *Red*. In the case of K2, the random value is 0.6080, and the value in the translation table is the last condition, *Blue*. Thus, values of 0.6000 up to, but not including, 1 will cause the return of *Blue*. The value 0.6080 meets this criterion. Incidentally, the cell colors in the *Translation* table are produced with conditional cell formatting.

Thus, if the distribution that we want to model is Discrete, as in the colorful stones example, we can simply partition the range of the RAND() proportionally, and then use a logical IF to determine the outcome. For example, if the proportion of Red, White, and Blue changes to 15%, 37%, and 48%, respectively, then the cell functions in the *Translation* table can easily be changed to reflect the new distribution—(=IF (K3<0.15, "Red", IF (K3<0.52, "White", "Blue")). Note that the second condition (K3<0.52) is the cumulative value of the first two probabilities (0.15 + 0.37 =0.52). If there are four possible discrete outcomes, then there will be a third cumulative value in a third nested IF in the translation table.

4. *We can use larger sample sizes, to achieve greater accuracy in outcomes.* In Exhibit 8.2 we also see a small table entitled *Table for Various Sample Sizes*. This table collects samples of various sizes (10, 20, 30, 50, and 100 observations) to show how the accuracy of an estimate of the population (the entire bowl) proportions generally increases as sample size increases. For example, for sample size 10, cells B14:K14 form the sample. This represents the top row of the translation table. As you can see, there are 3 red, 2 white, and 5 blue randomly selected colors. If we use this sample of 10 observations to make a statement about our belief about the distribution of colors, then we conclude that red is 30% (3/10), white is 20% (2/10), and blue is 50% (5/10). This is close, but not the true population color distribution.

What if we want a sample that will provide more accuracy; that is, a sample that is larger? In the table, a sample of 20 is made up of observations in B14:K14 and B15:K15. Of course, for any one sample, there is no guarantee that a larger sample will lead to greater precision, but if the samples are repeated and we average the outcomes, it is generally true that the averages for larger samples will converge to the population proportions of colors more quickly than smaller samples. It should also be intuitively evident that more data (larger sample sizes) leads to more information regarding our population proportions of colors. At one extreme, consider a sample that includes the entire population of 1 million colorful stones. The sample estimates of such a sample would estimate population proportions exactly—30% red, 30% white, and 40% blue. Under these extreme circumstances, we no longer have a sample; we now have a **census** of the entire population.

Note how the sample proportions in our example generally improve as sample size increases, although the sample size of 50 yields proportion estimates that are less accurate than the sample size of 30. This can and does occur, but in general, we will see better estimates with higher sample sizes. The sample size of 100 results in the exact values of the color proportions. Another sample of 100 might not lead to such results, but you can be assured that a sample size of 100 is usually better than a sample of 10, 20, 30, or 50.

To generate 100 new values of RAND() we recalculate the spreadsheet by pressing the F9 function key on your keyboard. This procedure generates new RAND() values each time F9 is depressed. Alternatively, you can use Calculation Group in the Formulas Ribbon. In Options, a tab entitled Calculation permits you to recalculate and also control the automatic recalculation of the spreadsheet. See Exhibit 8.3. You will find that each time a value or formula is placed in a cell location, all RAND() cell formulas will be recalculated if the Automatic button is selected in the Calculation Options subgroup. As you are developing models, it is generally wise to set Calculation to Manual. This eliminates the repeated, and

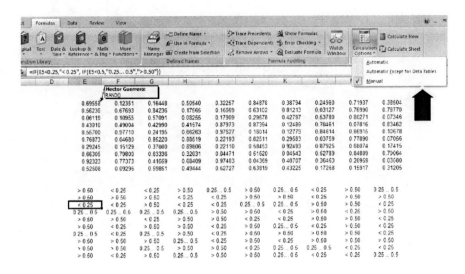

Exhibit 8.3 Control of worksheet recalculation

often annoying, recalculation of your worksheet that occurs as you are entering each cell value.

5. *Why do we need to perform many replications of an experiment?* In simulation, we refer to the repeated sampling of uncertain events as **replications**. The term refers to the replication, or repetition, of an experiment, with each experiment representing a single resolution of the uncertainties of the problem. For Fr. Efia, an experiment represents a single day of operation of his event, and each replication of the experiment results in observed daily, total revenue. In a complex problem, many individual uncertain elements will be combined to produce a complex distribution of a composite outcome value. The more experiments we conduct, the closer we approach the true behavior of complex models. In fact, in many cases it is impossible to understand what the combined distribution of results might be without resorting to large numbers of replications. The resulting distribution is the risk profile that the decision maker faces, and it becomes a tool for decision making. If we perform too few replications the risk profile is likely to be inaccurate. For example, in Exhibit 8.4 you can see the graphical results which we produced in Exhibit 8.2. The outcomes are attached to the graph in a data table below the graph. Although risk profiles are often associated with graphs that indicate the probability of some monetary result, the results in Exhibit 8.4 represents observed distribution of colors. Thus, the risk profile for the sample size of 10, the first of five, is 30% red, 20% white, and 50% blue. There are 5 risk profiles in the exhibit (sample size 10, 20, 30, 50, and 100), as well as the *Actual* distribution, which is provided for comparison. This exercise provides two important take-always: (a) an initial introduction to a *risk profile*, and (b) a demonstration of the value of larger sample sizes in estimating true population parameters, like proportions.

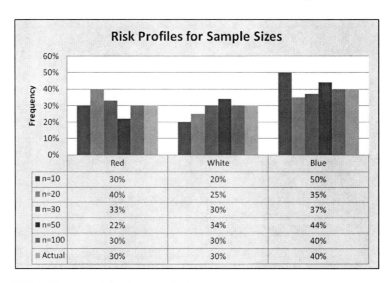

Exhibit 8.4 Risk profiles for various sample sizes

6. *In summary, the Monte Carlo methodology for simulation, as we will implement it, requires the following*: (a) develop a complete definition of the problem, (b) determine the elements of the model that are uncertain and the nature of the uncertainty in terms of a probability distribution that represents its behavior, (c) implement the uncertain elements by using the RAND() or other Excel functions, (d) replicate a number of experiments sufficient in size to capture accurate behavior, (e) collect data from experiments, (f) present the risk profiles resulting from the experiments, (g) perform sensitivity analysis on results, and (h) make the appropriate decisions based on results and the decision maker's attitude toward risk.

These steps represent a systematic approach for modeling processes and conducting simulation experiments.

At this point, it is wise for us to turn to a discussion of probability distributions (probability density functions for Continuous distributions) in a bit more detail. Since it will be of utmost importance that we incorporate uncertainty into our MCS, we will need a basic understanding of how we specify uncertain events. We will introduce the basics of a Poisson Arrival process in the next section, but this is just one way to deal with arrival uncertainty. There are many other ways to describe uncertain arrivals. In the discussion that follows, we will consider some commonly used probability distributions and density functions.

8.3.2 A Word About Probability Distributions

Obviously, we could devote an entire book to this topic, but in lieu of a detailed discussion, there are a number of issues that are essential to understand. First, there are three basic ways we can model uncertain behavior with probability distributions: theoretical distributions, expert opinion, or empirically determined distributions. Here are a few important characteristics to consider about distributions:

1. *Discrete Distributions*—Recall that distributions can be classified as either Discrete or Continuous. **Discrete distributions** permit outcomes for a discrete, or countable number of values. Thus, there will be gaps in the outcomes. For example, the outcomes of arrival of patients at an emergency room hospital during an hour of operation are discrete; they are a *countable* number of individuals. We can ask questions about the probability of a single outcome value, 4 patients for example, in a Discrete distribution, as well is the probability of a range of values, between 4 and 8 customers. The Poisson is a very important Discrete distribution that we will discuss later in one of our examples. It is restricted to having integer values.

2. *Continuous Probability Distributions or Probability Density functions*— **Continuous distributions** permit outcomes that are continuous over some range. *Probability density functions* allow us to describe the probability of events occurring in terms of ranges of numerical outcomes. For example, the probability of an outcome having values from 4.3 to 6.5 is a legitimate question to ask of a Continuous distribution. But, it is not possible to find the probability of a point value in a Continuous distribution. Thus, we *cannot* ask— what is the probability of the outcome 5 in a Continuous distribution? It is undefined.

3. *Discrete and Continuous Uniform Distribution*—Exhibits 8.5 and 8.6 show Discrete and Continuous Uniform distributions, respectively. As you can see in Exhibit 8.5, the probability of the outcome 7 is 0.2. In the case of the Continuous Uniform in Exhibit 8.6, we have a distribution from the outcome range of values 4 to 8, and the distribution is expressed as a *probability density function*. This relates to our discussion in 2, above. The total area under the Continuous distribution curve is equal to 1. Thus, to find the probability of a range of values, say the range 4 to 6, we find the proportion of the area under the distribution that

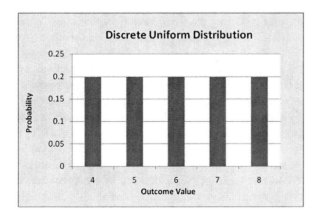

Exhibit 8.5   Discrete uniform distribution example

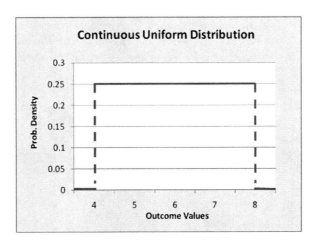

Exhibit 8.6   Continuous uniform distribution example

is implied by the range. In this case, the range 4 to 6 covers 2 units of a total interval of 4 (8-4). The area between each successive integer value (4 to 5, 5 to 6, etc.) represents 25% of the area of the entire distribution. Thus, we have 50% of the area under the curve ([6–4] * 0.25 = 0.5). The density is calculated as the inverse of the difference between the low and high values of the outcome range (1/[high range value – low range value]), in our case 0.25 (1/[8–4]). Although the Uniform describes many phenomena, there is one particular use of the Uniform that is very interesting and useful. It is often the case that a decision maker simply has no idea of the relative frequency of one outcome versus another. In this case, the decision maker may say the following—I just don't know how outcomes will behave relative to each other. This is when we resort to the Uniform to deal with our *total* lack of specific knowledge and attempt to be *fair* about our statement of relative frequency. For example, consider a family reunion to which I invite 100 relatives. I receive regrets (definitely will not attend) from 25, but I have *no* idea about the attendance of the 75 remaining relatives. A Discrete Uniform is a good choice for modeling the 0 to 75 relatives that might attend; any number of attendees from 0 to 75 is equally likely since we do not have evidence to the contrary.

4. *Specification of a distribution*—Distributions are specified by one or more parameters; that is, we provide a parameter or set of parameters to describe the specific form the distribution takes on. Some distributions, like the Normal, are described by a location parameter, the mean (Greek letter μ), and a dispersion parameter, the standard deviation (Greek letter σ). In the case of the attendees to our family reunion, the Uniform distribution is specified by a lower and upper value for the range, 0 to 75. The Poisson distribution has a single parameter, the **average arrival rate**, and the rate is denoted by the Greek letter λ. Exhibit 8.7 shows a Poisson distribution with a $\lambda = 5$. Note that the probability of obtaining an outcome of 7 arrivals is slightly greater than 0.1, and the probabilities of either

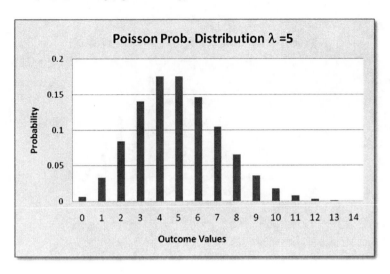

Exhibit 8.7   Poisson probability distribution with $\lambda = 5$

4 or 5 are equal, approximately 0.175 each. This must be taken in the context of an average arrival rate of $\lambda = 5$. It makes sense that if the average arrival rate is 5, the values near 5 will have a higher probability than those that are distant, for example 11, which has a probability of less than 0.01. Thus, the further away an outcome is from the average arrival rate, the smaller the probability of its occurrence for the Poisson.

5. *Similarity in distributions*—Many distributions often have another distribution for which they possess some similarity. Other distributions often represent a *family* of distributions. The Beta distribution family is one such Continuous distribution. In fact, the Uniform is a member of the Beta family. The Poisson is closely related to the Binomial distribution. Like the Poisson, the Binomial is also a Discrete distribution and provides the probability of a specific number of successes, k, in n trials of an experiment that has only two outcomes, and where the probability of success for an individual trial (experiment) is p. Thus, we could ask the probability question—what is the probability that I will get 9 heads (k) in 20 tosses (n) of a coin, where the probability of a head is 0.5 (p). In Exhibit 8.8 we see that the probability is approximately 0.12. In situations where the results of a trial can only be one of two outcomes, the Binomial is a very useful discrete distribution.

6. *Other important characteristics of distributions*—

   a. *Shape*. Most of the distributions we have discussed, thus far, are *unimodal*, that is, they possess a single maximum value. Only the Uniform is *not* unimodal. It is also possible to have bimodal, trimodal, and multimodal distributions, where the modes need not be equal, but are merely localized maximum values. For example, in Exhibit 8.9 we see a bimodal distribution where one local maximum is the outcome 4 and the other outcome is 12. This

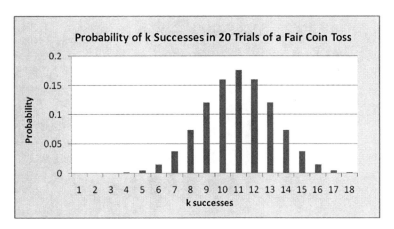

Exhibit 8.8 Binomial probability distribution example

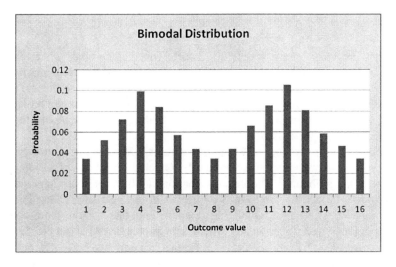

Exhibit 8.9 Bimodal distribution

terminology uses the definition of the mode (the maximum occurring value) loosely, but it is generally accepted that a bimodal distribution need not have equal probability for local modes. Additionally, some distributions are symmetrical in shape (Normal or Uniform), while others need not be (Poisson, Beta, or Binomial).

  b. *Behavior of uncertain events.* Often, we must assume particular behavior in order to justify using certain distributions. For example, in a Poisson arrival process, the arrivals are defined to be independent of one another, they occur at random, and the average number of events over a unit of time (or space) is constant. (More on the Poisson in the next section.) Even when we *bend* the strict nature of these conditions a bit, the approximation of a Poisson arrival process can be quite good.

c. *Empirically based distributions*. To this point our discussion has been about theoretically based distributions. In these distributions we assume a theoretical model of uncertainty, such as a Poisson or Normal distribution, but often decision makers can collect and record empirical data and develop distributions based on this observed behavior. Our distribution of colorful stones is such a case. We do not assume a theoretical model of the distribution of colors; we have collected empirical data, in our case, through sampling, which leads us to a particular distribution.

In the following section we will concentrate on an introduction to the Poisson distribution. It will be somewhat complex to create an approach that will allow Poisson arrivals.

8.3.3 Modeling Arrivals with the Poisson Distribution

Earlier, I mentioned that *arrivals* in simulations are often modeled with a Poisson distribution. When we employ the Poisson to describe arrivals, we refer to this as a **Poisson Arrival Process**. The arrivals can be any physical entity, for example, autos seeking service at an auto repair facility, bank clients at an ATM, etc. Arrivals can also be more abstract, like *failures* or *flaws* in a carpet manufacturing process or questions at a customer help desk. The arrivals can occur in time (auto arrivals during the day) or in space (location of manufacturing flaws on a large area of household carpet). So how do we sample from a Poisson distribution to produce arrivals?

Exhibit 8.10 serves as an example of how we will manage the arrival data for the Autohaus simulation we will perform later in the chapter; an advanced modeling example that is essentially a discrete event simulation. Describing a process as having Poisson arrivals requires that a number of assumptions should be maintained:

1. Probability of observing a single event over a small interval of time or within a small space is proportional to the length of time or area of the interval
2. Probability of simultaneous events occurring in time and space is virtually zero
3. Probability of an event is the same for all intervals of time and in all areas of space
4. Events are independent of one another, so the occurrence of an event does not affect another

Although it may be difficult to adhere to all these assumptions for a particular system, the Poisson's usefulness is apparent by its widespread use. For our purposes, the distribution will work quite well.

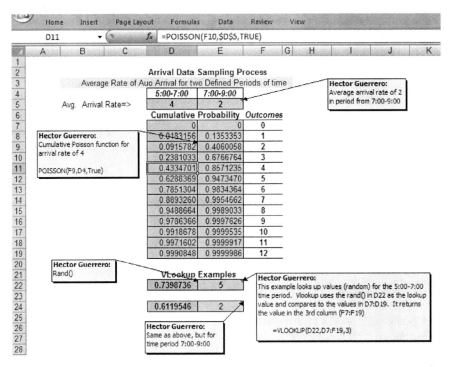

Exhibit 8.10 Auto arrival sampling area of brain

Now, let us examine in detail how we simulate Poisson arrivals. Recall how we introduced the RAND() function in Exhibit 8.2. We generated uniformly distributed random numbers and then assigned an outcome value (a color) to the random number depending on its value. If the random number value was between 0 and 0.3, a Red was returned; if between 0.3 and 0.6, a White was returned; if between 0.6 and 1.0 a Blue was returned. We will use the table of *cumulative* Poisson probability values in a similar fashion. Although we did not mention the cumulative nature of the comparison values before, the IF() used cumulative values to determine the colors for our sampling. To build a cumulative Poisson probability table, we will use the internal Excel cell function, **POISSON(x, mean, cumulative)**. Then we will use the table as the basis for sampling the number of arrivals for a unit of time. The arguments of the function, *x* and *mean*, are values that the user provides. By placing the term *true* in the third argument of the function, *cumulative*, a cumulative value will be returned; that is, the value x = 3 will return the probability of 0, 1, 2, and 3 arrivals for the Poisson distribution.

Now, consider the table in Exhibit 8.10 associated with the average arrival rate, λ, of 4 (in column D). In this exhibit we consider the arrival of autos at a repair facility in two distinct time periods: 5:00–7:00 and 7:00–9:00. The arrival rate of the later time period is 2; thus, on average more cars arrive earlier rather than later. Beginning in cell D7 and continuing to D19, the table represents the successive cumulative probabilities for a Poisson distribution with an average arrival rate of 4 (cell D5 value). Thus, the arithmetic *difference* between successive probability values represents the probability of obtaining a specific value. For example, the difference between 0.0183156 and 0.0915782 is 0.0732626, the probability of exactly 1 arrival for the Poisson distribution with an average arrival rate of 4 per unit of time. Similarly, the difference between 0 and 0.0183156 is 0.0183156, which is the probability of an outcome of exactly 0 arrivals. The numbers in cell F7 to F19 are the number of arrivals that will be returned by a lookup process of random sampling. Next we will discuss the details of the process of sampling through the use of lookup functions.

8.3.4   VLOOKUP and HLOOKUP Functions

To demonstrate how we can use the table to sample the number of random arrivals in a time period, let us turn our attention to cells E22 and E24 in Exhibit 8.10. These two cells form the heart of the sampling process and rely on the *vertical* lookup function, *VLOOKUP(value_lookup, table_array, col_index_num)*. To understand the use of the **VLOOKUP** and its closely related partner, **HLOOKUP** (horizontal lookup), we introduce Exhibit 8.11.

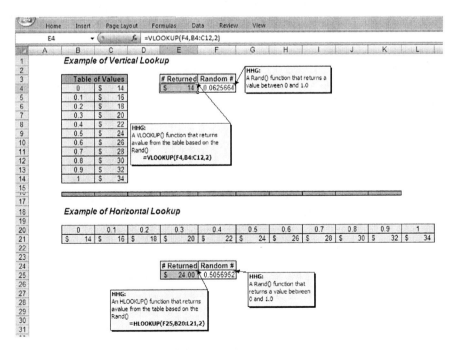

Exhibit 8.11   VLOOKUP and HLOOKUP example

The VLOOKUP and HLOOKUP are a convenient way to return a value from a table based on a comparison to another value. Except for the obvious difference in vertical and horizontal orientation, there is no difference in the cell functions. Consider the utility of such a function. In Chap. 4 we used the IF() to convert a dollar value into a *payment category*. In that example, we noted that depending on the number of categories, we might have more categories than the maximum allowed number of *nested* IF()s; thus, under these circumstances a lookup function should be used.

Now, let us take a look at the examples in Exhibit 8.11, lookups that convert a fractional value from 0 to 1 into a numerical dollar value. We will concentrate on the VLOOKUP due to the vertical nature of the Poisson cumulative probability table we used earlier. The function has three important arguments: *value_lookup*, *table_array*, and *col_index_num*. The lookup requires that a table be constructed with two types of values—a lookup value and a table value. The lookup value used in the example is a RAND() and represents a random sampling process, and this value is to be converted by the table to some associated table value.

We begin by generating a RAND(). It is compared to the leftmost column of a vertical table or the top row of a horizontal table. In Exhibit 8.11 a lookup value is in cell F4 and the table is located in B4:C12. The random number generated in the vertical example is 0.0625664. The last argument in the lookup function is the column index number which is to be returned, in this case 2, which will return a value in the 2nd column of the table. In simple two column tables, this number is always 2. If a table has more that two columns to permit return of other associated values, or to combine several tables into one, then a column number must be chosen to represent the table value of interest. Finally, the function *takes* the value lookup (0.0625664) and finds the region in the table that contains the value. For our example, 0.0625664 is found between 0 (B4) and 0.1 (B5). The convention used for returning a value from the 2nd column is to return the value associated with the topmost value (0) in the range. In this case, the return is $14 which is adjacent to 0. If the lookup value from the RAND() function had been exactly 0.1, the value returned would have been $16.

Now, let's return to our Poisson table example in Exhibit 8.10 to describe how we perform random sampling of a Poisson distribution. The VLOOKUP in cell E22 compares the value of a random number, D22, with the table of cumulative probability numbers. It then returns a value from 0-12 depending of the value of the RAND(). For example, the random number 0.7398736 is generated in cell D22 and the VLOOKUP searches values in D7 through D19. When it encounters a value higher than the random number, in this case 0.7851304, it returns the value in column F in the row *above* 0.7851304. The number returned, 5, is in the 3rd column of the table. If we repeat the process many, many times, the recorded values will have a frequency distribution that is approximately Poisson distributed, which is, of course, what we are attempting to achieve. Note that we can use the same approach for any Discrete distribution: (1) create a cumulative probability distribution and, (2) sample using a RAND() function to determine a randomly selected value from the distribution. This simple mechanism is a fundamental tool for generating uncertain events from a distribution.

8.4 A Financial Example—Income Statement

Not all simulations are related to discrete events. Let us consider a simple financial example that simulates the behavior of an income statement. I emphasize simple, because income statements can be quite complex. Our purpose in this exercise is to demonstrate the variety of simulation that is possible with the MCS method.

Table 8.1 shows the components of a typical income (profit or loss) statement and commentary on the assumptions that will be used to model each component. I have selected a variety of distributions to use in this example, including the **Normal distribution**, often referred to as the **Bell Curve** due to its shape. It is a very commonly used and convenient distribution that has a remarkable range of applications. It is distinguished by its symmetry, a central tendency ("peaked-ness"), and probabilities for lower and higher values that extend to infinity, but with very, very low probability of occurrence. The symmetry and central tendency of the Normal probability distribution (density function) is particularly useful to modelers. Such observable variables as individual's weight, shoe size, and many others, are often modeled with Normal distributions, although the infinitely extending low and high values could lead to a foot size that is miniscule or one that fills a soccer stadium. Fortunately, these extreme values have very, very low probability of occurrence.

A Normal distribution is described by two parameters—mean and standard deviation (or variance). The formula for generating a value from a Normal distribution in Excel is **NORMINV**(RAND(), mean, standard deviation). Note that the function uses the familiar RAND() function as its first argument. In our financial example,

Table 8.1 Income statement example data

Component	Assumptions
Sales Revenue	Sales Revenue = Units* Unit Price
	Distributions
	Discrete Units: 30%–75,000 units ; 70%–10,000 units
	Discrete Unit Price: 25%–$1.50 ; 50%–$2.00; 25%–$2.50
Cost of Goods Sold Expense	COGS = Percentage* Sales Revenue
(COGS)	Distribution
	Percentage = Normal Distribution—Mean–30 ; Std Dev–5
Gross Margin	= Sales Revenue-COGS
Variable Operating Expense	VOE = Sales Revenue * Percentage
(VOE)	Distribution
	Percentage = Continuous Uniform Distribution—10%–20%
Contribution Margin	= Gross Margin – VOE
Fixed Expenses (FE)	A constant value of $6000
Operating	= Contribution Margin – Fixed Expenses
Earnings (EBIT)	
Interest Expense (IE)	A constant value of $3000
Earnings before	= EBIT-Interest Expense
Income Tax	
Income Tax expense	Conditional Percentage of [EBIT-Interest Expense]
	35% < $20,000; 55% >=$20,000
Net Income (Profit)	= Earnings before Income Tax – Income Tax

the percentage of revenue used to calculate the cost of goods sold, COGS, is determined by a Normal distribution with mean 30 and standard deviation 5. See the COGS Expense in Table 8.1 for detail. It would be extremely rare that a value of 10 or 50 would be returned by the function, since this represents values that are 4 standard deviations below and above the mean. This is noteworthy since it is sometimes possible to return negative values for the Normal if the standard deviation is large relative to the mean; for example, a mean of 30 and standard deviation of 15 could easily return a negative randomly sampled value, since values 2 standard deviations below the mean are not difficult to obtain. If a Normal distribution is used with these types of parameter values, the modeler must introduce a function to truncate negative values if they are nonsensical. For example, a negative weight or height would be such a circumstance.

Now, let us examine the results of our simulation. I have placed the *Brain* on the same worksheet as the *Calculations*, and Exhibit 8.12 shows the general structure of the combined *Brain* and *Calculation* worksheets. The *Brain* contains the important values for sampling and also the constant values used in the calculations. In the lower part of Exhibit 8.12 you can see that the structure of the Profit or Loss statement is translated horizontally, with each row representing an experimental observation of the statement. Of the 13 observations visible for the model, only

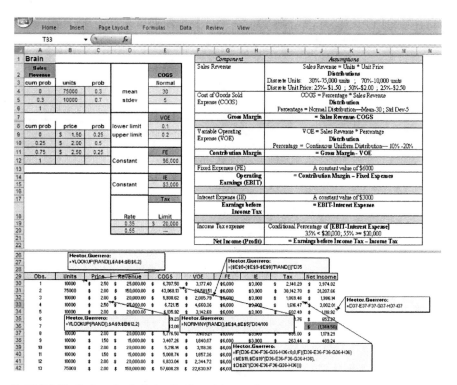

Exhibit 8.12 Brain and calculation for financial example

one, Obs. 7, results in a loss (–$1,348.50), and the others result in a profit. The advantages of this structure are: (1) there are many more rows visible on a worksheet than columns, and (2) since there are many more rows on a worksheet than columns, many more observations are possible. In this model, I will generate 500 observations of the Profit or Loss statement.

The calculations for this model are relatively straightforward. The calculations for *Units* and *Price*, which determine *Revenue*, are VLOOKUPs that sample from the *Sales Revenue* (cell A2) section of the *Brain* and are shown in Exhibit 8.12. The calculation of the *COGS* uses the NORMINV function discussed above, to determine a randomly sampled percentage. Variable Operating Expense, *VOE*, uses the values for lower and upper limits (10% and 20%) through a linear transformation to return continuous uniformly distributed values:

$$lower\ limit + (upper\ limit\text{-}lower\ limit) * RAND()$$

The formula leads to continuous values between the lower and upper limit, since RAND() takes on values from 0 to 1. For example, if RAND( ) is equal to the extreme upper value 1, then the value of the expression is simply the *upperlimit*. Conversely, if RAND() is equal to the extreme lower value 0, then the expression is equal to the *lower limit*.

In Exhibit 8.13 we see summary statistics (cell range I530:J535) for the simulation of 500 observations. The average profit or loss is positive, leading to a profit

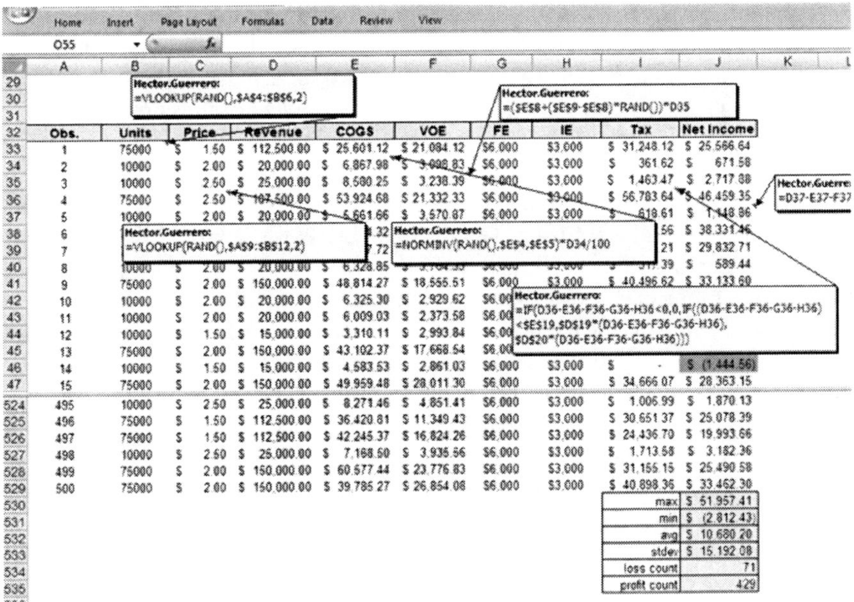

Exhibit 8.13 Summary statistics for financial model

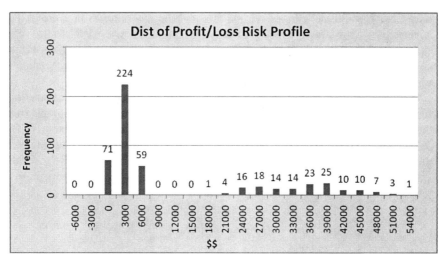

Exhibit 8.14 Risk profile for profit/loss statement

of $10,680.20 and a very substantial standard deviation of $15,192.08. The coefficient of variation for the 500 observations is greater than one (15,192.08/10,680.20). Thus, our model shows a very wide range of possible values with a max of $51,957.41 and a min of –$2,812.43. Also, note that 71 (cell J534) of 500 observations are losses, or 14.2% of the experimental total outcomes. This is valuable information, but the risk associated with this model is even more clearly represented by the risk profile in Exhibit 8.14. In this exhibit we see clearly a picture of the possible outcomes of the model. This is a classic risk profile, in that it presents the range of possible monetary outcomes and their relative frequency of occurrence.

Consider the calculations associated with row 33 of the worksheet in Exhibit 8.13, the first observation. The number of units produced is 75,000 and the price per unit is $1.50, both resulting from VLOOKUPs involving a random sampling of Discrete distributions. Cost of Goods Sold (COGS), $25,601.12, is a percentage of revenue determined by a Normal distribution outcome value divided by 100. The variable operating expense (VOE) is determined by sampling a Uniform distribution ($21,084.12). Finally, the fixed and interest expense (FE and IE) are constant values of $6,000 and $3,000, respectively. The calculation of net income is the summation of all expenses subtracted from revenue and results in a profit of $25,566.64.

Some of the results of the model are a bit unexpected. For example, the risk profile suggests a bi-modal distribution, one that is related to low or negative profits and has a relatively tight dispersion. The second mode is associated with higher profits and is more widely dispersed. This phenomenon is probably due to the distribution of demand, 70% probability of 10,000 units of demand and 30% of 75,000. In fact, the observations associated with the lower mode represent about 71% of the observations ([71+224+59]/500 = 70.8%).

8.5 An Operations Example—Autohaus

Let us now put into practice what we have learned about MCS methods on a more complex problem. We begin with item *6a* and *6b* from the *Implementing Monte Carlo Simulation Methods* subsection: develop a complete model definition of the problem (6a) and determine the uncertainty associated with model behavior (6b). Consider an example that models an automobile diagnostic service. Your sister, Inez, has decided to leave her very lucrative management consulting practice in business process modeling for a nobler calling—providing auto repair services for the automobile driving masses. She has purchased Autohaus, a well respected auto repair shop located in a small industrial park in her community, and she wants to expand the Autohaus service options to operators of large auto fleets and light-duty truck fleets—e.g. the local police department, the US Mail service, and corporate clients with auto fleets. The advantage that she foresees in this business model is a steady and predictable flow of contract demand. Additionally, Inez wants to limit the new services Autohaus provides to *engine/electrical* and *mechanical* diagnostics, and *oil change* service. Diagnostic service does not perform repairs, but it does suggest that repair may be necessary or that maintenance is needed. It provides fleet operators with a preventative maintenance program that can avoid costly failures. Inez wants to perform an analysis for the new services she is contemplating.

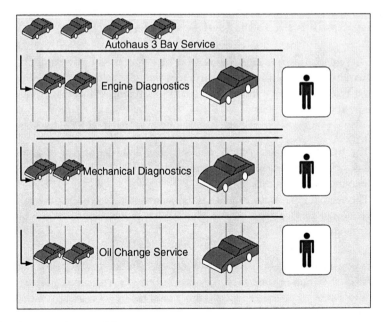

Exhibit 8.15 Autohaus model

The service facility she is planning has 3 service bays, each staffed by a single mechanic that works exclusively in each bay (see Exhibit 8.15). There is a large parking lot associated with the facility that can accommodate a very large number of parked vehicles. Inez has decided that the fleet operators will be required to deliver their autos to the lot every morning before the shop opens at 9:00 am. The autos are parked in the next available slot for the type of service required and then service is performed on a first-come-first-served basis. Although the hours of operation are 9:00 am to 7:00 pm, if at the end of the day an automobile has service started before 7:00 pm, service will be completed on that auto. Also, the autos need to be removed from the lot by fleet owners by the close of business; Inez wants to avoid the legal liability of insuring that the autos are safe overnight. As we have already stated, there are three types of service that are handled at Autohaus—*engine/electrical diagnostics*, *mechanical diagnostics*, and *oil change* service.

Wolfgang, her trusted head mechanic, has many years of experience, and he has kept mental records regarding the time that each service demand type requires. As a former modeling and simulation professional, Inez will find her expertise to be quite valuable for constructing a model of Autohaus. It is Inez's goal to understand the general behavior of this new business model in terms of the various types of service demand Autohaus will experience and the operations design that she is contemplating. This is a very reasonable goal if she is to organize the business for the highest possible return on investment. Inez decides that the first step in the assessment of the business model is to simulate the operation. She decides to have a conversation with Wolfgang to better understand the current business, and to get his input on important simulation design issues. The following are questions she wants to discuss with Wolfgang to assist in step 6a and 6b of the simulation process:

1. Are the general arrival rates (autos/hour) for the arrival of customers seeking service at Autohaus different for different hours? Wolfgang has noted that the arrival rate for the early time period, 5:00-7:00 am, is different from the later time period, 7:00–9:00 am.
2. She also wants to understand the uncertainty associated with the business model.

   a. The first type of uncertainty is associated with the arrival of autos. When and how many autos arrive and enter the facility? What type of services do arriving autos request?
   b. The second type of uncertainty is associated with service provision. Will the requested service type be available? What is the service time required for each auto arrival?

After considerable discussion with Wolfgang, she arrives at a flow diagram of the process, shown in Exhibit 8.16. The process begins with autos arriving *prior* to the facility's start of operation at 9:00 am. (This assumption greatly simplifies the operation of the simulation model as we will see in Table 8.2). The process flow elements in diamonds represent uncertainty. The *Autos Arrive with...* process element indicates that autos arrive in unknown quantities and within one of two contiguous

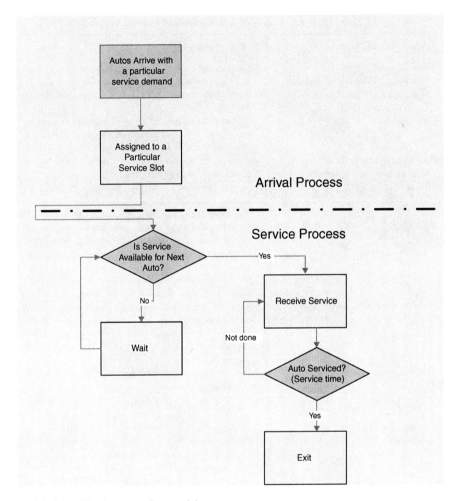

Exhibit 8.16   Simple process flow model

time horizons, 5:00–7:00 am or 7:00–9:00 am. We distinguish between these two
time horizons because they a have different arrival rate behavior. In general, more
demand arrives in the early period than the later due to traffic congestion issues
becoming more acute later in the morning. Next, the type of service demand that is
requested by an auto must also be determined. It will be restricted to one type of
service per auto, which is another simplifying assumption.

Once we have information on the number of autos that have arrived and the service type they have request, the model will direct the auto to one of three **queues**
(waiting lines) for each type of service—engine/electrical diagnostics, mechanical diagnostics, and oil change. This occurs in the process element *Assigned to
Particular Service Slot*. At this point, we exit the *Arrival Process* boundary and
enter the *Service Process*. The next step, although it might appear to be simple, is

Table 8.2 Details of Autohaus diagnosis model

Arrival process issue	Resulting model structure
When do autos arrive? 1) Strictly before the official start of service (9:00 am)? 2) All day long (5:00 am–7:00 pm)?	Review of the choices available: 1) This is the simplest choice to deal with as a simulation modeler, but maybe not very customer friendly. 2) This is a much more complex modeling situation. *Our choice—1) This assures that the days demand is available prior to starting service.*
How will the randomly arriving autos be assigned to service queues?	As an auto arrives, its demand for service must be determined, and only one type of service will be assigned per auto. Autos will be placed in one of 3 queues, each with a particular service demand type (Engine/electrical Diagnostics, etc.). The autos will be served according to a first-come-first-served service discipline. The distribution of arriving autos will be based on a *Poisson* (more on this later) distribution. *Our choice—Three service queues with first-come-first- served service discipline; random arrivals of autos; random service types.*
What happens to autos not served on a particular day?	This is also a relatively simple question to answer. They must leave the facility by close of business since they are there for diagnostic service only. This also eliminates the need to track vehicles that already in queues before the morning arrivals. *Our Choice—Cars are cleared from queues at the end of the day.*
Service process issues	**Resulting model structure**
How will service be initialized each day?	A queue discipline of first-come-first-served suggests that this is how service will be administered for the 3 types of service. Thus, we must keep track of vehicles in queues. These queues will also represent the 3 bays where a single mechanic is stationed. *Our Choice—In accordance to first-come-first served, the auto at the head of the queue will receive service next.*
How will service times be determined: empirical data or subjective opinion?	Empirical data is data that is recorded and retained over time. Subjective opinion comes from experts that supply their opinion. Wolfgang has a very good sense of the time required to perform the three service types. *Our Choice—In the absence of empirical data, we will use Wolfgang's subjective (expert) opinion.*

complex. The bays that are performing various types of service begin to operate on the available queues of autos. This requires that autos requesting a particular type of service be available (demand) and that service capacity (supply) for the service also be available, as noted in the diamond *Is Service Available for Next Auto?* As the mechanic in the bay administers service, the uncertain service time is eventually resolved for each auto, as shown in the process element *Auto Serviced?* When the simulation starts, the model will be operated for a predetermined period of time, simulating a number of days of operation, during which it will collect data on the servicing of autos.

How will we apply the MCS method to our problem; that is, how do we execute the steps *6c- 6f*? Although we have not yet specified every detail of Inez's problem, we now have a general idea of how we will structure the simulation. We can take the two processes described in Exhibit 8.16, simulate each, and use the results of the *Arrival Process* as input for the *ServiceProcess*. The steps for determining the remaining details of a model can be arduous, and there are numerous details that still need to be determined. Through interviews with Wolfgang, and through her own research, Inez must identify, to the best of her ability, the remaining details.

Table 8.2 is a compilation of some detail issues that remain for the model. This table takes us one step closer to the creation of an Excel simulation model by detailing the arrival and service processes. In the *Arrival* process we decide that arrival of demand is complete by 9:00 am. Permitting arrivals all day will make the model much more complex. Certainly model complexity should not suggest how she designs her service, but given that this business model will have *contract* demand as opposed to *drop-in* demand, she has incentives to manage demand to suit her capacity planning. Additionally, this procedure may suit the customer well by providing all customers with a structured approach for requesting service.

In the *Service* portion of the model, the arriving autos will receive a random assignment of a service type. Although the assignment is random, it will be based on an anticipated distribution of service type. Then service will be administered in a first-come-first-served manner at the corresponding bay, as long as capacity is available. Finally, we will use Wolfgang's subjective opinion on the service time distributions for the three services to consume service capacity at the bays.

We have resolved most of the issues that are necessary to begin constructing an Excel model. As usual, we should consider the basic layout of the workbook. I suggest that we provide for the following predictable worksheets: (1) introduction to the problem or *table of contents*, (2) a *brain* containing important data and parameters, (3) a *calculation* page for the generation of basic uncertain events of the simulation, (4) a *data collection* page for collecting data generated by the logic of the simulation, and (5) a *graph* page to display results.

8.5.1 Status of Autohaus Model

Now, let us take stock of where we are and add the final detail to our model of operation for Autohaus:

1. We have developed a sampling mechanism to determine the random Poisson arrivals of autos. Additionally, we have a similar discrete sampling mechanism for assigning the service type that is requested by each arrival.

2. The relative proportion of the type of service that is assigned to autos will remain constant over time and it is applied to autos arriving as follows: 40% of autos arrivals will request service type *Engine/electrical Diagnosis*; 25% will request *Mechanical Diagnosis*; and 35% will request *Oil Change*.

3. We will assume that the portion of the parking lot that is allocated to the new business is sufficiently large to handle any days demand without causing autos to **balk** (arrive and then leave). By not restricting the capacity of the parking lot, we reduce the complexity of the model, but we also eliminate the possibility of using the model to answer various questions. For example, we may want to consider how we would utilize a limited and costly lot capacity among the various services Inez provides. A *variable* lot capacity is clearly a more sophisticated modeling condition than we are currently considering.

4. Balancing the demand for service with the supply of service will be an important issue for Inez to consider. Customer demand that is greatly in excess of service supply can lead to customers spending inordinate amounts of time waiting and subsequently the potential loss of customers. Conversely, if demand is far less than supply, then the costs of operation can easily exceed the revenue generated leading to low profits or losses.

As we consider these issues, we can construct more flexible and sophisticated models, but at the cost of greater modeling complexity. The decision of how much complexity is needed should be made in light of Inez's goals for the simulation analysis.

8.5.2 Building the Brain Worksheet

Exhibit 8.17, the Brain worksheet, shows the results of our discussion of the arrival process. The *Brain* contains all the pertinent parameters and data to later supply our *Calculation* and *Data Collection* worksheets. Note that the worksheet has 5 major categories of information: *Arrival Data-Cumulative Probability Distribution, Selection of Arrival Order, Type of Service, Service Times Distributions,* and *Worker Assumptions.* As in Exhibit 8.10, 8.17 contains a table of calculations based on the cumulative Poisson distribution for the three customer categories: Corporate Client, Police, and US Mail. The arrival period is divided into two distinct periods of arrival—5:00–7:00 a.m. and 7:00–9:00 a.m. I have assumed Poisson average *hourly* arrival rates in the two time periods of [2, 2], [2, 1], and [1, 1] for Corporate Clients, Police, and US Mail, respectively. For example, cells B5 and C5 show the average arrival rates for *Corporate Clients* in the 5:00–7:00 period, 2 per hour, and in the 7:00–9:00 period, 2 per hour. The values in the table for the range B7:G18 are the corresponding cumulative Poisson probabilities for the various average arrival rates for all types of clients and times periods. The cell comment for G9 provides the detail for the formula used to determine the cumulative Poisson probability.

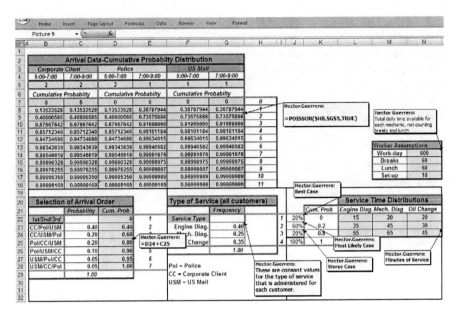

Exhibit 8.17 Brain worksheet for model parameters and assumptions

The order of customer arrival is also important in our model, and each customer arrives as a group in a period; that is, a caravan of Police autos might arrive at Autohaus in the 5:00–7:00 time period. The *Selection of Arrival Order* table provides the six possible combinations of arrival orders for the three clients, and is the result of some subjective opinion by Wolfgang. For example, there is a 40% chance that the precise order of arrival in a period will be *Corporate Client* first, *Police* second, and *USMail* third. As you can see, the *CorporateClient* is very likely to be the first to arrive in the morning, with a 60% (40+20%) chance of being first overall. Order will be important when Inez begins to examine which of the various clients does not receive service should a day's demand be greater than service supply.

The table entitled *Type of Service* provides the mix or service types for the arrivals. Notice it is deterministic: a fixed or non-probabilistic value. Thus, it is not necessary to resolve uncertainty for this model factor. If 20 autos arrive in a period of time, 8 (20*0.4) will be assigned to *Engine/electrical Diagnosis*, 5 (20*0.25) to *Mechanical Diagnosis*, and 7 (20*0.35) to *Oil Change*. If Inez anticipated a great deal of variation in the short term service types, then it might be wise to determine a distribution for the service types that can be sampled, as we did for arrival order. Service could also be seasonal, with certain types of service, for example engine/electrical diagnosis, occurring more frequently in a particular season. Our model handles service type quite simply and, certainly, more sophisticated models could be constructed.

Service Time Distributions are shown next to the *Selection of Arrival Order* table. The table suggests that service times are distributed with 3 discrete outcomes— 20% best case or shortest service time, 60% most likely case, and 20% worse case or longest service time. For the *Oil Change* service, the times are 20, 30, and 45 minutes, respectively, for the best, most likely, and worse case. This information is also the type of subjective information that could be derived from an interview with Wolfgang. The information gathering could be as simple as asking Wolfie to make the following determination: "If we define *Best Case* occurring 20% of the time, what value of oil change service time is appropriate for this case". Similarly, times for worse case and most likely can be determined. These estimates can be quite accurate for a knowledgeable individual, but great care must be taken in the process of eliciting these subjective values. The interviewer must make sure that the interviewee is fully aware of what is meant by each question asked. There are many excellent structured techniques that provide interviewers a process for arriving at subjective values.

Finally, the *Worker Assumptions* relate to the employee policies that Inez will set for the work force. As stated in Table 8.2, three mechanics, each in a single bay, provide service and it is assumed that they are available from 9:00 am to 7:00 pm (600 total minutes), with a one hour lunch break, four 15 minute breaks, and a changeover from one auto to another requiring 10 minutes of set-up time. The first three times are obviously a matter of workforce policy, but the set-up could be uncertain. This could include, placing the auto into the bay, selection of appropriate tools, and any number of other activities associated with the preparation for service of the next auto available. Set-up might also be dependent on the worker, the type of service, or other factors. It is possible that more experienced workers might require less set-up time than less experienced workers. Given the variety for types of automobile service, the total daily set-up time could be quite substantial, and this is time that Inez might want to reduce through special tools and standardization of work.

We have assumed a deterministic time for set-up to simplify our model. Also note that these numbers can be manually changed by the modeler to perform *what-if* or *sensitivity analysis*. For example, what if we could introduce training or equipment that might reduce set-up time significantly? We would want to know if the investment in the time reduction is worth the effort, or analyze competing technological changes to determine their cost/benefit. Having all these parameters on a single worksheet, the *Brain*, is beneficial to the modeler performing the analysis.

8.5.3 Building the Calculation Worksheet

Exhibit 8.18 provides a view of the calculation worksheet which simulates 250 days of auto arrivals. We will use the 250 days of randomly selected arrivals, along with their arrival order, to determine what autos the mechanics can service each day and which services they will provide. The Calculation worksheet will be used to determine some of the fundamental calculations necessary for the model. As you can see

250 days of Simulated Demand

Days	Corporate Client 5:00-7:00 Arrival (2 Order)		7:00-9:00 Arrival (2 Order)		Police 5:00-7:00 Arrival (2 Order)		7:00-9:00 Arrival (1 Order)		US Mail 5:00-7:00 Arrival (1 Order)		7:00-9:00 Arrival (1 Order)		Daily Totals	5:00-7:00 Rand #'s	7:00-9:00 Rand #'s
1	6	1	2	1	2	2	0	3	1	3	0	2	11	0.065493	0.544713
2	6	1	1	1	2	2	3	2	2	3	3	3	17	0.308855	0.286551
3	2	1	5	2	4	3	2	1	1	2	3	3	17	0.670286	0.743036
4	1	1	4	1	2	2	1	2	1	3	0	3	9	0.282767	0.374661
5	4	1	3	3	6	2	3	2	2	3	2	1	19	0.221381	0.914601
6	4	3	7	1	2	2	2	2	2	1	0	3	17	0.90818	0.234281
7	6	1	5	1	1	2	1	2	7	3	2		22	0.044499	0.118836
8	4		5	3	4	1						2	21	0.834824	0.890012
9	1	1	2	2	7	2						1	17	0.185505	0.990181
10	1	2	1	1	2	1						3	9	0.630309	0.123157
11	5	1	5	1	8	3						3	23	0.5765	0.059271
12	1	1	5	3	5	3						2	14	0.536398	0.863032
13	3	2	4	2	5	3						3	17	0.979834	0.614988
14	4	1	7	1	2	3	3	2	1	2	0	3	17	0.697775	0.384829
245	9	1	3	1	6	2	2	3	2	3	2	2	24	0.098176	0.406983
246	2	1	4	1	4	3	1	3	3	2	1	2	15	0.492138	0.416888
247	3	3	5	2	1	1	3	3	4	2	3	1	19	0.875559	0.967328
248	4	2	4	1	2	1	1	3	2	3	2	2	15	0.679523	0.504297
249	3										2	3	13	0.350389	0.7421
250	1										4	3	19	0.478435	0.279109

max	28
min	5
Avg	18.0
Std	3.9

Hector.Guerrero:
=IF(VLOOKUP(P9,Brain!$D$22:$E$28,2)=1,3,
IF(VLOOKUP(P9,Brain!$D$22:$E$28,2)=2,2,
IF(VLOOKUP(P9,Brain!$D$22:$E$28,2)=3,3,
IF(VLOOKUP(P9,Brain!$D$22:$E$28,2)=4,2,
IF(VLOOKUP(P9,Brain!$D$22:$E$28,2)=5,1,1)))))

Hector.Guerrero:
=VLOOKUP(RAND(),Brain!$B$7:$H$18,7)+VLOOKUP(RAND(),Brain!$B$7:$H$18,7)

Exhibit 8.18 Calculation worksheet for 250 days of operation

in the cell comment for B254, a pair of VLOOKUP()s with a lookup value argument of RAND() determines the hourly demand in each of the two hours. The cell formula uses the *Arrival Data* table in the *Brain*, B7:H18, to randomly select *two* one-hour arrivals. (Recall that each period is 2 hours in length and the arrival rate is hourly, thus the need for two lookups.) It is possible to calculate one hour's arrivals and multiply by two, but this will exaggerate a single hour's behavior.

There are three categories of demand (Corporate Client, Police, and US Mail) and hourly arrivals for each: columns B, D, F, H, J, and L. The arrival order is calculated for each client in a time period in the adjacent columns: C, E, G, I, K, and M; thus, in the Day 1 row (A5:P5), we find that in the 5:00–7:00 time period the number of arrivals for the Corporate Client is 6 (B5) and they are the 1st (C5) to arrive. For the US Mail arrivals in the 7:00–9:00 time horizon, there are 0 arrivals (L5) and they are the 2nd (M5) arrival. Thus the number and sequence of arrivals for the first simulated day is:

a) 5:00–7:00: Corporate Client=6. ...Police=2. ... US Mail=1 ....subtotal 9
b) 7:00–9:00: Corporate Client=2. ... US Mail=0. ...Police=0 ....subtotal 2
c) Thus, Total Day 1 demand =11

The logic used to determine the sequence of arrivals is shown in the cell M9 comment of Exhibit 8.18. The formula consists of an IF with four additional nested IFs, for a total of five. In the first condition, a random number is referenced in cell P9, as well as all of the nested IFs. This is done to make the comparisons consistent. Remember that every RAND() placed in any cell is independent of all other

RAND()s, and different values will be returned for each. The single random number in P9 insures that a single sequence (e.g. CC/Pol/USM) is selected, and then the IFs sort out the arrival sequence for each client type for that sequence. In this particular case, the P9 value is 0.914601. See the *Selection of Arrival Order* table in the *Brain* (Exhibit 8.17). The random number 0.914601 is compared to the values in D22:D28. Since the value falls between 0.90 (D27) and 0.95 (D28), the value 5 is returned in accordance to the lookup procedure. The value 5 indicates a sequence of USM/Pol/CC, and thus, the condition defaults to 1 and identifies the US Mail (USM) as the first position in the sequence. Cell I9 returns a 2 for the Police client (Pol) since it is second in the USM/Pol/CC sequence, etc. Although this may appear to be very complex logic, if you consider the overall structure of the cell formula, you can see that the logic is consistent for each of the 6 possible sequences. Finally, the calculation of *Daily Totals* is performed in column N by summing all arrival values. For Day 1, the sum of cells B5, D5, F5, H5, J5, and L5 is 11 arrivals in (N5).

8.5.4 Variation in Approaches to Poisson Arrivals—Consideration of Modeling Accuracy

Let us consider for a moment other possible options available to generate arrivals from the Poisson distributions of hourly arrivals. For the 7:00–9:00 time period, I have chosen a rather direct approach by selecting two randomly sampled values, one for the hour spanned in 7:00–8:00 and the other for 8:00–9:00. Another approach, which we briefly mentioned above, is to select a single hourly value and use it for both hours. This is equivalent to multiplying the single sample value by two. Does it really matter which approach you select? The answer is yes, it certainly does matter. The latter approach will have several important results. First, the totals will all be multiples of 2, due to the multiplication; thus, odd values for arrivals, for example 17 or 23, will not occur. Secondly, and related to the first, the standard deviation of the arrivals in the latter approach will be greater than that of the former approach.

What are the implications of these results? For the case of no odd values, this may not be a serious matter if numbers are relatively large, but this may also be a departure from reality that a modeler may not want to accommodate. In fact, the average for the arrivals sampled for several days will be similar for both approaches, as long as the sample size of days is large, for example 250 days. The second outcome is more problematic. If there is a need for the model analysis to study the variability of arrival outcomes, the latter approach has introduced variability that may not be acceptable or representative of *real* behavior. By accentuating (multiplying by two) extreme values, it is possible that larger than realistic extreme values will be recorded. In our case, this *is* an important issue, since we want to study the possible failure to meet extreme demand for daily arrivals.

To demonstrate the differences discussed above, consider the graph in Exhibit 8.19. In this simple example, the difference between the approaches becomes clear. The worksheet in Exhibit 8.19 contains two areas with different

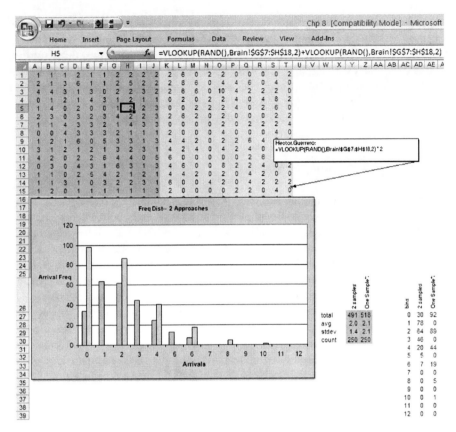

Exhibit 8.19 Two approaches to arrivals

approaches for simulating the arrivals in a 2 hour period. The range (A1:J25) has two separate VLOOKUP functions sampling a Poisson with an average arrival rate of one unit per hour. Why two VLOOKUPs? We have two because one is needed for each hour in the two hour period. This is the approach we used in Exhibit 8.18. The approach in range K1:T25 is a single sample which is then multiplied by two. Why a single sample? In this case we assume that the same value can be used for each of the two hours. Both approaches collect 250 samples of arrivals for a two hour period.

The summary statistics in Y27:Z30 show very similar behavior in terms of means and total arrivals for the 250 samples, approximately a mean of 2.0 and total arrivals 500 for both. Thus, both approaches appear to be similar in some summary statistics. The difference is seen in the standard deviation for the two VLOOKUP approaches. The approach with two randomly sample hours, has a lower value, 1.4, than the approach which used a single VLOOKUP multiplied by two, 2.1. The graph in Exhibit 8.19 shows the greater dispersion of results for the approach that generates a single VLOOKUP sample then multiplies by 2. It is far more likely that this approach will suggest a greater service capacity stock-out than the former, as evidenced by how the one sample graph extends far beyond the 2 sample. Additionally, the one sample graph has many more incidences of 0 arrivals. Thus, the distortion occurs for both extremes, high and low arrival values. Thus, the modeler must carefully consider the *reality* of the model before committing a sampling approach to a worksheet. The one sample approach may be fine, and certainly involves fewer excel functions to calculate, but it may also lead to exaggerated results.

8.5.5 Sufficient Sample Size

How many simulated days are sufficient? Obviously, simulating one day and basing all our analysis on that single sample would be foolish, especially when we can see the high degree of variation of *Daily Totals* that is possible in Exhibit 8.18—a maximum of 28 and minimum of 5. In order to examine the model behavior carefully and accurately, I have simulated a substantial number of representative[1] days. Even with a sample size of 250 days, another 250 days can be recalculated by depressing the F9 key. By recalculating and observing the changes in summary statistics, you can determine if values are relatively stable or if variation is too great for comfort. Also, there are more formal techniques for calculating an appropriate sample size for specific **confidence intervals** for summary statistics, like the mean. Confidence intervals provide some level of assurance that a sample statistic is indicative of the true value of the population parameter that the statistic attempts to forecast.

Without going any further, notice the substantial utility offered by our simple determination of demand over the 250 workday year in Exhibit 8.18. The worksheet provides excellent high level summary statistics that Inez can use for planning capacity. Note that there is a substantial difference between the minimum and maximum *Daily Totals*, 5 and 28. Thus, planning for peak loads will not be a simple matter given that the costs of servicing such demand will certainly include human capital and capital equipment investments. Also of great interest are the average number of daily arrivals, 18.0, and the standard deviation of arrivals, 3.9. An average of 18.0 is a stable value for the model, even after recalculating the 250 days many times. But what is the variation of daily arrivals about that average? There are several ways we can answer this question. First, we can use the average and the standard deviation to calculate a **coefficient of variation** of approximately 21.7% (standard deviation/mean= (3.9/18.0). It is difficult to make a general statement related to the variation of demand, but we can be relatively certain that demand varying one standard deviation above and below the mean, 13.9–21.9, will include

---

[1] We want to select a sample of days large enough to produce the diverse behavior that is possible from model operation. I have used 250 because it is a good approximation of the number of work days available in a year if weekends are not counted. After simulating many 250 day periods, I determine that the changes in the summary statistics (mean, max, min, and standard deviation) do not appear to vary significantly; thus, I feel confident that I am capturing a diversity of model behavior. If you are in doubt, increase the simulated days until you feel secure in the results.

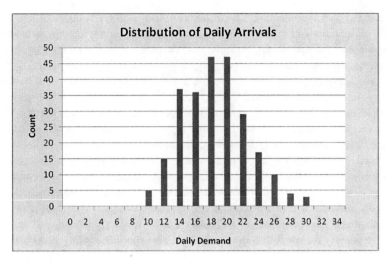

Exhibit 8.20   Distribution of daily arrivals for 250 simulated days

the majority of the 250 daily arrivals generated. In fact, for this particular sample, we can see in Exhibit 8.20 that the number of simulated days of arrivals between 14 and 22 is 196 of the 250 days of arrivals, or approximately 78%.

8.5.6 Building the Data Collection Worksheet

There is still an important element missing in our understanding of daily demand: we are unaware of the *types* of service that each of the arrivals will request and the related times for service. Although the analysis, thus far, has been very useful, the types of service requested could have a substantial impact on the demand for service time. What remains is the assignment of specific types of service to arrivals and the subsequent determination of service times. Both of these issues need to be resolved to determine the amount of daily demand that is serviced.

Should the calculations take place in the *Calculation* worksheet or the *Data Collection* worksheet? I suggest that when you are in doubt you should consider 2 factors: (a) is one worksheet already crowded and, (b) is the calculation important to collaboration, such that you would want it to appear on the same worksheet as other calculations. The location of some model processes is often a judgment call. In our case, the *Calculation* worksheet has been used to generate daily arrivals and the sequence of arrival. Thus, we will use the *Data Collection* worksheet to deal with service. This division of calculations will make the workbook more manageable to those using the workbook for analysis.

Now, let us consider the delivery of service for the model. There is an important issue to be aware of when considering service: simply because there is demand for service, this does not imply that the Autohaus service system can process all the demand that is requested. The simulation has resulted in approximately 4500 (18.0 * 250) autos seeking service during the 250 day year. Our goal now is to determine whether the operations configuration that Inez is contemplating can, in fact, manage this service demand. Depending on the availability of service capacity, the Autohaus service system may handle the demand easily, or it may be considerably short of the requested demand.

In our model, demand represents a *reservoir* of autos requesting service. To this point, we have generated demand without regard to Autohaus' capability to supply service. The question that Inez needs to answer is: given the demand that we have generated, how much of this demand can our service supply accommodate on a daily basis? Our simulation should be useful in a number of ways. Determining the amount of demand serviced will also depend on the contractual agreement that she has with customers. If she is contractually guaranteeing service over a period of time, then she may be forced to increase capacity to satisfy all demand or run the risk of a penalty for not fulfilling the contract. This should certainly be a concern for Inez as she begins to conceptualize her business model and as she finalizes plans for operation. The use of the simulation model to study service capacity will also help her understand the effects of various types of contractual agreements.

So how do we proceed? Consider Exhibit 8.21, the *Data Collection* worksheet. This is the worksheet where we perform the service analysis we just discussed. It is wise to first understand the broad layout of Exhibit 8.21, especially given the complex nature of the worksheet. This worksheet examines demand for each of the 250 days simulated in the *Calculation* worksheet (Exhibit 8.18) and returns the service that is provided. The *Data Collection* worksheet contains three major sections which we will detail in several exhibits due to their rather large size.

The first section, columns A:E, determines the percentage of each day's total demand that will be allocated to each of the three service types. In this example, day 1 demand is 11 arrivals (E4), and of this total, 4 (B4) are allocated to *Engine/electrical Diagnostics*, 3 (C4) to *Mechanical Diagnostics*, and 4 (D4) to *Oil Change*. Columns F to AX display the specific service arrivals and the service time required for each arrival. You can see in cell range F4:I4 the 4 *Engine/electrical Diagnostics* arrivals and their corresponding service times, 55, 35, 15, and 15,

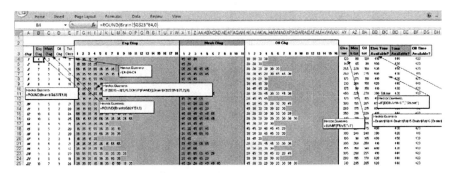

Exhibit 8.21 Data collection worksheet

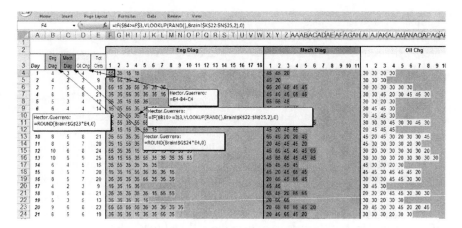

Exhibit 8.22 Allocation of service and service time determination

respectively. Finally, Columns AY:BF determine the availability of service, providing the modeler with an opportunity to detect a **service stock-out**; [2] that is, a day for which some demand is not satisfied by available service capacity. If a stock-out condition is observed, it is because the mechanic stationed in a bay is unable to service all daily demand for that particular service type. In summary, the *Data Collection Worksheet* has: (1) an initial allocation of overall demand to specific service types (*Engine/electrical Diagnosis*, etc.), (2) a determination of actual times, in minutes, for the service types, and lastly, (3) a comparison of the daily service time requested versus the daily service capacity available from a mechanic.

Now, let us take a closer look at the *Data Collection* worksheet. Exhibit 8.22 focuses on areas (1) and (2) in the summary above. As mentioned, columns B through D allocate the total daily demand over the three service categories. Cell E4 references the *Calculation* worksheet cell N5, the *Daily Total* for day 1. Cell range B4:D4 allocates the total, 11, on the basis of the deterministic percentages located in the *Brain* in the *Type of Service* table. Recall that *Engine/electrical Diagnosis* is constant at 40% of the arrivals. Therefore, 11 arrivals (E4) multiplied by 40% (G23 in the *Brain*) produces approximately 4 *Engine/electrical Diagnosis* service requests. This number is rounded to account for integer values for arrivals. Similarly, 3 *Mechanical Diagnosis* service requests and 4 *Oil Changes* are produced. Then *Oil Changes* are calculated as the difference of the *Total Clients* and the rounded *Engine/electrical* and *Mechanical Diagnostics* service requests, to insure that the sum of all service types requested is equal to the total, 11.

Next, column range F4:W4 generates 4 service times. Logic is needed to select only as many service times as there are arrivals. This is done by comparing the

---

[2] A service stock-out is defined as a daily period for which there is demand for service that is not met. It does not suggest the size of the stock-out. It is simply a stock-out occasion, without regard to the number of autos that do not receive service.

number of service requests to the index numbers in range F3:W3 (1–18). Note that 18 columns are provided to insure that a very large number of random arrivals are possible. It is very unlikely to ever experience 18 arrivals for any service type. For *Mechanical Diagnostics*, a maximum of 11 arrivals are permitted, and for *Oil Change*, a maximum of 16 are permitted. The cell function used to determine the service time value of *Engine/electrical Diagnostics* is a logical IF() with an embedded VLOOKUP. An example is shown in cell I10 of Exhibit 8.22. The IF() tests if B10 is greater than or equal to the index number in its column, 4. If this condition is true, which it is, then the cell formula in cell I10 will randomly sample the *Service Time Distributions* table in the *Brain* via a VLOOKUP and return a value.

If the index exceeds the total number of service requests, a 0 is returned. Note that cell J4 appears blank, although it is 0. This is accomplished by using conditional formatting for the cells. A logical test of the cell content—equal to 0—is made, and if the answer is to the test is true, then the cell *and* font colors are set to similar colors, thus resulting in a blank cell. This is done for clarity, and zeros could easily be allowed to appear without any loss of accuracy. The same is repeated in all other ranges for the various service types. By depressing the recalculate[3] key, F9, the contents of the three service areas change as new service arrivals are calculated in the *Calculation* worksheet.

Now, let us consider the analysis that is done in the *Data Collection* worksheet. In column range AY:BF of Exhibit 8.23, we perform the calculations necessary to determine if daily service demand is satisfied by available service capacity. The time available for service is the difference between the total time available from 9:00 am to 7:00 pm, 600 minutes, and breaks, lunch, and set-ups. Each row determines the sum of service time requested in a particular day for each service type (e.g. AY4:BA4) then compares the request to available service time. For example, for day 1 the available *Engine/electrical Diagnosis* time is 440 minutes, which is calculated by subtracting breaks (60), lunch (60) and set-up times for each service request (4*10 = 40) from a total of 600 minutes. These times are found in the *Worker Assumptions* table in the *Brain*. Thus, daily breaks and lunch always account for 120 minutes, while the set-ups sum will vary according to the number of service requests, 4 in this case requiring 10 minutes each. The total is 160 minutes, and this results in 440 minutes of available service capacity (600 − 160 = 440). Since 440 minutes of capacity is greater that 160 minutes of *Engine/electrical Diagnosis* requested, there is sufficient capacity to deal with demand, thus no stock-out occurs for that service type.

In row 14 of Exhibit 8.23, which corresponds to day 11 of the simulation, we see that a service stock-out occurs for *Engine/electrical Diagnosis*. The comment associated with the cell BC14 shows an IF() cell formula that compares available capacity of 400 minutes to a capacity demand of 440 minutes. Obviously, there is

---

[3] I would suggest setting the recalculation of formulas to manual in all worksheets. This gives you control and eliminates annoying and inopportune recalculations. The control can be set by using the Tools then Options menus. One of the available tabs is Recalculation and this is where you select manual.

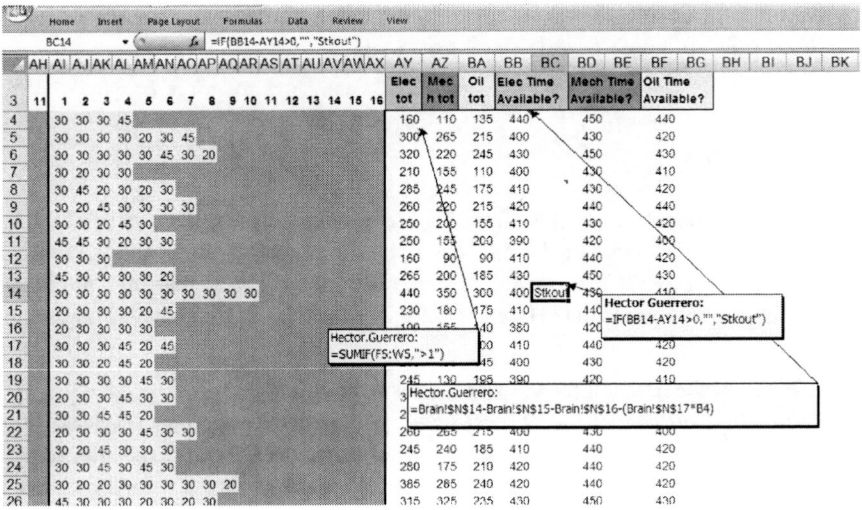

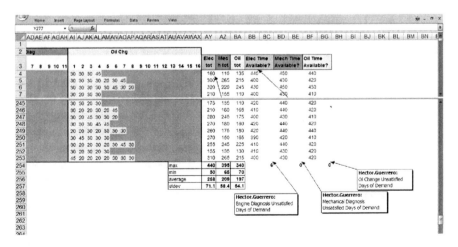

Exhibit 8.23 Analysis area of data collection

a service stock-out of 40 minutes, in which case the test results in a negative value and, thus, the cell value *Stkout*. There may be more detailed analysis that is possible for each stock-out. For example, is it possible to determine who will and will not receive service?

In Exhibit 8.24, we see that cell BC254 counts the number of *Engine/ electrical Diagnosis* stock-outs, 6. This is done by using a simple COUNTIF() function: COUNTIF(BC4:BC253,"Stkout"). We could easily count the number of stock-out minutes in column BC rather than simply determining that a stock-out

Exhibit 8.24 Determination of clients not receiving service

occurred. This would allow us to quantify the number of minutes that we are *under* demand capacity in each service area for 250 days.

8.5.7 Results

Inez has completed her analysis and we are now prepared to discuss some of the results. The major focus of the analysis is on the use of capacity and the level of service that is provided to clients. So what are the questions that Inez might ask about the simulation results? Here are a number that she might consider important:

1. How does the demand for the various services differ?
2. Will Autohaus be capable of meeting the anticipated demand for services?
3. Are there obvious surpluses or shortages of capacity? Given the revenue genera-tion capability of each service, do the model results suggest a reconfiguration of client services?
4. Can we be sure that the simulation model has sufficient sample size (250 days) to provide confidence in the results?

The first question does not have an obvious answer from the raw data avail-able in the *Brain*. Although the *Engine/electrical Diagnosis* represents the highest percentage of service (40%), it has shorter service times for worst, best, and most likely cases than *Mechanical Diagnosis*. The comparison to *Oil Change* also is not clear. The summary statistics near the bottom of Exhibit 8.24, row 256, show very clearly that *Engine/electrical Diagnosis* dominates demand by a substantial margin. The averages for demand are 258, 209, and 197 minutes, respectively, for *Engine/electrical, Mechanical,* and *Oil.* Thus, the average for *Engine/electrical* is about 23% ([258–209]/209) greater than the average for *Mechanical*, and 31% ([258–197]/197) greater than *Oil*.

What about the variation of the service demand time? Exhibit 8.25 shows that the distribution for *Engine/electrical Diagnosis* is wider than that of *Mechanical Diagnosis* or *Oil Change*. This is verified by the summary statistics in Exhibit 8.24, row 257, where the annual (250 days) standard deviations range from 71.1 for *Engine/electrical*, to 58.4 and 54.1 for *Mechanical* and *Oil*. Additionally, the range of values, max–min, for *Engine/electrical* (440–50=390) is substantially greater that *Mechanical* (395–65=330) or *Oil* (340–70=270). All this evidence indicates that *Engine/electrical* has much more volatile demand that the other services.

The answer to the second question has already been discussed. Again, we see in Exhibit 8.24 that the only area where there appears to be any significant demand that is not being met for a service type is for *Engine/electrical Diagnosis*. The simulation shows that there were 6 days of unsatisfied service. We need to be quite careful to understand what this suggests for capacity planning at Autohaus. It does not mean that demand was not met for all the demand that occurred for a service type in 6

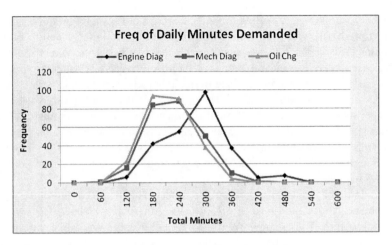

Exhibit 8.25   Graph of frequency of daily service minutes demanded

days; it does mean that some *portion* of the *Engine/electrical Diagnosis* demand for 6 days was not met. Although the model did not indicate the amount of service time that was not met, as we mentioned earlier, it is a simple matter to change the cell formulas in columns BC, BE, and BG to return the quantity rather than the indicator term *Stkout*. The *Stkout* in Exhibit 8.23 cell BC14 would then be –40 minutes (400–440= –40). By comparing this time (–40) to the time requested by the last demand in the queue we can determine if the auto receives service because it starts before the end of the day, or if the auto is not serviced because it has not yet started. Recall that if an arrival is being serviced at the end of the day, it will remain in service until it is finished. This would then not represent a *Stkout*, but would represent overtime to accommodate finishing the arrival.

Question three considers the appropriate use of capacity. Inez must consider the service stock-out for *Engine/electrical Diagnosis*. We have already suggested that some indications of a service stock-out are resolved by the *end-of-day* policy. She must also consider the cost associated with attempting to meet demand every day for every service type. A policy of 100% service could be a very costly goal. It may be far better for her to use overtime to handle excess demand and also accept an occasional service stock-out. Much will depend on alternative uses of the capacity she establishes for this new service. She may be able to accommodate other clients to handle the unused available capacity, in particular for *Mechanical* and *Oil* services since they appear to have substantial excess capacity. Additionally, she could consider making the 3 bays more flexible by permitting multiple services in a bay. This, of course, would depend of the capital investment necessary to make the flexibility available. These questions will require further analysis and the potential adaptation of the model to perform different types of modeling. It is not unusual for the results of a model to suggest the need to change the model to capture other behavior.

Finally, question 4, regarding the sample size, is always relevant in Monte Carlo simulation. It asks whether our simulation has a sufficiently large sample of model operation (250 days in our case) to accurately describe behavior. As a modeler gains experience, it becomes clear that relatively few sources of variation are sufficient to greatly affect results, even when data from a relatively large sample is collected. So how do we deal with this question? The obvious answer to this question is to repeat the experiments to produce results multiple times. In other words, simulate 250 days, simulate another 250 days, and repeat the process numerous times. Then examine the results of each simulation relative to the combined results of the simulations. This is easily done in our model by recalculating the spreadsheet and noting the changes with each recalculation.

In Exhibit 8.26 you can see *Stkout* data collected in 20 replications of the 250 day experiment. Notice the range of *Stkout* values for *Engine/electrical Diagnosis* in the first column. Values range from 2 to 15 and average 9.9, while the range for *Mechanical* and *Oil* are far less variable. A value of 2 appears to be a rare occurrence, but this example serves to suggest that some data elements will require substantial sample size to insure a representative result, and it may be necessary to increase sample size to deal with variation in *Engine/electrical Diagnosis*.

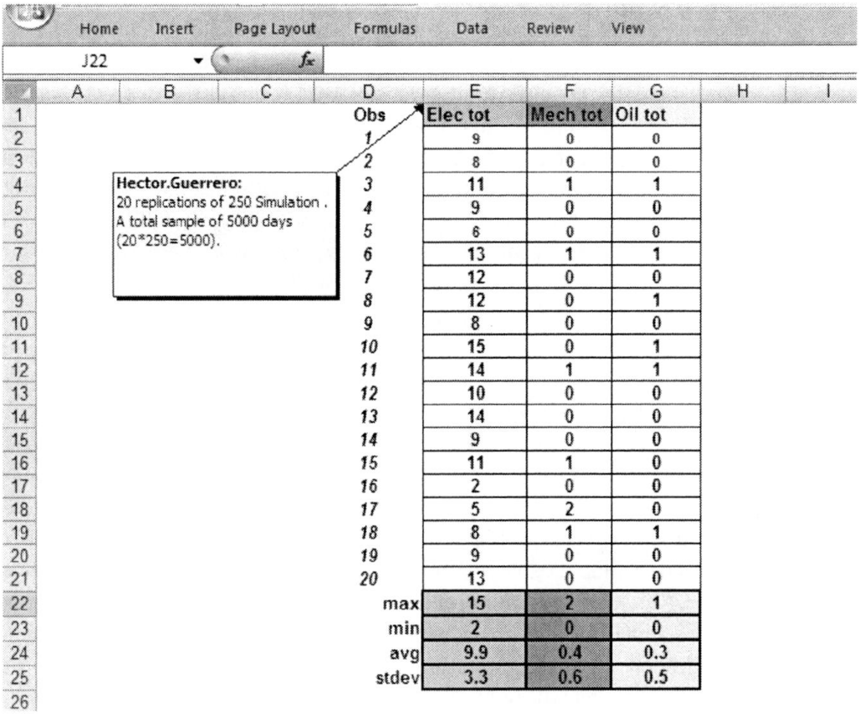

Obs	Elec tot	Mech tot	Oil tot
1	9	0	0
2	8	0	0
3	11	1	1
4	9	0	0
5	6	0	0
6	13	1	1
7	12	0	0
8	12	0	1
9	8	0	0
10	15	0	1
11	14	1	1
12	10	0	0
13	14	0	0
14	9	0	0
15	11	1	0
16	2	0	0
17	5	2	0
18	8	1	1
19	9	0	0
20	13	0	0
max	15	2	1
min	2	0	0
avg	9.9	0.4	0.3
stdev	3.3	0.6	0.5

Hector.Guerrero:
20 replications of 250 Simulation .
A total sample of 5000 days
(20*250=5000).

Exhibit 8.26 Multiple replications of 250 days simulation stock-out data

8.6 Summary

As we have seen in both examples, a relatively few sources of uncertainty are sufficient to produce results that are not easily predicted. Thus, the interactions of the uncertainties of a model are often unexpected and difficult to forecast. This is precisely why simulation is such a valuable business tool. It provides a systematic approach to understanding and revealing the complex interactions of models.

The value of a carefully conceptualized and implemented simulation can be great. Beyond the obvious benefits of providing insight into the risks associated with the problem, the process of creating a model can lead to far greater understanding of the problem. The process of creating a simulation requires a rigorous approach and commitment to very specific steps, not least of which is problem definition. Even the most clever simulation and sophisticated analysis is worthless if you are solving the "wrong" problem.

It is important to keep in mind that the goal of simulation is to determine the risk associated with a problem or decision; that is, what is the range of possible outcomes under the model conditions? One of the steps of simulation that we did not perform was *sensitivity analysis*. By changing the parameters of the problem, we can note the reaction of the model to changes. For example, suppose you find that a relatively small decrease in the set-up times used in Autohaus can lead to significant improvements in service stock-outs. You would be wise to investigate carefully the nature of step-ups and how you might be able to reduce them to take advantage of this leverage in service improvement.

In the next chapter we cover a number of tools that are extremely powerful and useful. They are available in the Data Ribbon—Solver, Scenarios, and Goal Seek. Some of these tools can be used in conjunction with simulation. The modeling and simulation we performed in Chaps. 7 and 8 has been **descriptive modeling**—it has provided a method by which we can describe behavior. In Chap. 9 we will introduce **prescriptive modeling**, where we are interested in prescribing what decisions *should be made* to achieve some goal. Both are important and often work together in decision making.

Key Terms

Model	Risk Profile
Simulation	RAND()
Point Estimates	Random Sampling
Monte Carlo Simulation	Resolution of Uncertain Events
Rapid Prototyping	Uniform Distribution
Continuous Event Simulation	Census
Discrete Event Simulation	Replications
Events	Discrete Distributions
Deterministic Values	Continuous Distributions

NORMINV()	Average arrival Rate
Empirical Data	Poisson Arrival Process
Balk	POISSON(x, mean, cumulative)
Confidence Intervals	VLOOKUP
Coefficient of Variation	HLOOKUP
Service Stock-out	Normal Distribution
Descriptive Modeling	Bell Curve
Prescriptive Modeling	Queues

Problems and Exercises

1. Name three categories of models and give an example of each.
2. Which of the following are best modeled by Discrete event or Continuous event simulation:

   a. The flow of water through a large city's utility system
   b. The arrival of Blue-birds at my backyard birdfeeder
   c. The number of customers making deposits in their checking accounts at a drive-up bank window on Thursdays
   d. The flow of Euros into and out of Germany's treasury
   e. The change in cholesterol level in a person's body over a 24 hour period
   f. The cubic meter loss of polar ice cap over a 10 year time horizon.

3. Monte Carlo simulation is an appropriate modeling choice when point estimates are not sufficient to determine system behavior-T or F?
4. Give three reasons why rapid-prototyping may be useful in modeling complex business problems.
5. Risk profiles always have monetary value on the horizontal axis-T or F?
6. Create a risk profile for the following uncertain situations:

   a. A $1 investment in a lottery ticket that may return $1,000,000
   b. A restaurateur's estimate of daily patron traffic through a restaurant where she believes there is 30% chance of 25 patrons, 50% chance of 40 patrons, and 20% chance of 75 patrons
   c. A skydiver's estimate of success or failure under particularly treacherous weather conditions, where the skydiver has no idea of the outcome (success or failure).

7. Create a simple simulation that models the toss of a fair coin. Test the results (% Heads/% Tails) for sample sizes of 5, 10, 30, and 100. Hint-Use the RAND() function.
8. Two uncertain events are related. The first event occurs and effects the second. The first event has a 35% chance of an outcome we will call *small* and 65% chance of a *large* outcome. If the first outcome is *small* then the second event will result in equal chances of 3, 4, 5, and 6 as outcomes; if the first event is

large then the second event has equal chances of 11, 13, 14, and 15 as outcomes. Create a simulation that provides a risk profile of outcomes. The simulation should replicate the experiment a minimum of 300 times.

9. Create a VLOOKUP that:

a. Allows a user to enter a percent (0 to 100%) and returns a categorical value based on the following data:

0–30%	31–63%	64–79%	80–92%	93–100%
A	B	C	D	E

b. For the same data above, create a VLOOKUP that returns a categorical value for a *randomly* generated %. Hint-Use the RAND() function.

c. Expand the table so that the category A and B is defined as *Good*, C as *OK*, and D and E as *Terrible*. With this new three row table return the new outcomes (*Good*, etc.) for exercise (a) and (b) above.

10. Create a simulation of a simple income statement for the data elements shown below and produce a risk profile and summary data for profit (average, max, min, standard deviation, etc.):

a. Revenue Normally distributed mean of $140 k and standard deviation of $26 k

b. COGS is a % of Revenue with outcomes of 24, 35, and 43%, all equally likely

c. Variable costs % of Revenue with outcome 35% one half as likely as outcome 45%

d. Fixed costs $20 k constant.

11. The arrival of email at your home email address can be categorized as *family*, *junk*, and *friends*. Each has an arrival rate: family—5/day; junk—4/day; friends—3/day.

a. Create a simulation that provides you with a profile of total, daily email arrival. What is the probability that you will receive 0 emails, 3 emails or less, 15 or more emails, and between 5 and 15 emails. Hint-Use the Poisson () function and replicate for 100 days.

b. If you read every email you receive and the time to read the categories is that shown below, what is the distribution of minutes spent per day reading email.

    i. Family email Normal distribution, mean 5 minutes and standard deviation 2

    ii. Junk email Discrete 0.5 minutes 80% and 3 minutes 20%

    iii. Friends Continuous Uniform from 3.5 to 6.5 minutes.

12. *An advanced problem*—Father Guido Aschbach, pastor of Our Lady of Perpetual Sorrow Church, is planning a weekly (Sunday after mass) charity

event, complete with food and games of chance. He has heard of Fr. Efia's success and would like to try his hand at raising money through games of chance. Each parishioner is charged $10 to enter the event and is given 3 tokens for each of the three games of chance: Wheel of Destiny, Bowl of Treachery, and the Omnipotent 2-Sided Die. See the table below for the odds and payouts of each game. The parishioners enjoy these events, but they are affected by the weather that occurs each Sunday—Rain produces the lowest turnout; Fair weather the largest turnout; Glorious sunshine results in the next largest turnout (see details below). Help Father Aschbach estimate the yearly (52 Sundays) returns for the events. Hint-You will have to decide whether, or not, you want to simulate the play of each individual parishioner, or if you will simply use the expected value of a player as we did before. If you can, and it will be much more difficult, attempt to simulate the play of each parishioner in the simulation.

Wheel of Destiny		Bowl of Treachery		Omnipotent 2-sided Die	
Type—Discrete		Type—Discrete		Type—Discrete	
Return	Prob.	Return	Prob.	Return	Prob.
$10*	0.4	$15	0.3	$100	0.35
−$10	0.6	−$20	0.7	−$40	0.65

*40% chance a parishioner will win $10.

Weather Attendance/Probability		
Rain	20	0.2
Fair	50**	0.55
Glorious	35	0.25

**55% chance 50 parishioners will attend.

Chapter 9
Solver, Scenarios, and Goal Seek Tools

Contents

9.1 Introduction

Chapters 1 through 8 have introduced us to some very powerful tools for formulating, solving, and analyzing complex problems. From the seemingly endless array of cell functions, to the sorting, filtering, and querying, to PivotTable and PivotChart reports, to the Data Analysis tools, there seems to be a tool for almost every analytical need. Yet, still there are a number of tools we have not exercised; in particular, the optimization tool Solver. **Solver** permits us to utilize a problem structure known as *constrained optimization*. Decision makers can search for a *best* solution for some desired *objective* under *constrained* conditions. We will learn more about this **prescriptive analysis** procedure later in this chapter.

Also of great interest will be the Scenarios and Goal Seek tools. The **Scenario** tool is used to implement what we have referred to as *what-if* analysis. Simply put, with Scenarios we have an efficient tool for automating the variation of inputs for a problem formulation, and subsequently recording the resulting output. Its function is to organize and structure. Without the formal structure of Scenarios it is very easy to lose important information in your analysis.

Goal Seek is also a member of the What-if sub-group in the Data Tools group. In situations where we know the outcome we desire from a problem, **Goal Seek**, as the name implies, is an approach that *seeks* to find the input that leads to our *desired* outcome. For example, if we are calculating the constant and periodic payments for a mortgage, we may ask the question—what interest rate will lead to a monthly payment of $1,000 for a loan with a term of 240 monthly periods and principal value of $100,000?

Before we become acquainted with these new tools let us take stock of where we have been thus far on our analytical journey. We began with a careful classification of data, from categorical to ratio, and we discussed the implications of the data type on the forms of analysis that could be applied. Our classification focused on *quantitative* and *qualitative* (*non-quantitative*) techniques for presenting, summarizing, and analyzing data. Through Chap. 6 we assumed that the data under consideration were available from the collection of primary data[1] or available from a secondary source. Additionally, our analysis was **descriptive** in nature, generally attempting to describe some characteristic of a population, by analyzing a sample from that population; for example, comparing the mean of a sample to the mean of the population, and then determining if the sample mean could have come from the population, at some level of statistical significance.

In Chaps. 7 and 8, Modeling and Simulation, we created models in which we generated data that we could then analyze. In these chapters, we began our modeling effort by using the descriptive procedures discussed above to define our model, and then we generated data from the model to use for prescriptive purposes. For example, in Chap. 8 we used our understanding of the operating behavior of Autohaus to create a Monte Carlo simulation. The data generated by the simulation allowed us to prescribe to Inez the possible design selections for the Autohaus business concept.

The models that we create with Solver have an even stronger prescriptive character than those encountered thus far. In using Solver, our goal is to determine the values of decision variables that will *minimize* or *maximize* some objective, while adhering to the technological constraints of the problem. Thus, the solution will *prescribe* very specific action about decision variables. As you can see from this

---

[1] Primary data is collected by someone in the role of collecting data for a specific purpose and comes from sources that are generally not available as a result of other studies. For example, a survey study performed by a company to determine their customers' satisfaction with their service is primary data, while data on similar industry-wide service that can be purchased from a consulting firm is secondary data.

discussion, our models have become ever more prescriptive in nature as we have progressed through our early chapters. Before we explore Solver, we begin with a discussion of constrained optimization.

9.2 Solver—Constrained Optimization

Constrained optimization is a subfield of numerical optimization, with the critical distinction arising from the term *constrained*. Without a doubt, the most commonly used type of constrained optimization is **Linear Programming** (LP). There are an abundance of software packages to provide solutions to LPs, including Solver. Additionally, there are numerous important applications of LP ranging from the distribution of electrical power, to scheduling, to menu planning. Some problems that can be formulated as LPs occur with such frequency, that they have received their own special designations; for example, the blending problem, the knapsack problem, the nurse or shift scheduling problem, and the capital budgeting problem. These are just a few of the generic formulations that have been developed over time as standard problem structures.

Linear programming, as a branch of constrained optimization, dates back to the 1940s. Over the years, as new innovations in LP techniques, computer programming, and digital computers have developed, the applications of LP have grown dramatically. Problems of such immense computational difficulty that a timely solution was once unimaginable can now be solved in fractions of a second by personal computers. Couple this with the availability of Solver, or other advanced Solvers that are add-ins to Excel, and now an ordinary personal computer user has a tool that was once the domain only of firms that could afford costly LP software and the experts to conduct LP analysis.

Yet, the adage that "a little knowledge is a dangerous thing" certainly applies to LP. Before we cavalierly rush into formulating LPs, we need to spend some time considering the structure of an LP. We could also invest time in understanding the underlying algorithms that provide a computational platform for solving LPs, but that is a very sophisticated topic and it will not be our focus. We will leave that for a text on the mathematical techniques of LP. Our focus will on *formulating* LPs; that is, structuring the important elements of a problem and then applying Solver to obtain a solution. Additionally, we will apply a special form of sensitivity analysis which is unique to LP solutions.

As we have discussed earlier, LPs consist of three major components:

1. **Decision Variables**—the continuous, non-negative variables that are selected to minimize or maximize an objective function which is subject to constraints. As continuous variables, they can take on fractional values. The objective function and constraints are comprised of *linear*, algebraic combinations of the decision variables. Thus, no powers, other than 1, of the decision variables (no values like X^2, $X^{1/2}$, etc) are allowed and only constants can be used as multipliers.

2. **Objective Function**—the *linear* combination of a decision maker's decision variables that are to be either minimized or maximized. For example, we may want to maximize revenue or profit, and we will want to minimize cost or a deviation from a desired target level. We will refer to the objective function as Z.

3. **Constraints**—the *linear* combination of decision variables that represent how the decision variables *can* and *can not* be used. Constraints are often referred to as the **technology of the linear program**.

As you can see, the term *linear* is important throughout the definitions above. If the conditions of linearity are not met, then you obviously do not have a LP. Yet, there are techniques for solving *non-linear* programs, and later we will briefly discuss the topic. It must be noted that it is far more difficult to solve non-linear programs, and it may also be very difficult to do so with Solver. There are software packages available for the specific purpose of solving non-linear programs. Now, let us consider a problem that can be formulated as a LP.

9.3 Example—York River Archaeology Budgeting

The principles of a thriving, but cash strapped business, York River Archaeology (YRA), are holding their quarterly budget meeting to decide upon the most critical budget that they have yet to produce. YRA does archaeological studies of sites that are to be developed for various uses—family housing, retail sales, public buildings, etc. Many cities and states require these studies. They are used to determine if important historical sites will be disturbed or destroyed due to development, and are quite similar to environmental impact studies. Thus, their services are in great demand in regions where significant building activity takes place.

In attendance at the meeting are the three principles, Nick, Matt, and Garrett, all accomplished Ph.D. archaeologists and historians, and their trusted business advisor, Elizabeth. Nick spoke to the group first—"The importance of this budget cannot be overstated. In the past 10 years we have not paid much attention to budgets, and it has gotten us into serious dilemmas with our cash flow. Back then, we could finance operations from our cash flow only, but now we have grown to the point where we need a substantial line of credit to keep operations running smoothly and also to finance our growth. Elizabeth has been telling us for years that it is important to create realistic budgets and to stick to them, but we haven't listened. Now, if we want our bank to provide us a reasonable line of credit, we have no choice but to do precisely as she has advised, and to show the bank that we can work *within* a budget."

YRA is in the business of managing archeological projects. Just prior to the beginning of each quarter, the principals and Elizabeth meet to select the projects they will accept for the forthcoming quarter. From the projects selected and the resources consumed in the process, they are able to produce a detailed quarterly budget. YRA has been quite successful, so much so, that they have more requests for projects than they are capable of accepting. Thus, their resource limitation is the only factor that restricts them from accepting all the available project contracts.

Elizabeth has convinced them that they can model the selection process with LP. The LP will select the projects that maximize revenue while maintaining feasible usage of resources; that is, the LP will *not* select a set of projects that consumes more than the limited resources available, and it will do so while maximizing the objective function (revenue).

In preparation for creating an LP, Elizabeth sets a project pricing policy that estimates the revenue and outside resource usage associated with each project. Although there are some variations in her estimates, the revenues and outside resource usage are very nearly constant over short spans of time (3–6 months), certainly well within the quarterly planning horizon. The outside resources needed are stated in terms of hours, and they represent hours of outside consulting that must be obtained to perform a project. These include services like fieldwork by contract archeologists, report writing, and other types of services. YRA has decided it is wiser to obtain outside services than it is to hire permanent staff; it permits them to be very agile and responsive to varying demand. We will assume four resource types—*Res-A*, *Res-B*, etc.

Exhibit 9.1 provides a convenient format for assembling the relevant data for YRA's project selection. For example, you can see that there are 25 projects of Type 1 available in the quarter, and that each project produces revenue of $45,000. The resources used in Type 1 projects are 6 hours of Res-A, 12 of Res-B, 0 of Res-C, and 5 of Res-D. Note that there are seven types of projects available—Project 1, Project 2, etc.

An examination of the data suggests that there is no possibility to accept all projects, because doing so would result in a total of 5,432 hours.[2] Since there are

Project Type	No. Projects Available	Rev.	Total Possible Rev	Res-A hrs. used	Res-B hrs. used	Res-C hrs. used	Res-D hrs. used
1	25	45,000	1,125,000	6	12	0	5
2	30	63,000	1,890,000	9	16	4	8
3	47	27,500	1,292,500	4	10	4	5
4	53	19,500	1,033,500	4	5	0	7
5	16	71,000	1,136,000	7	10	8	4
6	19	56,000	1,064,000	10	5	7	0
7	36	48,500	1,746,000	6	7	10	3
Total Available				Total Resource Hrs. Available in Quarter			
	226		9,287,000	800	900	700	375

Exhibit 9.1 Projects for YRA budget

---

[2] If we multiply the "No. Available" row by the individual "Res-A" through "Res-D" rows and sum the results, we find that a total of 4400 hours would be required to satisfy *all* the available projects. For example, for Res-A a total of 1338 hours would be required (6*25 + 9*30 + 47*4 + 53*4 + 16*7 + 19*10 + 36*6 = 1338). Similarly, 2022, 929, and 1143 are needed for Res-B through Res-D, respectively. Thus, 1338 + 2022 + 929 + 1143 = 5432.

only $2,775^3$ hours available, some projects will not be accepted. The LP will choose the best combination of projects that meets the constraints imposed by the resources, while maximizing the revenue returned. It is not obvious which projects should be accepted. To simply assume that the highest revenue projects will be selected until resources run out is not advisable. There may be lower revenue projects that consume fewer resources that are more attractive from a cost/benefit perspective. Fortunately for us, the LP will make this determination and guarantee an "optimal solution", or indicate that no solution is possible (**infeasibility**).

One of the features of an LP solution is the continuous nature of the resulting decision variables. Thus, it is likely that our solution will suggest fractional values of contracts; for example, it is possible that a solution could suggest 12.345 units of Project Type 1. Obviously, accepting fractional contracts is a problem; but if the numbers of projects that can be accepted are relatively large, rounding these fractional numbers up or down, while making sure that we do not exceed important constraints, should lead to a near optimal solution. Later we will see that by imposing a constraint on a variable to be either binary (0 or 1) or integer (1, 2, 3, etc.) we convert our LP into a non-linear program. As we suggested earlier, these are problems that are far more complex to solve, and will require careful use of the Excel Solver to guarantee that an optimal solution has been achieved.

9.3.1 Formulation

Now, let us state the problem in mathematical detail. First, the *decision variables* from which we can construct our objective function and constraints are:

$X_1, X_2, ..., X_7 =$ the number of each of the seven types of projects selected for the quarter

For example, X_4 is the number of projects of type 4 selected. The decision variables must all be non-negative values; selecting a negative number of projects does not make any practical sense.

Next, consider the *objective function*, Z, for the YRA in terms of the decision variables:

$$Z = 45,000X_1 + 63,000X_2 + 27,500X_3 + 19,500X_4 + 71,000X_5 + 56,000X_6 + 48,500X_7$$

The objective function sums the revenue contribution of the projects that are selected. If X_1 is selected to be 10, then the contribution of X_1 projects to the objective function, Z, is \$450,000 ($10* 45,000 = 450,000$).

---

[3] Total resource hours available in the quarter are 2775 ($800 + 900 + 700 + 375 = 2775$) for Res-A through Res-D, respectively.

Finally, we consider the *constraints* that are relevant to YRA. There is a number of constraints that must be met, and the first relates to project availability:

$$X_1 \leq 25;\ X_2 \leq 30;\ X_3 \leq 47;\ X_4 \leq 53;\ X_5 \leq 16;\ X_6 \leq 19;\ X_7 \leq 36$$

Note that these seven constraints restrict the number of project types that are selected to *not* exceed the maximum available. For example, $X_1 \leq 25$ insures that the number of type 1 projects selected cannot exceed 25, while permitting values less than or equal to 25. Although we also want to restrict variables to non-negative values, this can be easily and universally handled with an option available in Solver—*Assume Non-Negative*. This condition is particularly important in minimization problems, since values for decision variables that are negative can contribute to the minimization of an objective function. Thus, in such a case, unless we set the non-negative condition, the LP will attempt to make the values of decision variables more and more negative to achieve a lower and lower Z value.

But we are not done with the constraints yet; we still have a set of constraints to consider that relate to the consumption of resource hours. For example, there is a maximum of 800 Res-A hours available in the quarter. Similarly, there are 900, 700, and 375 available hours of Res-B, Res-C and Res-D, respectively. The consumption of Res-A occurs when the various projects are selected. Thus, if we multiply each of the decision variables by the number of hours consumed, the resulting linear constraint relationships are:

Res-A constraint.... $6X_1 + 9X_2 + 4X_3 + 4X_4 + 7X_5 + 10X_6 + 6X_7 \leq 800$
Res-B constraint.... $12X_1 + 16X_2 + 10X_3 + 5X_4 + 10X_5 + 5X_6 + 7X_7 \leq 900$
Res-C constraint.... $0X_1 + 4X_2 + 4X_3 + 0X_4 + 8X_5 + 7X_6 + 10X_7 \leq 700$
Res-D constraint.... $5X_1 + 8X_2 + 5X_3 + 7X_4 + 4X_5 + 0X_6 + 3X_7 \leq 375$

Let's take a close look at the first inequality, *Res-A constraint*, above. The **coefficients** of the decision variables represent the *technology* of how the maximum available hours, 800, are consumed. Each unit of project type 1, X_1, that is selected results in the consumption of 6 hours of resource A; each unit of project type 2, X_2, consumes 9 hours, etc. Also recall that since LP's permit continuous variables, it is possible that a fraction of a project will be selected. As we mentioned above, later we will deal with the issue of continuous variables by imposing integer restrictions on the decision variables.

Our LP is now complete. We have defined our *decision variables*, constructed our *objective function*, and identified our *constraints*. Below you can see the **LP formulation**, as it is called, of the problem. This format is often used to provide the complete structure of the LP. The problem can be read as follows: Given the decision variables which we have defined, maximize the objective function Z, subject to the constraint set. Next we will see how to use the Solver to select the optimal values of the decision variables. As usual, dialogue boxes will prompt the user to input data and designate particular cells and ranges on a spreadsheet for data input and calculations.

9.3.2 Formulation of YRA Problem

Maximize:

$$Z = 45{,}000X_1 + 63{,}000X_2 + 27{,}500X_3 + 19{,}500X_4 + 71{,}000X_5 + 56{,}000X_6 + 48{,}500X_7$$

Subject to:

$$X_1 \leq 25; X_2 \leq 30; X_3 \leq 47; X_4 \leq 53; X_5 \leq 16; X_6 \leq 19; X_7 \leq 36$$
$$6X_1 + 9X_2 + 4X_3 + 4X_4 + 7X_5 + 10X_6 + 6X_7 \leq 800$$
$$12X_1 + 16X_2 + 10X_3 + 5X_4 + 10X_5 + 5X_6 + 7X_7 \leq 900$$
$$0X_1 + 4X_2 + 4X_3 + 0X_4 + 8X_5 + 7X_6 + 10X_7 \leq 900$$
$$5X_1 + 8X_2 + 5X_3 + 7X_4 + 4X_5 + 0X_6 + 3X_7 \leq 375$$

All X_i, where $i = 1$ to 7, are non-negative.

9.3.3 Preparing a Solver Worksheet

Let us begin the process of using Solver to solve our LP. First, we create a worksheet that enables Solver to perform its analysis. Exhibit 9.2 appears to be quite similar to Exhibit 9.1 in terms of the data contained, but there are some important differences. The differences are essential to the calculations that will be performed, and the worksheet must be prepared in a manner that Solver understands.

There are two areas in the worksheet that must be *set* in order to execute Solver: a *target cell* and *changing cells*. The **target cell** is the cell in which we provide Solver with algebraic instructions on how to calculate the value of the objective function. This is the value of Z in our formulation, and as you recall, it is the sum of the number of projects of each type multiplied by the revenue that each project returns (coefficients). In Exhibit 9.2 it is cell E10, and it is the summation of E2 through E8, the individual contributions by each project type. The values shown in the *target* and *changing* cells are calculated by Solver and will not contain these values until you have executed Solver. Once a solution is obtained, you can perform sensitivity analysis by changing the parameters currently in the problem. The values in the *changing cells* and the *target cell* will be recalculated after the next execution of Solver.

Although we have yet to perform the procedure, a solution is shown in Exhibit 9.2, and we can see that Solver uses the **changing cells** as the cell locations for storing the values of the decision variables, in this case the optimal. Similarly, the optimal value of Z is stored in the *target cell*. This is quite convenient and makes reading the solution simple.

Note below that Solver has produced a solution that selected non-zero, positive values for project types 1, 2, 4, 5, 6, and 7, and a value of 0 for project type 3:

$$X_1 = 5.3; X_2 = 19.7; X_3 = 0; X_4 = 2.6; X_5 = 16.0; X_6 = 19.0; X_7 = 36.0$$

	Home	Insert	Page Layout	Formulas	Data	Review	View	Add-Ins

| A1 | | fx | Project Type |

	A	B	C	D	E	F	G	H
1	Project Type	No. Projects Available	Rev. per Project	Value Selected	Contribution to Rev			
2	1	25	45,000	5.3	$ 239,491.53	hector.guerrero: Changing Cells-- D2:D8		
3	2	30	63,000	19.7	$ 1,244,250.00			
4	3	47	27,500	0.0	$ -	Each cell contains the values selected by Solver to maximize the formulation		
5	4	53	19,500	2.6	$ 51,228.81			
6	5	16	71,000	16.0	$ 1,136,000.00			
7	6	19	56,000	19.0	$ 1,064,000.00			
8	7	36	48,500	36.0	$ 1,746,000.00	hector.guerrero: The formula in this		
9		Total Available		Total Selected	Sum	cell is-- C7*D7		
10		226		98.70	$ 5,480,970.34			
11						hector.guerrero: Target Cell-- E10		
12								
13	Res-A hrs. used	Res-B hrs. used	Res-C hrs. used	Res-D hrs. used		The sum of all projects' contribution to revenue		
14	6	12	0	5				
15	9	16	4	8				
16	4	10	4	5				
17	4	5	0	7				
18	7	10	8	4				
19	10	5	7	0				
20	6	7	10	3				
21		Total Resource Hrs Available in Quarter						
22	800	900	700	375		hector.guerrero: Resource Hours Computation--		
23	738.19	900.00	700.00	375.00				
24						=D14*$D$2+D15*$D$3+D16*$D$4+D17*$D		
25						$5+D18*$D$6+D19*$D$7+D20*$D$8		
26								

Exhibit 9.2 Spreadsheet structure for solver

Value of the maximum Z. . . . . $ 5,480,970.34
Total Projects selected . . . . 98.7.

The solution resulted in fractional values for project types 1, 2, and 4. How we deal with this situation depends on how strongly we feel that the constraints must be strictly maintained. For example, we could simply round every variable in the solution *up* to the next highest value: $X_1 = 6$; $X_2 = 20$; $X_4 = 3$. Rounding these fractional values *up* will require more resource hours. This will violate our constraints, but may or may not create a problem for YRA. To determine the effect on the use of resource hours, insert these values into the worksheet area containing the values of the decision variables, D2:D8.

It is easy to assume a *false-precision* when devising the constraints of a formulation. The constant values of the **right-hand side** of the constraints (which we call RHS) may simply be estimates. For example, the RHS of the Res-A hours constraint, 800, could be 812 or 789. So there may be some latitude in our ability to deal with fractional results. Conversely, if the constraints are *hard and fast*, a solution could be implemented that rounds all, or some, of the fractional values *down* to insure strict adherence to the constraints. YRA should have an understanding of the precision of the data used in the formulation, and they will have to decide how to deal with this issue. Regardless, they have available the formulation to test the adjusted solutions that they develop. We will see later that one of the reports that is produced by Solver will provide information on the **slack** (unused resources) for each constraint. If there is no slack in a constraint, then the entire RHS (maximum amount of the resource) is consumed.

Let us now turn to the use of the Solver dialogue boxes to input our formulation. In Exhibit 9.3 we can see the *target* (E10) and *changing cells* (D2:D8) in the *Solver Parameters* dialogue box. Now we introduce the constraints of the formulation. In row 23 (Exhibit 9.2) we calculate the use of resource hours. The cell comment in D23 shows the calculation for the resource D hour usage. Generally, the formula sums the products of the number of projects selected times the hours consumed by each project type. These four cells (A23:D23) will be used by Solver to determine the use of the RHS (800, 900, 700, and 375) and to insure that the individual RHS's are not exceeded. We must also account for the maximum available projects for each type in B2:B8 (Exhibit 9.2).

How do we enter this information into our formulation? We begin by selecting the Solver in the Analysis group in the Data ribbon. See Exhibit 9.3. Solver should appear in the group, but if it does not, it is because you have not enabled the Solver add-in. To enable the add-in, select the Office button and Options. One of the options is Add-ins, as we noted before with the Data Analysis tools.

The *Solver Parameters* dialogue box is relatively simple, but does require some forethought to use effectively. The entry of the *target cell* and the *changing cells* is straightforward; the entry of constraints is slightly more complex. You can see in the *Subject to the Constraints* area a number of constraints that have been entered

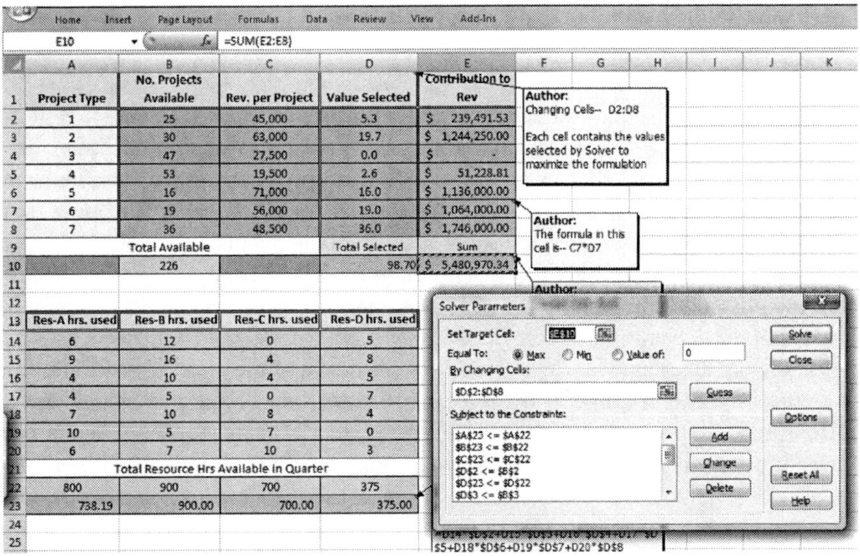

Exhibit 9.3 Solver parameters dialogue box

	Home	Insert	Page Layout	Formulas	Data	Review	View	Add-Ins	

| E10 | ▾ | fx | =SUM(E2:E8) |

	A	B	C	D	E	F	G	H	I
1	Project Type	No. Projects Available	Rev. per Project	Value Selected	Contribution to Rev	Author: Changing Cells-- D2:D8			
2	1	25	45,000	5.3	$ 239,491.53	Each cell contains the values selected by Solver to maximize the formulation			
3	2	30	63,000	19.7	$ 1,244,250.00				
4	3	47	27,500	0.0	$ -				
5	4	53	19,500	2.6	$ 51,228.81				
6	5	16	71,000	16.0	$ 1,136,000.00				
7	6	19	56,000	19.0	$ 1,064,000.00	Author:			
8	7	36	48,500	36.0	$ 1,746,000.00	The formula in this cell is-- C7*D7			
9		Total Available		Total Selected	Sum				
10		226		98.70	$ 5,480,970.34				
11						Author:			
12						Target Cell-- E10			
13	Res-A hrs. used	Res-B hrs. used	Res-C hrs. used	Res-D h	Add Constraint				
14	6	12	0						
15	9	16	4		Cell Reference:		Constraint:		
16	4	10	4		D2 ▦	<= ▾	B2 ▦		
17	4	5	0						
18	7	10	8		OK	Cancel	Add	Help	
19	10	5	7	0					
20	6	7	10	3					
21		Total Resource Hrs Available in Quarter							

Exhibit 9.4 Constraint entry for $X_1 \le 25$

manually. Imagine that we are beginning the process of entering the two types of constraints in our formulation. The first is the maximum number of project types that can be selected in a quarter. By selecting the *Add* button in the constraint area of the dialogue box, the *Add Constraint* dialogue entry box appears. Exhibit 9.4 shows the entry of $X_1 \le 25$. The *Cell Reference* entry is represented by the *changing cell* that contains the value for the project type 1 units selected, D2, and the *Constraint* is the RHS of the constraint, B2 (25). Rather than enter 25 in the *Constraint* area, I have provided a cell location, B2. This will permit you to simply change the cell location value at any time, and the constraint will be automatically updated to the new value when Solver is executed.

The entry of the resource hour constraints is a more complex task. Above we noted that cells A23:D23 are the summation of the products of the decision variables and hours consumed by the use of each project type. Thus, each of these four sums represents the use of each resource hour type due to the selected number of projects. We will use these cells, A23:D23, as the *Cell Reference* entries and the corresponding RHS entries (*Constraint*) are found in A22:D22. In Exhibit 9.3 you can see the first entry (A23 <= A22) represents the constraint for resource A. The creation of the formula in A23 greatly simplifies data entry. Yet, it is possible to enter the constraints directly into the *Constraint* section of the *Add Constraint* dialogue box. We would do so by typing the same formula that appears in cell A23 and then entering the RHS, A22, in the *Cell Reference*. Note that you will have to change the sense of the inequality between these quantities to >= to preserve the proper inequality of the constraint.

9.3.4 Using Solver

Now let us consider the application of Solver and examine the resulting solution. Assuming we have entered all the necessary formulation data and correctly prepared the worksheet for Solver, we are now ready to execute the tool. Exhibit 9.5 shows the dialogue box called *Solver Options*. This dialogue box is available via a button in the *Solver Parameters* box. The top five numerical parameter settings (*Max Time*, etc.) are defaults and the recommended values for most problems. I suggest that you not change these settings, unless, you have a well-informed justification to do so. These settings control the technical aspects of the solution algorithms associated with Solver, and they are important to its precision and functioning.

In Exhibit 9.5 we can also see that the *Assume Linear Model* and *Assume Non-Negative* boxes are checked. Both these conditions apply to our LP model and result in improved calculation performance. When Solver can assume LP characteristics, it simplifies and streamlines the solution procedure and also affects the solution output format. As with the numerical parameter settings, leave the *Estimates*, *Derivatives*, and *Search* settings as they appear. You can now return to the *Solver Parameters* box by selecting the *OK* button, and you are ready to launch Solver by selecting the *Solve* button.

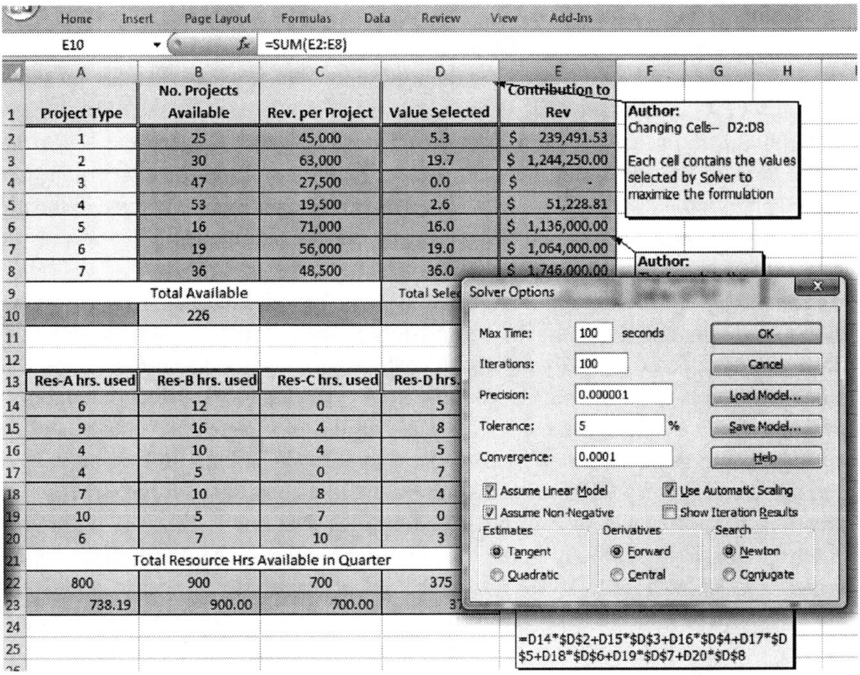

Exhibit 9.5 Running the solver tool

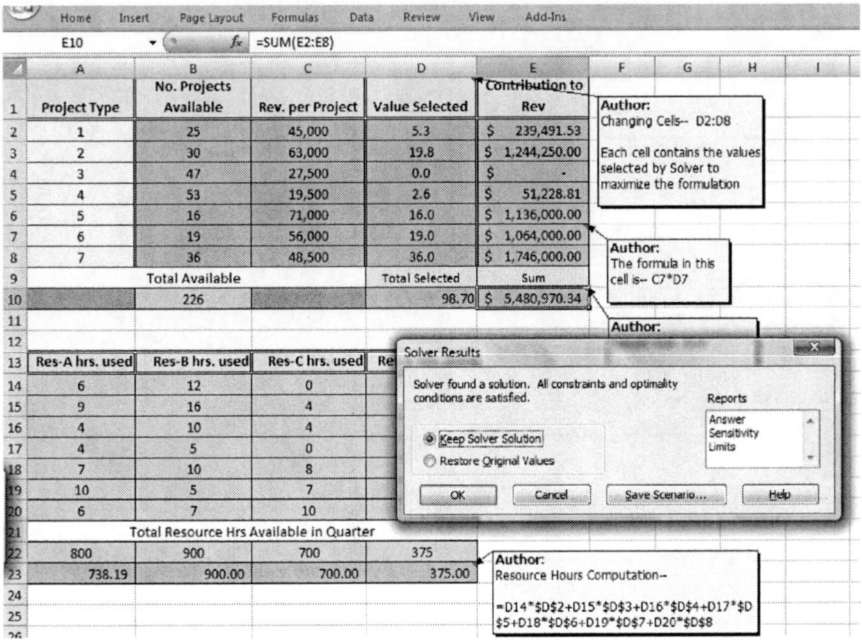

Exhibit 9.6 Implementing solver solutions

Finally, make sure you have selected the *Max* button in Solver Parameters dialogue box, since we are maximizing our objective function. The moment of truth arrives when you receive verification that Solver has produced a feasible solution: a solution that does not violate the constraints and meets the optimality condition of the problem. This appears in the *Solver Results* dialogue box in Exhibit 9.6. It is possible that your formulation may not have a solution, either because the constraints you have imposed simply do not permit a solution, or you have made a mistake in declaring your constraints. In both cases, no combination of feasible decision variable values can be found that satisfy your constraints.

9.3.5 Solver Reports

There are three reports that can be selected in the *Solver Results* box: *Answer*, *Sensitivity*, and *Limits*. We will focus on the *Answer* and *Sensitivity* reports. Exhibit 9.7 shows the *Answer Report*. It provides the basic information for the optimal solution in three output sections:

1) *Target Cell (max)* section: the optimal value of the objective function (Z) determined by Solver—$5,480,970.34
2) *Adjustable Cells* section: the values of the seven decision variables that lead to the optimal solution—$X_1 = 5.3, X_2 = 19.8$, etc.

| Home | Insert | Page Layout | Formulas | Data | Review | View | Add-Ins |

K20   ▾   *fx*

	A	B	C	D	E	F	G
6		Target Cell (Max)					
7		Cell	Name	Original Value	Final Value		
8		$E$10 Sum		$ -	$ 5,480,970.34		
9							
10							
11		Adjustable Cells					
12		Cell	Name	Original Value	Final Value		
13		$D$2 Value Selected		0.0	5.3		
14		$D$3 Value Selected		0.0	19.8		
15		$D$4 Value Selected		0.0	0.0		
16		$D$5 Value Selected		0.0	2.6		
17		$D$6 Value Selected		0.0	16.0		
18		$D$7 Value Selected		0.0	19.0		
19		$D$8 Value Selected		0.0	36.0		
20							
21							
22		Constraints					
23		Cell	Name	Cell Value	Formula	Status	Slack
24		$A$23 Res-A hrs. used		738.19	$A$23<=$A$22	Not Binding	61.80932203
25		$B$23 Res-B hrs. used		900.00	$B$23<=$B$22	Binding	0
26		$C$23 Res-C hrs. used		700.00	$C$23<=$C$22	Binding	0
27		$D$23 Res-D hrs. used		375.00	$D$23<=$D$22	Binding	0
28		$D$2 Value Selected		5.3	$D$2<=$B$2	Not Binding	19.6779661
29		$D$3 Value Selected		19.8	$D$3<=$B$3	Not Binding	10.25
30		$D$8 Value Selected		36.0	$D$8<=$B$8	Binding	0
31		$D$5 Value Selected		2.6	$D$5<=$B$5	Not Binding	50.37288136
32		$D$6 Value Selected		16.0	$D$6<=$B$6	Binding	0
33		$D$7 Value Selected		19.0	$D$7<=$B$7	Binding	0
34		$D$4 Value Selected		0.0	$D$4<=$B$4	Not Binding	47

Exhibit 9.7 Solver answer report

3) *Constraints* section: information relating to the constraints and the amount of resource consumed by the optimal solution.

The information relating to constraints comes in the form of a constraint designation found in the *Status* section, as either **Not Binding** or **Binding**. You can see that there are 11 items in the *Constraint* section corresponding to the formulation's 11 constraints. The first constraint is the resource A usage. This constraint is *Not Binding* and has a slack of 61.81. What does this mean? If we sum the *Slack* and the *Cell Value*, the result is 800 hours, which are the maximum available hours for resource A. Thus, we can interpret this result as follows: the optimal solution consumed 738.19 hours of resource A and therefore had 61.81 hours of unused capacity, or *Slack*. The next constraint is for resource B hours, and the *Status* of the constraint is *Binding*. For this constraint all the available capacity, 900 hours, is consumed and there is no unused capacity, so *Slack* is 0. The same analysis can be applied to the decision variable maximum limits, e.g. $X_1 \leq 25$, $X_2 \leq 30$, etc. Project type 1, which can be found in cell D28, has a value of 5.3. This is less than the maximum value of 25, thus the *Slack* is approximately 19.7. We can state that 19.7 units of project type 1 were not utilized in the solution.

How might we use this information? Let us consider Resource A. If more hours of resource A could be found and added to the RHS, would the objective function benefit by the addition? Why? Currently you have unused hours of A; the constraint has *slack*. The addition of a resource that is currently underutilized cannot be of any value to the objective function. The solution algorithm sees no value to the objective function by adding an additional unit of A. We are far wiser to acquire additional hours of resource B, C, and/or D since their constraints are *binding* and have no *slack*. To demonstrate the point, I will change the formulation to increase the number of resource B hours by 1 hour. Although this is a very minor change, it does lead to different decision variable values and to a higher objective function value as seen in Exhibit 9.8.

The new solution increases the number of project type 1, from 5.3 to 5.4, and reduces the number of project type 4 from 2.6 to 2.5. All other decision variables remain the same, including project type 2. Recall it was not at its maximum (30) in the previous solution and the addition of a single unit of resource B has not caused it to change. The new value of the objective function is $5,484,656.78, which is $3,686.44 greater than the previously optimal solution of $5,480,970.34. In essence, the value of an additional hour of resource B is $3,686.44, and although the changes in decision variables are minor, the change has been beneficial to the objective function. We could perform this type of analysis with all our resource hours to determine the *marginal value* of an additional unit of resource (RHS) for binding constraints.

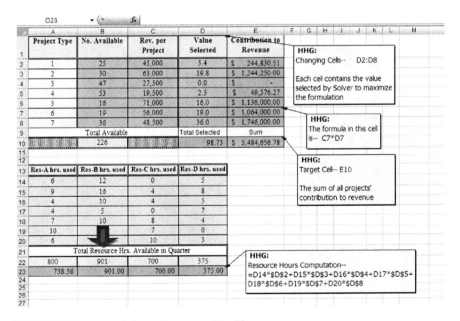

Exhibit 9.8 Incremental change in resource B to 901

Although we do not perform the analysis here, if you increase resource A to 801 hours, I assure you that the solution will not change since it is non-binding and has slack of 61.81 hours.

These types of changes in the problem formulation are a form of sensitivity analysis. It would be convenient to have access to the results of these analyses without having to perform each individual change in the formulation. As you might guess, Excel does provide this type of analysis. It comes in the form of one of the three *Report* options in the *Solver Results* dialogue box (see Exhibit 9.6)—the *Sensitivity Report*. Exhibit 9.9 shows the Sensitivity Report for the YRA formulation. Let us focus on the *Constraints* section of the report, and particularly the column entitled *Shadow Price*. **Shadow prices** are associated with constraints and result from the solution algorithm used to determine an optimal solution. Each of our four resource hour constraints can have a shadow price and it can be interpreted as the benefit to the objective function if the RHS of a constraint is increased by one unit. Additionally, only binding constraints have shadow prices; thus, the resource A constraint will not have a shadow price since it is non-binding. Resource B, C, and D constraints do have a shadow price due to their binding condition, as can be seen in Exhibit 9.9.

As we examine the shadow prices, we can see a very interesting similarity. The shadow price of an additional hour of resource B is 3,686.44, which is precisely the change in objective function value we obtained earlier when we *manually* increased the RHS to 901. The shadow prices of resources C and D are less, 699.15 and 152.54, respectively. Thus, if I am allowed to increase any RHS of the four resource

Microsoft Excel 12.0 Sensitivity Report
Worksheet: [Chapter 9.xls]Budget Problem
Report Created: 12/16/2007 1:58:37 PM

Adjustable Cells

Cell	Name	Final Value	Reduced Cost	Objective Coefficient	Allowable Increase	Allowable Decrease
$D$2	Value Selected	5.3	0.0	45000	1800.000004	6160.958905
$D$3	Value Selected	19.8	0.0	63000	6098.305086	2796.610169
$D$4	Value Selected	0.0	-12.923.7	27500	12923.7288	1E+30
$D$5	Value Selected	2.6	0.0	19500	10312.5	750.0000016
$D$6	Value Selected	16.0	27,932.2	71000	1E+30	27932.20339
$D$7	Value Selected	19.0	32,673.7	56000	1E+30	32673.72882
$D$8	Value Selected	36.0	15,245.6	48500	1E+30	15245.76271

Constraints

Cell	Name	Final Value	Shadow Price	Constraint R.H. Side	Allowable Increase	Allowable Decrease
$A$23	Total Resource Hrs. Available in Quarter	738.19	-	800	1E+30	61.80932203
$B$23	Res-B hrs. used	900.00	3,686.44	900	31	44.85714286
$C$23	Res-C hrs. used	700.00	699.15	700	17.44444444	64.5
$D$23	Res-D hrs. used	375.00	152.54	375	62.8	12.91666667

Exhibit 9.9 Sensitivity report for YRA

hour constraints, the most valuable hour is for resource B, followed by C, then D. Of course, it is important to consider the cost of each extra unit of resource. For example, if I can obtain an additional hour of resource D for $155, it would not be wise to make the investment given that the return will only lead to $152.54 in benefit the objective function. The results would be a net loss of $2.46.

In the last two columns of the *Constraints* area, additional valuable information is provided regarding the **Allowable Increase** and **Allowable Decrease** for the change in the RHS. As the titles suggest, these are the allowable changes in the RHS of each constraint for the shadow price to remain at the same value. So, we can state that if our increase of resource B had been for a 30 hour increase (the *Allowable Increase* is 31), our return would be 110,593.20 (30*3686.44). Beyond 31 units, there may still be a benefit to the objective function, but it will not have the same shadow price. In fact, the shadow price will be lower.

In the *Adjustable Cells* section of the *Sensitivity Report*, we also have information regarding the *estimated* opportunity cost for variables. These opportunity costs are found in the **Reduced Cost** column, and they apply to variables in our solution that are currently "0"; they relate to the *damage* done to the objective function if we force a variable to enter the solution. We have one such variable, X_3. If we force the type 3 project variable to equal 1, that is, we require our solution to accept exactly one of these contracts, the reduction in the objective function value will be approximately $12,923.90. Why do we expect a reduction in the objective function value Z? The value of Z will be reduced because forcing the entry of a variable that was heretofore set to "0" by the optimization algorithm will obviously damage, rather than help, our solution. The reduced costs for the X_5, X_6, and X_7, although appearing to have a value, cannot be interpreted as the positive effect on the objective function if we increase the RHS by one unit, e.g. $X_5 = 17$.

One last, but important, use of the *Sensitivity Report* is related to the *Allowable increase* and *Allowable decrease* in the coefficients of the objective function variables (revenues for projects). These increases and decreases represent changes that can be made, one at a time, without changing the solution values for the decision variables. The objective function value will change, but not the selected decision variable values. As often is the case with complex problems, uncertainty will play a part in our analysis. The objective function coefficients of our decision variables are likely to be estimates that are dependent on many factors. This form of analysis will therefore be very useful in understanding the range of possible coefficient change, without affecting the current decision variable values.

9.3.6 Some Questions for YRA

Now, let us put our understanding of LP to the test. Elizabeth has received an urgent message from Nick. He wants her to meet him at a local coffee shop, the Daily Grind, to discuss some matters related to the quarterly budgeting process they have been working on. The following discussion occurs at the Daily Grind:

Elizabeth: Nick, your phone call sounded like something related to the budget needs urgent attention.

Nick: It does. I have been talking to a couple of our big clients and they are upset with the idea that we may not choose to accept their projects. They are threatening to take their business elsewhere.

Elizabeth: That is a problem. What project types are they talking about?

Nick: The problem is that they want us to take on 2 projects of type 3, and as you know, the solution that you determined suggested that we should not take on *any* type 3 projects. They are terrible projects from the standpoint of the low revenue they return, and the high resource hours they consume.

Elizabeth: Yes I know, they are a disaster for YRA, but if we want to keep these clients I guess we have to bend a bit and accept a couple of type 3 projects. It's the price of being customer focused.

Nick: There is also another issue we have to talk about. We need to reconsider the revenue we assumed for project type 2 and 4. I believe these are coefficients in the objective function of the formulation. Well, I think that we may have been inaccurate for both. In the case of project type 2, the revenue is more accurately between $60,500 and $68,000. For project type 4, the range is likely to be between $19,000 and $32,000.

Elizabeth: Nick, this could be a serious issue. I'll have to take a look at the sensitivity analysis for the solution. We may be fine with the current selection of projects, but with the requirement that we must select 2 projects of type 3, we will clearly have to make changes.

Nick: Also, while you are looking, can you also consider an increase in resource C of 12 hours. You know that many of the numbers we used in the formulation were point estimates of some range of values, and I have a feeling I was incorrect about that one.

Elizabeth: Nick, I want to use the current *Sensitivity Report* wherever I can, but I may have to reformulate and solve the problem again. Don't worry; these things happen all the time. Solver can answer many of the questions.

Nick: Thanks for stopping by; we really need to answer these questions quickly.

Elizabeth: You are right. I should have answers for you by tomorrow.

Elizabeth considered the tasks ahead of her. She decided to start with the issue of accepting 2 of the unprofitable type 3 projects. Depending on the results with this analysis, she then will consider the issues related to the revenues (coefficients of the objective function), and finally, the change in the resource hours.

She begins by returning to Exhibit 9.9 where she finds that the *Reduced Cost* of the *Project Type 3* is approximately −12,924 (see cell E11). As you recall, this number represents the penalty associated with forcing a unit of a decision variable into the solution that is currently set to 0. Therefore an estimate of the *reduction* of the objective function by including 2 units of type 3 projects is approximately $25,848. Exhibit 9.10 verifies that this is approximately the change to the objective function if we re-solve the LP with a change in the constraint for resource type 3.

	Home	Insert	Page Layout	Formulas	Data	Review	View	Add-Ins

J15   ▾   fx

	A B	C	D	E	F	G	H
1	Microsoft Excel 12.0 Answer Report						
2	Worksheet: [Chapter 9.xls]Budget Problem						
3	Report Created: 12/16/2007 6:29:44 PM						
4							
5	Target Cell (Max)						
6	Cell	Name		Original Value	Final Value		
7	$E$10 Sum			$ 5,480,970.34	$ 5,455,122.88		
8							
9	Adjustable Cells						
10	Cell	Name		Original Value	Final Value		
11	$D$2 Project Type 1			5.3	6.2		
12	$D$3 Project Type 2			19.8	17.7		
13	$D$4 Project Type 3			0.0	2.0		
14	$D$5 Project Type 4			2.6	2.8		
15	$D$6 Project Type 5			16.0	16.0		
16	$D$7 Project Type 6			19.0	19.0		
17	$D$8 Project Type 7			36.0	36.0		
18							
19							
20	Constraints						
21	Cell	Name		Cell Value	Formula	Status	Slack
22	$A$23 Res-A hrs. used			734.50	$A$23<=$A$22	Not Binding	65.50423729
23	$B$23 Res-B hrs. used			900.00	$B$23<=$B$22	Binding	0
24	$C$23 Res-C hrs. used			700.00	$C$23<=$C$22	Binding	0
25	$D$23 Res-D hrs. used			375.00	$D$23<=$D$22	Binding	0
26	$D$8 Project Type 7			36.0	$D$8<=$B$8	Binding	0
27	$D$2 Project Type 1			6.2	$D$2<=$B$2	Not Binding	18.76271186
28	$D$3 Project Type 2			17.7	$D$3<=$B$3	Not Binding	12.25
29	$D$5 Project Type 4			2.8	$D$5<=$B$5	Not Binding	50.16949153
30	$D$6 Project Type 5			16.0	$D$6<=$B$6	Binding	0
31	$D$7 Project Type 6			19.0	$D$7<=$B$7	Binding	0
32	$D$4 Project Type 3			2.0	$D$4=2	Not Binding	0
33							

Exhibit 9.10   Answer report for YRA with 2 units project type 3

The new solution, appearing in the *Final Value* column, is shown next to the old solution values, in the *Original Value* column. The inclusion of an *Original Value* and a *Final Value* is a convenient mechanism for comparison. The change from the previous objective function value, $5,480,970.34, is to a value $25,847.46 lower, $5,455,122.88. The reduction is very close to $25,848. The new decision variables selected look familiar, but with a few significant changes:

$$X_1 = 6.2; X_2 = 17.7; X_3 = 2.0; X_4 = 2.8; X_5 = 16.0; X_6 = 19.0; X_7 = 36.0$$

Elizabeth's reply to Nick's first issue is that complying with the customers' request for selecting 2 units of project type 3 will lead to a $25,848 loss, and the number of the other projects selected will change slightly. All in all, this is not a terribly disruptive outcome given the necessity to satisfy important clients.

Now for the remaining questions, consider the action Elizabeth has taken to re-solve the problem under the new constraint condition. If she is to answer the question related to the change in revenue, she must be careful with the project type 3 changes imposed on the solution, above. In the case of the small change Nick suggested (2 projects), we can see from comparing the *Sensitivity Reports* in Exhibits 9.9 and 9.11, *pre* and *post* inclusion of project 2, that the *Adjustable Cells* area is unchanged (the Constraints area is changed). The reduced costs for both reports are identical, but if the change to the project type 3 was greater, for example 29 units, the two reports would not be the same. Thus, she can answer the questions without regard to the change (2 required) in type 3 projects. Since it is difficult to know how large a change is necessary to change the *Reduced Costs* and related *Allowable Increase* and *Decrease*, it is wise to re-solve the problem and not take the risk of missing the change. And now for the answer to Nick's question about changes in revenue:

1) The allowable change in the revenue of type 2 projects is from $60,203.39 to $69,098.31. This range includes the $60,500.00 to $68,000.00 that Nick has proposed; thus, the current solution will not be changed by Nick's change in type 2 revenue.

2) The allowable change for type 4 projects is $18,750.00 to $29,812.50, which includes Nick's lower boundary of $19,000.00, but the upper boundary, $32,000.00, exceeds the allowable upper boundary. Therefore, if Nick believes that the revenue can indeed be greater than $29,812.50, then the decision variables in the current optimal solution will be changed. The current revenue of $19,500.00 appears to be quite low given Nick's new range, and he should probably revisit his point estimate and attempt to improve his estimate.

Finally, Nick has asked Elizabeth to determine the effect of increasing the RHS of the resource C by 12 hours from 700 to 712. We can see from Exhibit 9.11 that the *Shadow Price* for the resource is 699.15, and we are within the allowable *Constraint R.H. Side*, 638.5 (700–61.5) and 720.44 (700+20.44). The results will be an increase to the objective function of $8389.8 (12*699.15).

Microsoft Excel 12.0 Sensitivity Report
Worksheet: [Chapter 9.xls]Budget Problem
Report Created: 12/6/2007 6:29:44 PM

Adjustable Cells

Cell	Name	Final Value	Reduced Cost	Objective Coefficient	Allowable Increase	Allowable Decrease
$D$2	Project Type 1	6.2	0.0	45000	1800.000001	6160.958904
$D$3	Project Type 2	17.7	0.0	63000	6098.305085	2796.610169
$D$4	Project Type 3	2.0	-12,923.7	27500	12923.72881	1E+30
$D$5	Project Type 4	2.8	0.0	19500	10312.5	750.0000004
$D$6	Project Type 5	16.0	27,932.2	71000	1E+30	27932.20339
$D$7	Project Type 6	19.0	32,673.7	56000	1E+30	32673.72881
$D$8	Project Type 7	36.0	15,245.8	48500	1E+30	15245.76271

Constraints

Cell	Name	Final Value	Shadow Price	Constraint R.H. Side	Allowable Increase	Allowable Decrease
$A$23	Res-A hrs. used	734.50	-	800	1E+30	65.50423729
$B$23	Res-B hrs. used	900.00	3,686.44	900	33.4	52.57142857
$C$23	Res-C hrs. used	700.00	699.15	700	20.44444444	61.5
$D$23	Res-D hrs. used	375.00	152.54	375	73.60000001	13.91666667

Exhibit 9.11 Sensitivity report for new solution

The value of LP is not just the determination of a set of decision variables that optimize an objective function. As important as the optional solution, are the uses of the sensitivity analysis associated with the solution. We have seen the power of the *shadow price, reduced cost,* and *allowable changes of coefficients* and *RHS's* in our YRA example. Since LP is a deterministic procedure (all parameters are point estimates), sensitivity analysis permits the consideration of ranges of parameter values, even though we have represented the ranges with point estimates. Thus, the reality of the uncertainty and variation associated with parameters can be recognized and investigated.

Solver is not restricted to the solution of linear programs. A related class of problems is known as **Non-Linear Programs** (NLP) that, broadly defined, contain non-linear relationships in the objective function and/or in constraints. These can be very difficult problems to solve. Additionally, there are **Integer Programs** (IP) where the decision variables are restricted to integer values, **Mixed Integer Programs** (MIP) where decision variables can be both continuous (fractional value) and integer values, and **0–1 Integer Programs** where variables are binary (having two states). Again, these conditions require substantially more complicated solution algorithms than LP.

Earlier, we ignored the integer nature of the decision variables by assuming that we could simply round the variables and not worry about the potential violation of constraints, as long as they were not severe. What if the integer condition is important? We can impose integer values on the decision variables by adding constraints to the problem. Exhibit 9.12 demonstrates how to use the *Add Constraint* dialogue box to declare the number of project type 1 to be restricted to integer. By selecting the *int* designation in the pull-down menu between the *Cell Reference* and *Constraint*, we can convert the decision variables to integers. The *Add Constraint* dialogue box shows the integer constraint for cell D2, project type 1. Exhibit 9.13 shows the new solution (after imposing integer conditions on all decision variables) where we can clearly see that all variables result in integer values, and the value of the associated objective function is $5,477,500.00, not a great deal below the continuous variable solution. The projects selected are:

$$X_1 = 7.0; X_2 = 19.0; X_3 = 0.0; X_4 = 1.0; X_5 = 16.0; X_6 = 19.0; X_7 = 36.0$$

Note though, that the *Sensitivity Report* for integer and binary programs are no longer valid due to the algorithm used to solve these problems. This of course is unfortunate and a weakness of the approach since we will no longer have the sensitivity analysis available to answer questions. Now, let us move on to another applications tool—Scenarios.

9.4 Scenarios

The Scenarios tool is one of the *what-if analysis* tools in Excel. It also has been incorporated into Solver as a button on the *Solver Results* dialogue box. The basic function of Scenarios is to simplify the process of management, record keeping, and

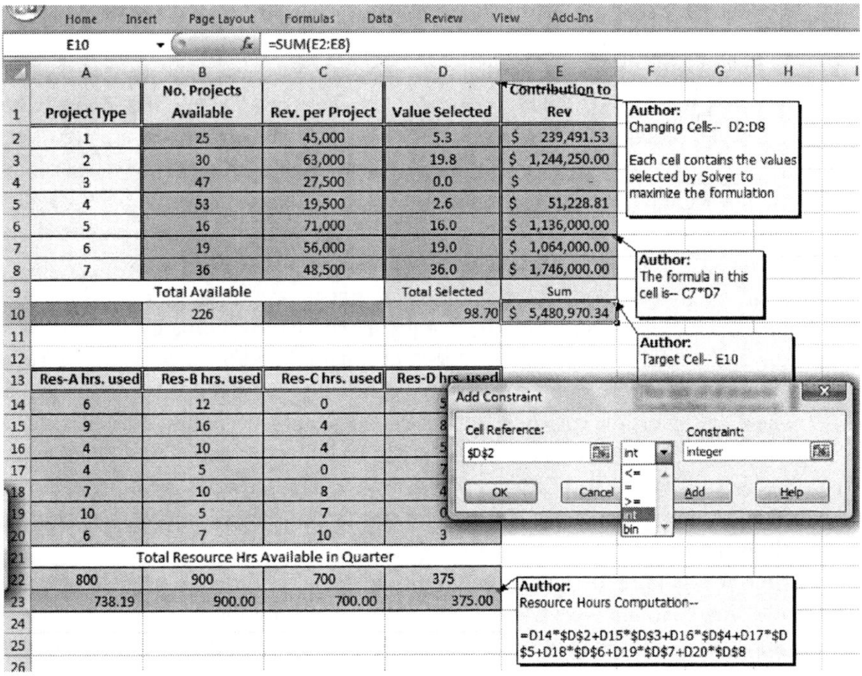

	E10		▾	f_x	=SUM(E2:E8)			

	A	B	C	D	E	F	G	H	I
1	Project Type	No. Projects Available	Rev. per Project	Value Selected	Contribution to Rev				
2	1	25	45,000	5.3	$ 239,491.53				
3	2	30	63,000	19.8	$ 1,244,250.00				
4	3	47	27,500	0.0	$ —				
5	4	53	19,500	2.6	$ 51,228.81				
6	5	16	71,000	16.0	$ 1,136,000.00				
7	6	19	56,000	19.0	$ 1,064,000.00				
8	7	36	48,500	36.0	$ 1,746,000.00				
9		Total Available		Total Selected	Sum				
10		226		98.70	$ 5,480,970.34				
11									
12									
13	Res-A hrs. used	Res-B hrs. used	Res-C hrs. used	Res-D hrs. used					
14	6	12	0	5					
15	9	16	4	8					
16	4	10	4	5					
17	4	5	0	7					
18	7	10	8	0					
19	10	5	7	0					
20	6	7	10	3					
21		Total Resource Hrs Available in Quarter							
22	800	900	700	375					
23	738.19	900.00	700.00	375.00					
24									
25									
26									

Author:
Changing Cells-- D2:D8

Each cell contains the values selected by Solver to maximize the formulation

Author:
The formula in this cell is-- C7*D7

Author:
Target Cell-- E10

Add Constraint

Cell Reference: Constraint:
$D$2   int   integer

OK   Cancel   Add   Help

int
bin

Author:
Resource Hours Computation--

=D14*$D$2+D15*$D$3+D16*$D$4+D17*$D $5+D18*$D$6+D19*$D$7+D20*$D$8

Exhibit 9.12 Imposition of integer decision variables

entry of data for the repeated calculation of a spreadsheet. It is often the case that we are interested in asking repeated *what-if* questions of a spreadsheet model. The questions are generally of the form—what if we change the inputs to our model to *this*, then to *this*, then to *this*, etc.

You will recall that we dealt with this question when we introduced Data Tables. Data Tables display the value of a particular calculation as one or two inputs are varied. Although this is a powerful tool, what if we have many more than two inputs to vary? We may need to construct many Data Tables, but the comparison between tables will be difficult at best. Scenarios permit you to determine the changes in a calculated value while varying as many as 32 inputs and each different set of input values will represent a scenario.

9.4.1 Example 1—Mortgage Interest Calculations

After many years of hard work, Padcha Chakravarty has experienced great success in her import-export business. So much so that she is considering the purchase of a yacht that she can claim as a second home. It meets the United States Internal Revenue Service criteria for a second home by being capable of providing "sleeping, cooking, and toilet facilities", and it is a very convenient way to reduce her

	A B	C	D	E	F	G
1	Microsoft Excel 12.0 Answer Report					
2	Worksheet: [Chapter 9.xls]Budget Problem					
3	Report Created: 12/16/2007 7:52:21 PM					
4						
5	Target Cell (Max)					
6	Cell	Name	Original Value	Final Value		
7	$E$10 Sum		$ 5,480,970.34	$ 5,477,500.00		
8						
9						
10	Adjustable Cells					
11	Cell	Name	Original Value	Final Value		
12	$D$2 Project Type 1		5.3	7.0		
13	$D$3 Project Type 2		19.8	19.0		
14	$D$4 Project Type 3		0.0	0.0		
15	$D$5 Project Type 4		2.6	1.0		
16	$D$6 Project Type 5		16.0	16.0		
17	$D$7 Project Type 6		19.0	19.0		
18	$D$8 Project Type 7		36.0	36.0		
19						
20						
21	Constraints					
22	Cell	Name	Cell Value	Formula	Status	Slack
23	$A$23 Res-A hrs. used		735.00	$A$23<=$A$22	Not Binding	65
24	$B$23 Res-B hrs. used		900.00	$B$23<=$B$22	Binding	0
25	$C$23 Res-C hrs. used		697.00	$C$23<=$C$22	Not Binding	3
26	$D$23 Res-D hrs. used		366.00	$D$23<=$D$22	Not Binding	9.00000001
27	$D$8 Project Type 1		36.0	$D$8<=$B$8	Binding	0
28	$D$2 Project Type 2		7.0	$D$2<=$B$2	Not Binding	18
29	$D$3 Project Type 3		19.0	$D$3<=$B$3	Not Binding	11
30	$D$5 Project Type 4		1.0	$D$5<=$B$5	Not Binding	52
31	$D$6 Project Type 5		16.0	$D$6<=$B$6	Binding	0
32	$D$7 Project Type 6		19.0	$D$7<=$B$7	Binding	0
33	$D$8 Project Type 7		36.0	$D$8=integer	Binding	0.0
34	$D$2 Project Type 1		7.0	$D$2=integer	Binding	0.0
35	$D$3 Project Type 2		19.0	$D$3=integer	Binding	0.0
36	$D$4 Project Type 3		0.0	$D$4=integer	Binding	0.0
37	$D$5 Project Type 4		1.0	$D$5=integer	Binding	0.0
38	$D$6 Project Type 5		16.0	$D$6=integer	Binding	0.0
39	$D$7 Project Type 6		19.0	$D$7=integer	Binding	0.0
40	$D$5 Project Type 7		1.0	$D$5<=$B$5	Not Binding	52

Exhibit 9.13   Answer report for integer variables

tax burden in the coming years. The mortgage interest deduction is one of the few remaining personal income tax deductions available in the US tax code.

Padcha has decided that a short term mortgage of 4–6 years (these are the shortest terms she can find) is in her best interest since she may sell the yacht soon (2–3 years) after the capture of the initial tax advantages. Knowing that mortgage payments consist overwhelmingly of interest in early years, she is interested in finding a loan structure that will lead to a beneficial interest tax deduction while satisfying other criteria.

Padcha decides to construct a spreadsheet that calculates the cumulative interest paid for two years for numerous scenarios of principal, term, and interest rate. She has discussed the problem with a yacht broker in Jakarta, Indonesia, and he has provided six yacht options for her to consider. He is willing finance the purchase, and has forwarded the following scenarios to Padcha in Table 9.1:

A spreadsheet for the calculation of the scenarios is shown in Exhibit 9.14. In Exhibit 9.14 we introduce a new cell formula (see C18 and C19) that is part of the financial cell formulas contained in Excel—**CUMIPMT** (rate, nper, pv,

Table 9.1 Scenarios for Yacht purchase

Yacht	A	B	C	D	E	F
Interest (%)	7	6.75	6.5	6.25	6	5.75
No. of Periods	72	72	60	60	48	48
Principal		160,000	150,000	140,000	180,000	330,000	360,000

start_period, end_period, type). It calculates the cumulative interest paid over a specified number of time periods and contains the same arguments as the PMT cell formula. There are also two additional inputs, start_period and end_period; they identify the period over which to accumulate interest payments.

For Padcha's mortgage problem, the periods of interest are the first year (1–12) and the second year (13–24). This suggests the first payment will begin in January and the last will be in December. Since income taxes are paid annually, it makes sense to accumulate over a yearly time horizon. Of course, if payments do not begin in January, we must select the end_period to reflect the true number of payments

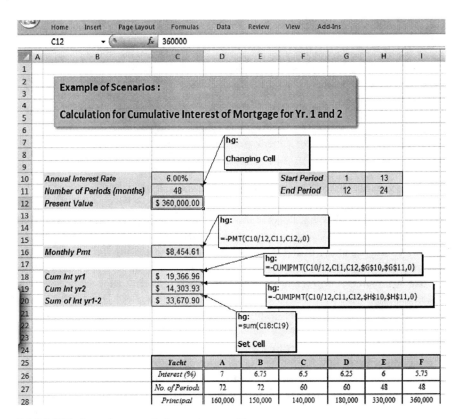

Exhibit 9.14 Scenarios example for mortgage problem

Exhibit 9.15   Creating scenarios

of interest in the initial year. For example, if we begin payment in the month of September, the start_period is 1 and the end_period is 4, indicating that we accumulated interest payments for four months, September through December. At the bottom of Exhibit 9.14 are the values of the six scenarios, A through F, for Padcha's model.

So how do we create a scenario? The process of creating scenarios is shown in Exhibit 9.15a, 9.15b, and 9.15c and is described as follows:

1. We begin by engaging the What-If Analysis tools in the Data Ribbon and Data Tools Group. Section *a* of Exhibit 9.15, *Scenario Manager*, shows the first dialogue box encountered. As you can see, no scenarios are currently defined.
2. In section *b* of Exhibit 9.15 we depress the *Add* button, and the *Edit Scenario* dialogue box becomes available. Here we name scenarios and identify *Changing Cells*: the cells that contain the data inputs for the calculations of interest.
3. Next, the *Scenarios Values* dialogue box permits entry of the individual values for the cells, as shown in section *c* of Exhibit 9.15. Note that Excel recognizes the cell name for C10:C12—C10 as *IntRate*, C11 as *NumPeriods*, and C12 as *Principal*. The cell ranges were *named* in the spreadsheet for ease of identification.
4. The process is repeated for each scenario by selecting the *Add* button on the *Scenarios Values* dialogue box.
5. When you return to the *Scenario Manager* dialogue box by selecting *OK*, the named scenarios will appear in the window.
6. Finally, we are able to select the *Summary* button to generate a report, as either a *Scenario summary* or a *Scenario PivotTable report*, as shown in the *Scenario Summary* dialogue box in section *c* of Exhibit 9.15.

The resulting *Scenario summary* report is shown in Exhibit 9.16. In this report, I also have named the results cells: (1) *MntlyPmt* is monthly payment for the mortgage, (2) *CumIntyr1* and *CumIntyr2* are cumulative interest payments in years 1 and 2, respectively, and (3) *SumIntyr1_2* is the sum of year 1 and 2 cumulative interest. The report provides a convenient format for presenting comparative results. If

	Current Values:	A	B	C	D	E	F
Scenario Summary							
Changing Cells:							
IntRate	6.00%	7.00%	6.50%	6.50%	6.25%	6.00%	5.75%
NumPeriods	48	72	72	60	60	48	48
Principal	$ 360,000.00	$ 160,000.00	$ 150,000.00	$ 140,000.00	$ 180,000.00	$ 330,000.00	$ 360,000.00
Result Cells:							
MntlyPmt	$8,454.61	$2,727.84	$2,521.49	$2,739.26	$3,500.87	$7,750.06	$8,413.41
CumIntyr1	$ 19,366.96	$ 10,495.50	$ 9,127.87	$ 8,378.87	$ 10,363.36	$ 17,753.05	$ 18,550.64
CumIntyr2	$ 14,303.93	$ 8,887.87	$ 7,712.75	$ 6,738.58	$ 8,317.13	$ 13,111.94	$ 13,685.15
SumIntyr1_2	$ 33,670.90	$ 19,383.38	$ 16,840.62	$ 15,117.46	$ 18,670.49	$ 30,864.99	$ 32,235.79

Notes: Current Values column represents values of changing cells at time Scenario Summary Report was created Changing cells for each scenario are highlighted in gray.

Exhibit 9.16 Scenario summary for mortgage problem

Padcha believes she would like to generate the highest interest deduction possible, she may consider either scenarios E or F. If more modest interest deductions are more appealing, then scenarios B and C are possible. Regardless, she has the entire array of possibilities to choose from, and she may be able to generate others based on the results she has observed, for example the Current Values shown in column D. This ability to manage multiple scenarios is a very attractive feature in spreadsheet analysis.

9.4.2 Example 2—An Income Statement Analysis

We now consider a slightly more complex model for scenario analysis. In this example, we consider a standard income statement and a related set of scenarios that are provided by a decision maker. The decision maker would like to determine the *bottom-line* (net profit) that results from various combinations of input values. In Exhibit 9.17 we can see that we have 7 input variables and each variable has two possible values. This is not a particularly complex problem, but with a greater number of possible input values, this problem could easily become quite cumbersome. The 7 input values represent standard inputs that are often estimated in *proforma* Income Statement analysis:

- Sales Revenue = (Volume)(Price)
- COGS = (percentage[4])(Sales Revenue)
- Variable Operating Expense = (percentage)(Sales Revenue)
- Fixed Operating Expenses
- Depreciation Expense
- Interest Expense

---

[4] The estimation of Cost of Goods Sold (COGS) and Variable Operating Expense as a percentage (%) of Sales Revenue is common approach to estimation of Income Statements, but not an approach without its detractors.

	A	B	C	D	E	F	G	H	I
						f_x			
1	**Brain**								
2				**Uncertain Events**					
3		Changing Cells							
4		2.25	4000000	25%	35%	300000	300000	250000	
5									
6		Possible Scenario Combinations				Fixed Expenses			
7		Price	Volume	COGS	VarExp	Operating	Deprtn	Interest	
8		2.25	2000000	45%	25%	300000	250000	170000	
9		1.75	4000000	25%	35%	450000	300000	250000	
10									
11				**Income Tax-Progressive**					
12				500000	or below	23%			
13				500001	or above	34%			
14	**Calculation**								
15				**Income Statement**					
16									
17		Sales Revenue					9,000,000.00		
18		Cost of Goods Sold Expense					2,250,000.00		
19				Gross Margin				6,750,000.00	
20		Variable Operating Expenses					3,150,000.00		
21				Contribution Margin				3,600,000.00	
22		Fixed Expenses							
23				Operating Expenses			300,000.00		
24				Depreciation Expense			300,000.00		
25				Operating Earnings (EBIT)				3,000,000.00	
26		Interest Expense					250,000.00		
27				Earnings before income tax				2,750,000.00	
28		Income Tax expense					880,000.00		
29				Net Income				1,870,000.00	
30									

Exhibit 9.17   Income statement analysis example

Obviously, we cannot use a two variable Data Table for this type of analysis; there are too many variables to consider simultaneously. This example is an excellent use of the Scenarios tools. Exhibit 9.18 shows the results of the 7 scenarios. They range from a loss of $300,000 to a gain of $1,870,000.

9.5 Goal Seek

There is another tool in Excel's *What-If Analysis* sub-group, Goal Seek. It is similar to Solver, except that it functions in reverse: it determines the value of an input that will *result* in a specified output. While Solver can manipulate numerous variables and has a generalized goal to maximize or minimize the objective function, Goal Seek knows *a priori* the goal and must find a single variable value, among several, to arrive at the goal. For example, assume that you want to have a payment of exactly $1000 for a loan. There are three inputs in the PMT function—interest rate, number

		Current Values:	Base	S1	S2	S3	S4	S5	S6
Scenario Summary									
Changing Cells:									
	Price	2.25	2.25	2.25	1.75	2.25	1.75	1.75	1.75
	Volume	2000000	2000000	2000000	2000000	4000000	4000000	2000000	2000000
	COGS	45%	45%	25%	45%	25%	45%	45%	45%
	VarExp	25%	25%	25%	35%	35%	35%	25%	35%
	FxExOpng	300000	300000	450000	450000	300000	450000	450000	450000
	FxExDptn	250000	250000	300000	300000	300000	300000	300000	250000
	Interest	170000	170000	170000	250000	250000	250000	250000	170000
Result Cells:									
	BtmLine	470,800.00	470,800.00	932,800.00	(300,000.00)	1,870,000.00	308,000.00	38,500.00	(170,000.00)

Notes: Current Values column represents values of changing cells at
time Scenario Summary Report was created. Changing cells for each
scenario are highlighted in gray.

Exhibit 9.18 Income statement scenarios

of periods, and present value of the loan principal. Goal Seek will allow the user
to select one of the three inputs, such that it will result in a payment of $1000 per
period. It is a limited tool in that it will permit only a single variable to be changed to
arrive at the goal. Thus, it is not possible to vary interest rate and number of periods
and present value simultaneously.

In the next section we will examine two examples that demonstrate the power
and the pitfalls of Goal Seek. The first example is relatively simple and relates to the
calculation of Padcha's loan, in particular the PMT function. The second example
is a more complex application related to Padcha's problem of accumulating interest
in years 1 and 2, and it utilizes the CUMIPMT cell function. Although the PMT
function is similar to the CUMIPMT function, the application of Goal Seek to the
latter cell function is somewhat problematic.

9.5.1 Example 1—Goal Seek Applied to the PMT Cell

Consider the mortgage example we introduced in the Scenarios section. Imagine
that Padcha has determined the yacht that she will purchase, the Queen of Malacca,
along with its price, $240,000. The broker for the yacht has agreed to finance at an
interest rate of 7%; he is anxious to sell the Queen of Malacca due to some rather
unfortunate history of the yacht's previous owners—pirates and gun runners. He is
not concerned with the *term* of the loan as long as he gets an agreement to purchase.
Padcha sees an opportunity to *set* a loan payment and determine the term that will
be implied given the broker's interest rate and the principal of the loan. She decides
that $5000 per month is a very manageable loan sum for her. Exhibit 9.19 shows the
Goal Seek dialogue box for Padcha's problem. There are three entries:

1. *Set cell* entry is the cell that she will set as a goal—*Monthly Pmt*, C16.
2. *To value* is the value she selects for the *Set cell*—$5000.
3. *By changing cell* is the cell where changes will be permitted—*Number of periods*
   (months), C11.

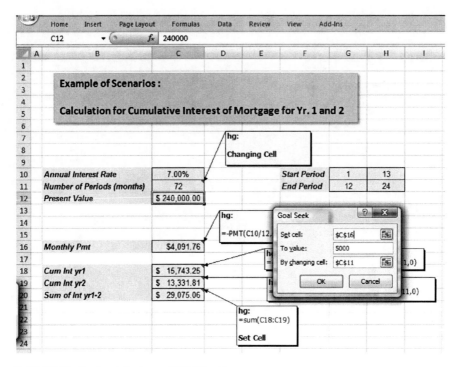

Exhibit 9.19   Goal seek for term of PMT function

Exhibit 9.20 shows the results of the Goal Seek. The term that will lead to a loan payment of $5000 per month is 56.47907573 months, or approximately 56. The solution is found in a fraction of a second; thus, you could perform many *what-if* scenarios with little effort and in a minimal amount of time. Now, let us move to the next example to see how we might run into problems with Goal Seek in more complex situations.

9.5.2   Example 2—Goal Seek Applied to the CUMIPMT Cell

Suppose that Padcha, after some consideration, has decided that she would like the sum of two years of cumulative interest to be exactly $25,000: this is her new goal. As before, she has decided on the level of investment she would like to make, $240,000, and the interest rate that the yacht broker will offer on financing the purchase is 7%. Thus, the variable that is available to achieve her goal is the term of the loan. This appears to be an application of the Goal Seek tool quite similar to Example 1. As before, the tool seeks to obtain a goal for a calculated value, by manipulating a single input. Note that the calculated value is much more complex than before (CUMIPMT), but why should that make a difference? In fact, this more

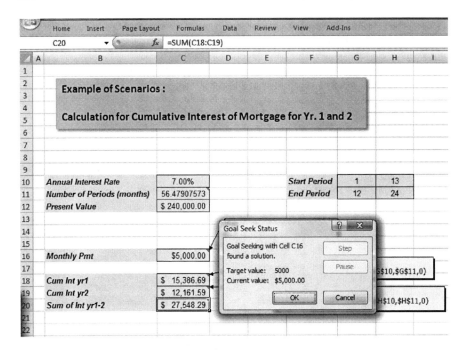

Exhibit 9.20 Goal seek solution to PMT of $5000

complex calculation may make a very significant difference in the application of
Goal Seek.

We will repeat the Goal Seek for a new set of inputs, and now we will change
the *Set cell* entry to C20, the sum of two years of accumulated interest, and
the *To value* entry to $25,000. The *Changing cell* entry will remain C11. In
Exhibit 9.21 we see the new Goal Seek entry data, and in Exhibit 9.22 the results
of the Goal Seek analysis. The results are a bit troubling in that the dialogue box
indicates that the tool "*may not have found a solution.*" How is this possible? The
algorithm used to find solutions is a search technique that does not guarantee a solution in all cases. Additionally, these types of algorithms are often very sensitive to
where the search starts, i.e. they use the value that is currently in the cell to begin
the search for the goal. In the case of Exhibit 9.21, the changing cell contained 48
periods, so this is where the search began. The search terminated at 24 periods and a
cumulative sum of $26,835.08, but the tool was unsure of the solution. The problem
we face is that it is impossible to achieve a $25,000 in a term of greater than or
equal to 24 months and the problem required that 24 months be used in the calculation period. But, some experimentation shows that the end period in cell H11 can
be changed to 18 and 19 months to achieve a value very near $25,000, $24,890.93
and $25,443.72 respectively. Obviously, this is a complex condition and may take
considerable experience before it is easily identified by an Excel analyst.

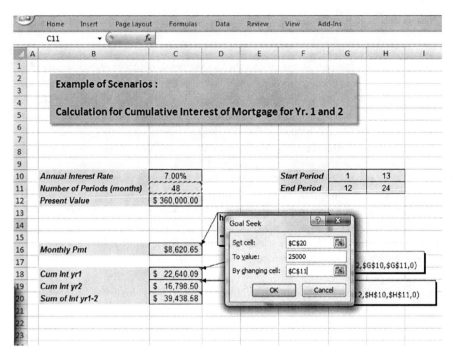

Exhibit 9.21   Goal seek for cumulative interest payments

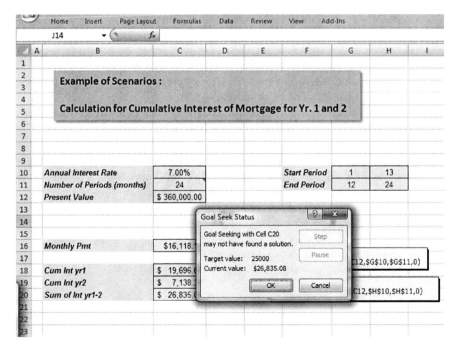

Exhibit 9.22   Uncertain goal seek status of cumulative interest

9.6 Summary

Solver, Scenarios, and Goal Seek are extremely powerful tools for quantitative analysis. Yet, we must be careful to use these tools with caution. In the words of US President Reagan—*Trust, but verify*. We have seen how a complex goal seek function can lead to problems if some forethought is not applied to our analysis. The nature of the search algorithms that are used in the Solver and Goal Seek tools, and the possible non-linear nature of the problem structure can baffle the search algorithm; it can lead to uncertainty in the veracity of the answer, or it can also lead to wrong answers. Although we did not spend much time discussing non-linear programs in our Solver section, other than to say they were very difficult to solve, it is not wise to assume that an optimal solution is always optimal. If the objective function and/or the constraints of a problem are non-linear, you might experience a solution that is a **local optimum**. A local optimum occurs when the search algorithm assumes that it need not search any further for a better solution, but in doing so, it has actually ignored other regions of the function where better solutions are possible.

How are we to know when we have a local optimum, or that a solution that has been identified as optimal is possibly not optimal? A little common sense is invaluable in making this determination. Here are a few tips that might help you avoid accepting the claim of an optimal solution when it is not, or help you verify whether an uncertain solution is in fact optimal:

1. If you have a non-linear target cell or objective function for a formulation in a *single* variable, attempt to plot the function by using successive values of inputs to see if the function might be a candidate for a local optimum. You can do this by copying the function to a long column of cells and placing consecutive values of input in an adjacent column. Then plot the results and note the shape of the curve. Of course, this is only possible for a single variable and in most problems we have far more that one input variable.
2. In the case of multi-variable problems, you may want to resort to simulation of inputs and to see if you can find some combination that outperforms the so-called optimal solution.
3. If a solution is uncertain, but appears to be correct, investigate by examining values near the solution that is proposed. Be careful to consider a local optimum condition.
4. Be careful to note any odd solutions—negative values where none are possible and values that are either too large or too small to accept as possible.
5. Verify that the constraints that are imposed on a formulation are satisfied.
6. Remember that in spite of your best efforts, you may still, on rare occasions, have problems dealing with these issues.

There is nothing more embarrassing than presenting a solution that contains a clear inconsistency in a solution which you have overlooked. Verification of an analysis is much like editing—it is not a pleasant task, but it is foolhardy to avoid it.

Key Terms

Solver	Changing Cell
Prescriptive Analysis	Right-Hand Side (RHS)
Scenario	Slack
Goal Seek	Not Binding
Descriptive Analysis	Binding
Constrained Optimization	Shadow Price
Linear Programming	Allowable Increase
Decision Variables	Allowable Decrease
Objective Function	Reduced Cost
Constraints	Non-Linear Programs (NLP)
Technology of LP	Integer Programs (IP)
Infeasibility	Mixed Integer Programs (MIP)
Coefficients	0-1 Integer Programs
LP Formulation	CUMIMTP
Target Cell	Local Optimum

Problems and Exercises

1. Name 2 types of Prescriptive Analysis and 2 types of Descriptive Analysis.
2. Simulation is to Linear Programming as Descriptive is to _____?
3. Constrained optimization optimizes an objective function without regard to factors that constrain the selection of decision variables—T or F?
4. Decision variables in Linear Programming are always integer valued—T or F?
5. Identify the following relationship as either linear of non-linear:

   a. $2X + 3Y = 24$
   b. $4/X + 3Y^2 = 45$
   c. $3XY - 8Y = 0$
   d. $4X = 6Y$

6. For the following linear programs, what is the solution? Do not use Solver; use strict observation:

   a. Maximize: $Z = 4X$; Subject to: X<=6
   b. Maximize: $Z = 2X + 5Y$; Subject to: X=5 and X + Y <=12
   c. Minimize: $Z = 12X + 2Y$; Subject to: X>=3 and Y >= 4
   d. Minimize: $Z = X - Y$; Subject to: X>=0 and Y<= 26

7. *Knapsack Problem* - Consider a number of possible investments in contracts for various projects that can fit into a budget (knapsack). Each investment in a contract has a cost and a return in terms of millions of dollars, and the contracts

can be purchased in multiple quantities. This problem is known as the Knapsack Problem due to its general structure—selecting a number of items (projects in this case) for a limited budget (knapsack).

Available contract types	# of contract investments available	Cost per contract	Value of contract
Project 1	4	1	2
Project 2	3	3	8
Project 3	2	4	11
Project 4	2	7	20

Budget $<= 20$; X_j = # of contracts of project j with possible values 0, 1, 2, ...

a. Formulate the LP for this problem.
b. Solve the problem with Solver.
c. What is the marginal effect on the objective function of adding one unit of each project to *# of contract investments available*? Recall this sensitivity analysis is done by looking at each decision variable, one at a time.
d. How does the solution change if the budget constraint is reduced by 5 units to 15?

8. *Nutrition Problem* - A farmer raises hogs for profit. They are organic pigs of the highest caliber. The farmer is a former nutritionist and has devised the following table of possible nutrients and minimum daily requirements for *3* nutritional categories. For example, a Kilogram of corn provides 90 units of carbohydrates, 30 of protein, and 10 of vitamins. Also, there are 200 units of carbohydrates required daily, and corn costs $35 per kilogram. The farmer is interested in knowing the kilograms of Corn, Tankage, and Alfalfa that will minimize his cost.

Daily nutritional requirements

	Kg. of Corn	Kg. of Tankage	Kg. of Alfalfa	Min. daily requirements
Carbohydrates	90	20	40	200
Protein	30	80	60	180
Vitamins	10	20	60	150
Cost $	$35	$30	$25	

a. Formulate the LP for this problem.
b. Solve the problem with Solver.
c. What is the marginal effect on the objective function of adding one unit of each ingredient (Min. daily requirements)? Recall this sensitivity analysis is done by looking at each decision variable, one at a time.

d. How does the solution change if an additional 15 units of RHS are added for each of the nutrition constraints? Add them one at a time; for example, change 200 to 215, then 180 to 195, etc.

9. You have joined a CD club and you are required to purchase two types of music CD's in a year: Country Music (X) and Easy Listening (Y). Your contract requires you to purchase a minimum of 20 Country CDs. Your contract also requires you to purchase at least 20 Easy Listening CDs. Additionally, you must purchase a minimum of 50 CDs (both types—Country Music and Easy Listening) yearly. If the Country CDs cost $7 per CD and the Easy listening cost $10 per CD, what is the solution that minimizes your yearly investment in CDs?

   a. Solve this LP with Solver.
   b. Which constraints are binding?
   c. Will the optimal solution (number of each CD type) change if the cost of Country CD's increases to $9?

10. You are interested in obtaining a mortgage to finance a home. You borrow a principal of $150,000 for 30 years. If you would like to have a monthly payment of $700, what is the interest rate that will permit this payment?

11. The CUMPRINC() function is similar to the CUMIPMT() function, except rather than calculate the cumulative interest paid, the function calculates the cumulative principal paid in a period of time. You have a mortgage loan at 6% over 30 years for $150,000.

   a. Use Goal Seek to find the approximate period in which $5,000 in principal payments have been accumulated.
   b. Use Goal Seek to find the approximate period in which $75,000 in principal payments have been accumulated.
   c. If the use of Goal Seek is problematic for problem 11b, what do you suggest as an alternative to finding the approximate period that still uses the CUMPRINC() function?

12. Advanced Problem—*Shift Scheduling Problem* - I run a call center for customers using my *personal sensitivity training* tapes. I guarantee that an understanding and caring voice (UCV) will be available to customers 24 hours a day. In order to satisfy this promise I must schedule UCV's based on historical demand data shown in the table below. I must determine how many UCV's must report to work at the beginning of each *Period*. Once a worker reports for duty they will work an 8-hour shift. There is no restriction in which of the periods a worker starts a shift. For example, if a worker begins at 3 pm, then they will work until 11 pm.

   a. Formulate and solve the LP that will minimize the total workers needed to cover the overall UCV historical demand. Note that the assignment of a partial worker is not a realistic situation. Therefore you may have to consider a constraint that guarantees integer UCV's.

Period	Time	UCV historical demand
1	3–7 am	3
2	7–11 am	12
3	11am–3 pm	16
4	3 pm–7 pm	9
5	7–11 pm	11
6	11 pm–3 am	4

b. What happens to the solution if the demand for UCV's in period 6 changes to 8? What is the new solution?

c. How would you handle varying cost for the time periods? For example, what if the cost of the period 5 (7 pm–11 pm) and period 6 (11 pm–3 am) time period is twice as high as other time periods. How does the objective function change if you want to cover the UCV demand at the minimum cost under the new condition that costs of UCV's are not equal?

CPSIA information can be obtained at www.ICGtesting.com
Printed in the USA
LVOW101415150512

281850LV00001B/2/P

9 783642 108341